Labor and the State in Egypt

Labor and the State in Egypt

Workers, Unions, and Economic Restructuring

Marsha Pripstein Posusney

Columbia University Press

New York

Columbia University Press
Publishers Since 1893
New York   Chichester, West Sussex

Library of Congress Cataloging-in-Publication Data
Posusney, Marsha Pripstein.
Labor and the state in Egypt : workers, unions, and economic restructuring / Marsha Pripstein Posusney.
p. cm.
Includes bibliographical references and index.
ISBN 0-231-10692-0. — ISBN 0-231-10693-9 (pbk.)
1. Working class—Egypt—History—20th century. 2. Industrial relations—Egypt—History—20th century. 3. Trade-unions—Egypt—History—20th century. 4. Corporate state—Egypt—History—20th century. I. Title.
HD8786.P67 1997   97-3929
331'.0962'09045—dc21   CIP

∞

Casebound editions of Columbia University Press books are printed on permanent and durable acid-free paper.
Printed in the United States of America
c 10 9 8 7 6 5 4 3 2 1
p 10 9 8 7 6 5 4 3 2 1

For Tom

Contents

Acknowledgments

This book incorporates material from earlier research, where I focused on the Egyptian labor movement and union/state relations from the Free Officers coup in 1952 through 1987, based on 1987–88 fieldwork conducted in Egypt. I have updated it with extensive archival research on developments since that time, ongoing written communication with several key Egyptian contacts, and a follow-up trip to Cairo over the 1994/95 semester break.

I am especially grateful for the friendship and assistance of Joel Beinin, who generously shared his insights into labor and the left in Egypt. In the initial stages of my research, Joel helped me to identify and locate essential books and periodicals and introduced me to a number of individuals who proved critical to my investigation. Later, he faithfully read and commented on numerous drafts of original manuscript and manuscript-in-process chapters.

At the University of Pennsylvania Frederick W. Frey, my dissertation supervisor, helped me to understand and prepare for the difficulties and limitations of conducting research in a foreign culture and an authoritarian political environment, and shaped my thinking about labor/state power relations. Peter Swenson inspired me to relate my work to theoretical debates in the study of European workers, and Thomas Callaghy pointed me to the comparative implications of my findings for the study of Third World political economy. The reader familiar with their work will find evidence of each of their influences here; the final argument is, of course, my own responsibility.

I am also grateful to John Waterbury, who served informally as an outside reader of that project, sharing his knowledge of Egyptian political economy.

I would like to thank Leslie Elliott Armijo, Ragui Assaad, Eva Bellin, Nancy Brooks, Corinna-Barbara Frances, John Gerring, Ellis Goldberg,

Miriam Golden, Rick Harbaugh, Antoine Joseph, Audie Klotz, Zachary Lockman, Dwayne Oldfield, Wendy Schiller, Diane Singerman, and Robert Vitalis for helpful comments on drafts of various portions of this book. Some of these individuals, as part of my Egyptian "family," also provided ongoing support, encouragement, and numerous insights about Egypt during my conversations with them over the years. For these latter contributions, I would also like to thank Marilyn Booth, Ken Cuno, Rebecca Foote, Ann Lesch, Dirk Vandewalle, and Carrie Rosefsky-Wickham.

Preliminary drafts of some chapters were presented at a conference on Middle East labor at the Harvard University Middle East Center, organized by Zachary Lockman, the Northeastern University conference on Privatization in Egypt and the Middle East, organized by Denis Sullivan, and the conference on Privatization in Socialist Countries of the Middle East at Banz Castle, Germany, organized by Hans Hopfinger. I would like to acknowledge the enjoyable and helpful interchanges I shared with the other participants at these conferences.

In Egypt I was fortunate to have the intellectual support and friendship of a group of scholars associated with Cairo University who helped me to understand the general political terrain of their country and directed me to individuals and institutions that were critical to my research. I am especially grateful in this regard to Samya Sa'id, Amani Qandil, and Ahmed Abdalla, and to Hamdi Mahmud al-Dawi who facilitated my introduction to the latter two individuals. Ragui Assaad helped me to make sense of the maze of available statistics on Egyptian labor, and ultimately provided me with the wage data on the 1970s and 1980s that is reported here. I would also like to thank Taha 'Abd al-'Alim Taha and Muhammad al-Sayyid Sa'id at the Al-Ahram Center for Strategic Studies for numerous discussions of Egyptian political economy, and for providing me with access to the facilities of the Center. In the company of all these individuals, I met dozens of other scholars too numerous to name; I am grateful to all of them for welcoming me to forums, dinners, and parties, and making me feel like part of their intellectual community.

I was also made welcome at the headquarters of the Egyptian Trade Union Federation, where much of my research was conducted. I am grateful to Ahmad al-'Amawi, president of the ETUF in 1987–88, for granting me permission to conduct interviews with the union officials based there, and especially to Niyazi 'Abd al-'Aziz of the Engineering, Electrical, and Metal Workers' Federation, who made time to answer all my questions and ensured that I always felt comfortable in the offices of that federation. 'Aisha 'Abd al-Hadi and Ga'far 'Abd al-Mun'im Ga'far of the Pharmaceutical and Chemical

Workers' Federation were likewise exceptionally accommodating. I would also like to thank Ahmed Harik, then editor of the ETUF newspaper *al-'Ummal*, for granting me permission to use the archives of the paper and providing me with an office to conduct that research.

Similarly, I am grateful to al-Sayyid al-Tahari, editor of the Ministry of Labor magazine *al-'Amal*, for allowing me to use that magazine's archives, for the office space I was given there, and for all of the assistance I received from his staff. I would also like to thank the librarians, archivists, and staff of *Al-Ahram* newspaper, the National Planning Institute, the Workers' University, the National Library Dar al-Kutub, and the Cairo offices of USAID for access to their research facilitities.

At all of these institutions my stay was made more comfortable by the secretaries who provided companionship and administrative assistance, and by the *farashin* who made sure that I was never without a cup of tea. The hospitality of Egyptians is legendary, but nevertheless deserves yet another note of appreciation.

The above-named institutions were critical sources of information about the institutional history of the Egyptian trade union movement, which forms a major part of this book. Another of its components, and what I feel is one of its key contributions to the study of Egypt's political economy, is its emphasis on rank-and-file struggles and the role of leftists within the workers' movement. For helping me to obtain the relevant information I would especially like to thank the Darwish family and the members and friends of the Voice of the Workers group. I am grateful as well to 'Abd al-Hamid al-Shaykh, head of the Tagammu' Party's labor bureau, for giving me permission to attend their meetings, and to all the members of that bureau for their assistance. In addition to welcoming me at other Tagammu' functions, officials of the party permitted me to conduct research in the archives of their newspaper, *al-Ahali.* I would also like to thank the staff of that newspaper for their help and hospitality. I hope that I have adequately repaid all of the time these leftists gave me, and the personal risks they took, by telling their stories here.

Excellent Arabic instructors at the Center for Arabic Study Abroad in Cairo, Middlebury College, and the University of Pennsylvania helped me to develop the necessary language skills to carry out my research. I would also like to thank Tawfiq Michael for his special assistance with the vocabulary and dialect of the Egyptian workers I interviewed.

Funding for my dissertation research and writing was provided by the American Research Center in Egypt, Inc., Fulbright-Hays, and the Social Sciences Research Council; I thank all for their generous support. I would also like to thank Bryant College for funding my follow-up trip to Egypt, and for a summer stipend to support completion of this book. I am particularly grateful to Conny Sawyer for providing proofreading assistance and secretarial support.

Finally, I would like to thank Kate Wittenberg and her staff at Columbia University Press for guiding me through the manuscript preparation process.

My son Eric was born while I was completing my dissertation, and has been a joyous counterweight to the rigors of writing ever since. Usually unsung and underpaid, the directors and teachers of the day care centers he attended while I was writing are the hidden heroes and heroines of this project; by providing him with a safe, loving, and nurturing environment, they gave me the peace of mind necessary to carry out my work. My husband Tom, himself a manual worker for much of our married lives, supported my graduate studies, and then risked the loss of his job to accompany me to Egypt, struggling to master Arabic both before and during our trip. His unswerving faith in my ability to complete this project saw me through occasional bouts of self-doubt, though of course I hated to admit it when he was right about something.

A Note on Transliteration

The Arabic bibliography for this dissertation follows the transliteration system adopted by the Library of Congress, minus diacritical marks and using an apostrophe to represent both the *hamza* and the letter *'ayn*. Where an author is commonly known in Egypt by a different last name, I have noted this in parenthesis after the official listing. The Library of Congress system is generally used as well in the text, but with some modifications to reflect commonly used Egyptian pronunciation and romanizations. Thus the Arabic "j" (*jim*) is rendered as a "g" in the text. Similarly, I use the common Egyptian spellings for cities such as Helwan, and periodicals based on a romanized name, like *Rose el-Youssef*.

Labor and the State in Egypt

Introduction

On August 12, 1952, just a few weeks after the Free Officers military group, led by Gamal 'Abd al-Nasir, had overthrown the British-backed monarchy in Egypt, workers at the Misr Fine Spinning and Weaving plant in Kafr al-Dawwar went on a sit-down strike in demand of wage increases and an end to company unionism. The army quickly crushed the strike, and the new regime tried 29 workers before a military tribunal as instigators. Two, accused of membership in illegal communist organizations, were publicly hanged.

Some ten years later, the same plant was again occupied by workers. It had since been nationalized, becoming part of a rapidly expanding public industrial sector in a mixed economy the Nasir regime called "Arab socialism." Workers were demanding repayment of deductions that had been taken from their pay prior to nationalization. They electrified the factory fence to prevent company officials from entering but, to demonstrate their loyalty to the government and its development strategy, they continued to operate their machines.

While the 1963 incident was resolved peacefully, bloodshed returned to Kafr al-Dawwar in 1994. Parastatals like the Misr Spinning and Weaving Company were now being restructured as part of the government's pledge to privatize the economy. Workers again occupied the factory, protesting the threat of layoffs and benefit cutbacks initiated by a new manager trying to make the plant more profitable prior to sale. State security forces opened fire on the workers and their supporters outside the plant, killing six and wounding more than a hundred.

Volumes have been written about the history of Egypt since the military came to power, but in most of this scholarship informal workers' protests like these in Kafr al-Dawwar, and often the formal activities of the trade union leadership as well, are noticeably absent and thus implicitly rendered insignificant. Combining institutional and interpretive analysis, this book explores—and restores—the role of labor in the unfolding political economy of Egypt after the 1952 coup. It focuses centrally on how labor has been, sometimes simultaneously, both an important basis of support and legitimacy for Egypt's authoritarian regimes and a force for challenging the government and restricting its prerogatives. Special emphasis is placed on the period since the early 1970s, when Presidents Anwar Sadat and Husni Mubarak attempted, with different degrees of success, to pursue economic liberalization measures in the context of limited and oscillating political liberalization. I argue that factors inherent to the labor movement and its relations with the state can account for much of the variation in the success or failure of these reform measures.

In this introductory chapter, I first present an overview of the role of labor in the political economy of Egypt after the 1952 coup. The subsequent sections lay out in detail the theoretical arguments which inform this analysis. These are followed by a discussion of the methodological issues surrounding research into labor/state relations in Egypt, and then a brief outline of the empirical chapters of the book.

Labor in the Political Economy of Egypt

From the standpoint of labor/state relations, the political economy of Egypt since 1952 can be divided into three historical periods, although the lines of demarcation between them are gradual rather than abrupt. The first begins with the coup and extends, roughly, through 1955. During this period, as Nasir individually was struggling to consolidate political power within a regime of military officers with conflicting ideological orientations, the government's central economic development strategy was to maintain and expand the existing foreign investment in the country. The repression of a labor movement that had been an active force in struggling against the British-backed monarchy was a cornerstone of this strategy. Shortly after the first Kafr al-Dawwar incident, the military outlawed strikes and indefinitely postponed plans by a core of union activists to found a national trade union confederation.

The second period, marked by a shift toward an autarchic development plan, coincided with Nasir's undisputed hegemony. Relying first on private capital to stimulate growth, the regime moved in the late 1950s and early 1960s to claim ownership of the country's largest banks, insurance companies, utilities, and industries. Other populist measures were enacted as workers came to be seen by the regime as a base of legitimacy for the government, and trade unions as an important vehicle for mobilizing their support as well as their efforts in production. The establishment of a confederation was now encouraged, with the understanding that its leaders would be screened for their loyalty to the regime.

Political participation was channeled through a single, mass party known as the Arab Socialist Union, membership in which was made a condition for solicitation of any union office. But despite the regime's socialist rhetoric and etatist policies, Nasirism is not considered here to constitute any genuine kind of socialism. It is instead held here, following Waterbury and others, to be a form of state capitalism.[1] The latter is defined by a large public sector in an environment where private enterprise continues to exist, the market remains the principal means of distribution, and state ownership signifies little in the way of actual workers' *control* over the means of production.

Sadat's ascension to power in May 1971, following a brief period of uncertainty after Nasir's death at the end of 1970, marks the beginning of the third period; it extends into the mid-1990s when this manuscript was in preparation. Engineering a shift in Egypt's strategic orientation from the East to the West, Sadat undertook a program of limited political and economic liberalization; Husni Mubarak, who succeeded Sadat after his 1981 assassination, essentially continued these policies. The political reforms entailed the disbanding of the ASU and the creation of political parties which, except for the ruling party, have enjoyed only limited ability to criticize the government and contest a series of highly manipulated parliamentary elections. At the same time, the economic liberalization programs, increasingly entailing the involvement of the International Monetary Fund (IMF), the World Bank (WB), and other Western multilateral agencies, threatened the various forms of government economic intervention which Egyptians had come to expect. As workers, traditional leftists, and adherents of Nasir's philosophies began to mobilize against these policies, the government's demands on union leaders changed: now they were expected to provide political support for the president and his party, and to forswear labor protest and preempt any independent actions by workers. With these demands never fully realized, labor opposition to reform became a frequent government justification for its repeated failure to live up to IMF and WB targets.

Each of these periods holds important lessons for the comparative study of developing countries. At its broadest level, the central argument of this book is the ability of societal forces, in this case labor, to affect governmental policies even in the context of repressive, authoritarian regimes. It is thus set against the state-centered approach which has characterized much recent scholarship in political science and much of the nonanthropological, social scientific work on the Middle East since the 1950s. The intent here is not to claim, like the earlier, Western-based "pluralist" literature justifiably rejected by scholars of authoritarian regimes, that societal pressures will uniquely determine policy outcomes.[2] Rather, I seek to contribute to the development of balanced inquiries which focus on the dynamic interplay between state and society. Two analytical undertakings are required to achieve this balance: first, the disaggregation of the categories of "state" and "labor" into the different organizational and individual players who have agency on a given issue, in order to accurately expose the interaction between them; and second, the combination of an institutional focus on how existing rules and structures shape the options available to the relevant actors with an interpretive examination of the beliefs and motives which influence the choices that these actors make. These latter dimensions reflect, respectively, issues of *institutional capacity* and *individual will*.

Thus my argument about the early 1950s focuses on ideological clashes and competing power motivations among the military leaders in power. With the labor movement similarly divided, several separate clusters of unionists each found different patrons in the regime. Official policy toward labor was shaped by the interplay of these forces, but at the same time, implementation of these policies was impeded by the weak institutional capacity of the new regime.

The formation of the trade union confederation in 1957 did not resolve the conflicts within either the government or the union movement over the proper structure of the unions and the nature of union/state relations. Rather, contradictions within each agency, and the interplay between them, spilled over into the next two periods and resulted in repeated revisions to the laws governing union behavior. The formation, and subsequent disbanding, of the ASU represented a new arena for power struggles among aspirants for better positions in the government or union hierarchy.

At the same time, the inability of the regime in the early 1950s to affect measures seemingly intended to better the lot of workers contributed to its later turn toward more populist/socialist policies. Egypt's embrace of etatism in the second period is the backdrop against which the struggle over economic reforms in the third takes place. I hold here that this etatism cemented, among workers, belief in a moral economy in which their wages and benefits came to be viewed as entitlements in exchange for their contribution to the cause of national economic development. The persistent repression of labor, principally in the form of the ongoing ban on strikes and the repeated interference of the government in the union leadership selection process, reinforced workers' beliefs that the government placed a high value on their productive efforts. In promoting this view of reciprocal rights and responsibilities between workers and the state, the Nasir regime staked its own legitimacy on meeting its part of the bargain, and was constrained to do so.

The subsequent efforts of the Sadat and Mubarak regimes to abandon etatist policies in the face of severe macroeconomic distortions can thus be seen as attempts to extricate the government from this moral economy. These efforts have met with steadfast resistance from labor. The level of spontaneous rank-and-file protest increased dramatically with Sadat's embrace of economic liberalization, while Marxists and Nasirists in the union movement pushed the confederation to adopt and maintain a firm stance against the government's retraction from the economy.

In the 1990s, after successive stabs at devaluing the currency, raising domestic interest rates, eliminating price controls, and cutting public spending, especially by modifying subsidies on basic necessities, the Egyptian government appeared to have finally stabilized its longstanding fiscal deficit. The IMF gave Egypt a successful review of a standby credit agreement in 1993; it was Egypt's first positive review from the Fund after thirty years of relations punctuated with broken negotiations and canceled agreements. But both the Fund and the World Bank remained dissatisfied with Egypt's progress on trade liberalization and, especially, the privatization of public sector enterprises. Although the government had enacted a sweeping privatization bill in 1991, prior to signing the IMF accord, few major sales had since occurred. Egypt's Western creditors complained that the terms of the 1991 legislation, which subject privatized firms to the existing labor laws that prohibit mass layoffs without government permission and sharply restrict individual firings, make the parastatals unattractive to potential purchasers. New labor legislation which would remove these impediments to the rationalization of labor markets in Egypt, first proposed in 1991, was being held up by the insistence of unionists that it include legalization of strikes. Although representatives from the government, businessmen's associations, and the labor unions reached final agreement at the end of 1994 on "exchanging the right to strike for the right to fire,"[3] the new legislation had still not been passed a year and a half later.

This impasse over privatization underscores the centrality of the labor movement to the structural adjustment process. I will show here that previous governmental attempts to implement privatization, as well as the three successive Egyptian presidents' efforts to initiate stabilization measures, were thwarted to different degrees by opposition from labor. Yet the degree of this opposition has varied over time, as well as according to the reform in question itself. These variations can be traced to the intersection of workers' *attitudes* toward the different issues, shaped by the moral economy, and the organizational *capacity* of the union movement, shaped by the outcomes of the institutional struggles described above. Union leaders' motivations, which are intertwined with both of these variables, form an intermediary factor.

This book is thus intended in part as a contribution to the recent spate of studies on economic reforms in developing countries, but it is also meant as a corrective to the state-centered focus of much of this literature.[4] Scholars in the statist tradition have identified numerous variables affecting the success or failure of economic liberalization attempts, revolving around the commitment of regime elites to the reforms and the capacity of the state to implement them.[5] Labor opposition to reform is not completely neglected in this approach—commitment in part reflects the willingness of the rulers to confront societal resistance, and capacity in part the ability to shield technical personnel from societal pressures—but it is mediated through statist variables. This emphasis on the state obscures potential differences in the strength of labor objections to different components of a reform program, and to variations in its capacity to effectively obstruct them. An approach which focuses greater attention on labor can therefore add to our understanding of cross national variations in the success or failure of different reforms programs, as well as to diachronic differences within individual countries.[6]

Because the findings of this study may prove useful to practitioners seeking ways to circumvent or preempt labor opposition to economic liberalization, it is important to clarify here that its normative underpinnings lie in the opposite direction. Many of the advocates of orthodox reform, among both scholars and economic practitioners, see organized and especially public sector labor as an elite class in the third world, whose privileges come at the expense of the unorganized workers, the rural poor, and the informal sector. The reform process must thus, they maintain, entail the abrogation of these privileges.[7]

I hold this view to be both impolitic and immoral. From the normative perspective, even if urban workers are not the poorest of the poor, many nevertheless live in appalling conditions of overcrowding and poor sanitation. Some Egyptian workers I visited while conducting this research lived in apartment buildings lacking any access to light, with rudimentary bathroom facilities shared by all residents of the dwelling, on garbage-laden streets too narrow for an automobile to drive down. What is the rationale for lowering the living standards of such workers, when no similar sacrifices are demanded from the country's well-to-do?

Politically, it is always difficult for a government to deprive its citizens of something to which they have become accustomed. Selling sacrifice to workers for the sake of economic liberalization is especially problematic because the logic of the argument is inherently flawed. The orthodox reform strategy rests of the assumption of a market rationality in which every individual acts in his/her own self-interest; its heroes are private sector entrepreneurs, formerly constrained under etatism, whose unleashed drive for profit is expected to foster growth and employment. But how can those who promote the selfishness of capitalists as the road to economic salvation then propose that workers be self-sacrificing?

Because it is difficult to win popular support for orthodox reforms, there is an apparent contradiction between economic and political liberalization. Governments which are withdrawing from the economy, with deleterious consequences for the lower classes, are unlikely to fare well in free and fair elections. Some advocates of orthodox reforms therefore hold that authoritarian regimes should delay democratization until the government's retraction from the economy is complete.[8] In Egypt, President Husni Mubarak himself advanced this argument in the early 1990s as justification for stalling political liberalization. The mid-1990s impasse over the new labor legislation in part reflected the government's hesitancy to relinquish its control over the unions, which could exacerbate demands for an end to repression from other quarters as well.

Western economists and policy makers, while doubting Mubarak's commitment to the World Bank and IMF programs, have not publicly questioned his rationale for delayed democratization. Whether successful economic reform is sufficient justification for prolonging authoritarian rule is a normative issue. But the orthodoxy of the market is not necessarily the best alternative to the import-substituting, etatist economic policies formerly pursued by Egypt and many other developing countries. The litany of economic imbalances and inefficiencies associated with these policies cannot be denied, but neither have the virtues of neoliberalism been proven.[9] The "trickle-down" effect of enriching the elite has seldom materialized. Rather, numerous countries that have embraced orthodoxy, from Russia to Argentina, have seen income gaps widen while unemployment and inflation depress the living standards of the lower classes.

For these reasons, labor opposition to economic liberalization, vilified by the reform advocates, is viewed here with sympathy. Even when it takes the form of defending an unviable status quo, opposition to orthodoxy is the best way for workers to ensure a compromise outcome under which the state continues to regulate private capital and provide protection against the joblessness that inevitably results from "market failures" under capitalism.[10] Battling reforms also helps to build the organizational experience and skills that public sector workers will need to protect their interests if privatization takes hold and a freer market for labor is established in Egypt.

The dismantling of parastatals and their transfer to private hands, combined with withdrawal of the government from the sphere of price-setting, means that workers' earnings and their purchasing power will be determined much more by localized negotiations between labor and employers than by nationally set economic policies. The organizational capacity of local unions, where they exist, thus becomes more meaningful to workers as privatization is enacted, or even threatened. In Egypt, this inevitably focused greater attention on government interference in union affairs and the ban on striking, and increased pressure for democratic unionism.

Economic Issues and Labor Capacities: State Corporatism Reconsidered

While this book departs normatively from those advocates of economic liberalization who decry the efforts of labor to impede reform efforts, it takes issue theoretically with other scholars who deny its ability to do so. Several decades of scholarship have disparaged the potential for third world workers to influence policy making, at least in the latter stages of industrialization. Proposing that states can completely control the activities of labor if they so desire, this approach leads to the implication that failed reforms must reflect the weakness of government will rather than the strength of societal forces.

Theorists of the "New International Division of Labor" (NIDL) school focus on the consequences for labor of export-led strategies of industrial development. They argue that the successful initiation of export-led growth requires capital and technology that can only come from multinational corporations (MNCs). At the same time, with the recent advances in communications and transportation technology, it has become both possible and desirable for advanced industrial firms to disaggregate their production processes, locating the manufacture of different components of a final product in different areas according to their resource endowment. The MNCs will seek to place the most labor-intensive production processes in the areas where workers are both cheap and plentiful, that is, the third world. Developing countries must therefore compete to attract these foreign firms, and creating the appropriate conditions to attract foreign capital means, first and foremost, ensuring a supply of cheap labor. The low wage environment is achieved by the outlawing, or suppression, of labor unions by authoritarian rulers. Because of the surplus labor conditions prevailing in the LDCs, those countries which deviate too far from the MNCs' standards will be punished by the withdrawal of foreign investors. Furthermore, those regimes which are initially reticent to pursue the export-led growth path will eventually be brought to heel by Western-dominated international lending agencies.[11]

While NIDL theory is based largely on the experience of the East Asian "tigers," similar arguments were made earlier about Latin America. The bureaucratic-authoritarian (BA) literature posited that military regimes there, bent on attracting foreign firms to produce capital goods for the domestic market, successfully tamed unions that had previously been nurtured and incorporated by the state.[12] The theoretical impact of this literature was to focus attention on the state as the locus of change, contributing to a larger paradigmatic shift away from society-oriented scholarship.

Both NIDL and BA theories cited union incapacity as the explanation for labor's weakness. Capacity is an institutional issue, reflecting the organization of the union movement in relationship to the economic policy-making structure, and the presence or absence of legal restrictions on union behavior. The Latin American variants of the literature emphasized in particular the emasculating effects of state corporatism, in which labor is organized into noncompetitive, centralized and hierarchically run "peak organizations" whose leaders are screened, if not chosen, by the state; workers are required to join these unions.[13] The authoritarian Latin American regimes supplemented interference in union leadership selection with legislation restricting popular protest, including strikes, and intense repression of the left from whose ranks many labor militants are customarily drawn.

Thus, Mericle wrote that "the ability of the state to control labor protest depends primarily on two variables: the extent to which the corporatist system has been imposed on the labor organization, and the will and capacity of the state to utilize the control apparatus. . . . When employed by an authoritarian government, this control system has an amazing capacity to prevent the effective articulation of working class demands and to suppress open conflict behavior."[14] Similarly, Cohen argued that the establishment of corporatist labor arrangements by government elites in Brazil "preempted the formation of a more autonomous labor movement and minimized industrial conflict. To this day, they have maintained urban workers virtually powerless and utterly dependent on the protection of the state."[15]

This study will demonstrate that a system resembling the state corporatist ideal type developed in Egypt under Nasir's rule. During the second period described above, a singular, hierarchically ordered labor confederation was consolidated. Strikes remained illegal, with fines and jail terms specified in the criminal code for workers who refused to perform their jobs. Moreover, emergency laws in effect virtually continually since 1952 gave the state the right to prosecute participants in other forms of labor protest. Although repression of labor and the left in Egypt has been comparatively milder than it was under the Latin American bureaucratic-authoritarian regimes, the generally arbitrary exercise of police power has offered few protections against the routine torture of detainees.

Arguments similar to those made for Latin America have thus been applied to Egypt. Galenson blamed corporatism for the timidity of Egyptian unions in the first decade after the coup: "The conception of labor market organization . . . in countries with corporate systems such as Egypt and Spain, regards the interests of labor, as well as of other social groups, as subordinate to the interests of the state . . . Trade unions, or similar bodies, are regarded accordingly as administrative arms of the state, charged with the primary responsibility of maintaining discipline and furthering productivity. . . . They are not permitted to exert any real pressure for shorter hours or higher wages."[16] Works by Elsabbagh, Shaaban, and Kamel hold that this situation continued throughout the 1960s.

For John Waterbury as well, labor was largely irrelevant to Egyptian political economy under Nasir, and has remained so thereafter. One reason for labor's weakness, he argues, is that the privileges of the public sector workforce were granted from above rather than won through actual struggle. Acknowledging that labor protests have occurred in Egypt and other developing countries with similar economic structures, Waterbury nevertheless discounts the possibility of "these actions truly driving the state to make concessions to labor. When organized labor has chosen to confront the state . . . it has almost invariably been smashed." Thus he attributes the failure of structural adjustment in Egypt to the government's lack of commitment to reform.[17]

Here I maintain, in contrast, that labor has been able to pursue economic demands and wring concessions from the state, in spite of corporatist controls. Its ability to do so, however, is contingent on the specific issue at hand and how policy around that issue is made. Rather than uniformly strengthening or weakening labor, corporatism differentially affects workers' capacity to respond to issues at the local, industrial, regional, and national levels.

In developing this argument I uphold, and marshal new evidence in support of, two critiques of the "pessimistic approaches" to the study of labor in developing countries cited above.[18] The first emphasizes the ability of workers to break out of the corporatist straitjacket and fight for their interests outside of the formal union framework, by forming alternative organizations and/or through wildcat protests. Such actions may directly impede implementation of economic policies and influence political developments as well.[19] Below, and more fully in chapter 3, I describe the nature of informal workers' protests in Egypt and analyze the motivations behind them.

The resilience of workers in the face of corporatist controls and episodes of severe repression may explain the rapid resurgence of labor militancy after the demise of Latin America's bureaucratic-authoritarian regimes. O'Donnell himself predicted this, and Kaufman and Haggard confirm that unions have been able to successfully challenge economic stabilization programs in the fledgling "transitional democracies" which the collapse of authoritarianism has in some cases produced.[20] Yet, as Roxborough notes, it is unlikely that labor movements which had been so thoroughly suppressed could regroup and regain their influence so quickly, suggesting activity beneath that surface that the pessimistic theorists overlooked.[21]

Putting the spotlight on ordinary workers reveals that labor quiescence around specific issues, or during particular time periods, may stem from attitudinal rather than capacity factors; such attitudinal variations are ignored by the pessimists' emphasis on institutional paralysis. Youssef Cohen has argued that the Brazilian state historically, and successfully, portrayed itself to the masses as a "benevolent leviathan" which would provide for their basic needs. He attributes the smoothness of the transition to bureaucratic-authoritarianism there to the internalization of this philosophy by Brazilian workers. Thus, while corporatism effectively diminished the *capacity* of the workers to resist government policies, its imposition was facilitated by an initial lack of *propensity* of workers to challenge the regime.[22] This implies the potential for workers' struggles to erupt, without any change in the legal framework for labor organizing, if the attitude of workers is transformed. Such underlying changes in workers' beliefs may account for the sudden outbursts of protest which broke out in Korea and the Philippines during the 1980s.[23]

At the same time, this book also departs from those scholars, primarily on the left, who would dispense entirely with the study of corporatist unions and their officers, presuming them to be merely extensions of the state.[24] While exclusive studies of the informal activities of workers may be valuable in their own right, they overlook the potential for unions to marshal resources behind workers' demands in spite of government regulation and interference in their affairs. In this regard it is noteworthy that the Asian countries highlighted in NIDL theory initially sought to avoid establishing corporatist systems.

Especially at the early stages of export-oriented manufacturing, rather than imposing large and hierarchical federations, laws forbid unions altogether, or required separate locals in each individual plant, with company officials heavily involved. In some cases, such provisions remained in effect even as an advanced stage was reached. Thus in East Asia, the capacity of labor to organize was diminished not by the presence of a centralized and hierarchical union structure, by rather by the unavailability of the organizing capacity that such large unions can provide.[25]

Haggard and Kaufman, somewhat bemoaningly, find considerable potential for labor movements to effectively challenge stabilization programs in countries "where unions or informal sector workers possess sufficient resources for defensive mobilization but are still vulnerable to periodic repression and lack secure access to decision making or clear rights to organize."[26] Egypt in fact fits nicely into this category, and the negative assessments of the Egyptian labor movement's influence cited above have been challenged by other scholars of the country's political economy. Mustapha Kamel al-Sayyid and Amani Qandil, in particular, both argued convincingly that while Egypt's union movement has been subservient to the state on political issues, it has been oppositional with regard to economic policies. Bianchi, drawing on their work, found labor to be quite effective in resisting public sector reform in Egypt through the mid-1980s.[27] Chapter 4 reviews and augments their evidence in this regard, while focusing on how workers' and unionists' responses to the different components of economic liberalization have varied.

The discrepancies between these contradictory arguments about corporatism's effect can be somewhat reconciled when one considers the focus of the inquiry. My intervention here is to establish the importance or relating arguments about organizational structures to the nature of specific issues. Where Bianchi, and Haggard and Kaufman found labor serving as an impediment to reform, it was in relation to policies made or attempted by national governments. It is at this national level where corporatist union leaders have the greatest potential to influence government policy, since centralization grants the greatest freedom of maneuver to unionists at the top of the organizational hierarchy.

Most of the pessimistic arguments, in contrast, were focused on wages and job regulations, and the presumed absence of struggles at the individual plants where these are at least partially determined. If unions are to respond to local conditions, the initiative must come from local officials. It is these cases, however, where the corporatist structure serves to decrease rather than increase labor's capacity to react effectively. In a singular confederation structure like

Egypt's, leaders of an industrial federation may be hamstrung in addressing sectoral issues until they have gotten the approval of confederation leaders, a process that requires time, bargaining and perhaps compromise with other federations who are not affected by the reform scheme. By the same token, local leaders may be straitjacketed when it comes to individual plant issues, if they require the consent of federation officials before taking any action. In other words, provisions for hierarchy, which facilitate the influence of top union leaders with national policy makers, serve simultaneously as restraints on lower level union leaders and the workers they represent.

Wages, of course, are also influenced by national policies, such as minimum wage legislation and, potentially, wage indexing agreements. Local struggles are nevertheless critically important in determining take home pay as well as working conditions. In Egypt, where the state sector embodies virtually all of the largest industrial firms and basic salaries for parastatal employees are set by national policy, there is still tremendous scope for individual plant managers to determine discretionary wage components and working conditions, and thus the potential for *local* unions to be involved in these issues.

This distinction between national, industrial, and individual plant issues is relevant to economic liberalization because the variety of issues it encompasses involve different policy-making configurations which affect labor, and tap into the union structure, in different ways.[28] Privatization, for example, can be implemented in what I call "one fell swoop" schemes where the entire parastatal sector is put up for sale at once via subscription sales. Such a plan naturally invites a response by unionists at the national level. In Egypt, legislation to this effect was proposed and defeated several times before the most recent initiative. But there are a variety of ways in which ownership transfers can be staggered by industry, region, or the decisions of conglomerate holding companies with mixed portfolios. In this case, which characterizes the privatization project in Egypt after the 1991 legislation, plants are affected one at a time, making the local union's initiative more important to the response. Similarly, cutbacks in government spending can be effected by changes in price controls or subsidies which have a national impact, but they can also come about through workforce reductions in individual plants or agencies. Trade liberalization and public sector rationalization are also examples of policies that may be decided at the national level, but whose implementation may affect only circumscribed groups of workers.

In sum, while state corporatism creates the potential for bringing the weight of the entire union movement to bear on local struggles, it simultaneously renders this less likely by subjecting such decisions to the will of the uppermost layer of union leadership—the leaders closest to, and most carefully screened by, the state. Thus the argument here is that union centralization best enhances the capacity of labor to respond to centralized issues. Where decisions are made, or their effect is only felt, at the industrial or local level, it is more likely that the hierarchical structure will serve as a barrier to industrial or local initiatives.

Although I make this claim here only in the context of authoritarian countries, it should be noted that some scholars have criticized the stifling of local initiative under European "societal corporatism" as well.[29] The charge that peak association leaders have "sold out" in advanced industrialized countries is based on the association between European corporatism and union cooperation with wage moderation programs. For some rational choice theorists, such wage moderation contributes to economic growth and is therefore in the longer-run interests of workers. Corporatist union structures, in this view, can be seen as providing workers with a public good, i.e. the organizational ability to overcome militancy at individual plants, which causes inflationary spirals and thereby damages national economies. Whether or not wage moderation is really in workers' interests, however, depends on a number of factors.[30] Arguably, therefore, the real public good that European corporatism provides is not moderation per se, but rather the ability for workers to formulate, implement, and evaluate *any* national plan for dealing with capital.

It is exactly this kind of democratic strategizing that authoritarian varieties of corporatism prohibit. Nevertheless, this study will show that intensive government screening of senior union leaders does not prevent the latter from challenging the government's economic policies under appropriate conditions, to be explored below. First, though, I explain why the neglect of rank-and-file concerns by government-sponsored federation and confederation leaders does not stop workers from taking action outside of the union structure.

Economic Issues and Workers' Attitudes: Moral Economy Theory and Egyptian Labor

It is a central argument of this book that the structure of the union movement, and the legal environment in which it interacts with the state, are critical variables which shape the capacity of workers to respond to different economic issues. But the study of labor cannot be reduced to the activities of unions. Egypt's trade unions have historically excluded large segments of the working class, who are nevertheless influenced by government policies and able to communicate their sentiments to the state. By the same token, unionized workers sometimes resort to organizing outside of the union structure when it fails to address their concerns. For these reasons, as well as because pressures from below form one of the constraints under which union leaders act, the starting point for any analysis of labor's influence on policy must be the sentiments of ordinary workers.

The term "workers" is used here to refer to nonagricultural laborers who own no means of production and must therefore sell their labor power in order to earn a living. Workers should also be understood to be those who, by virtue of lack of education and/or family background, have little opportunity for career advancement; their jobs correspondingly carry little status in society. The core of the working class consists of industrial laborers, blue-collar service workers, and those occupying unskilled clerical positions. Workers are thus defined here as an objective category, independent of the subjective presence or absence of class consciousness.[31] The focus of this study is on workers in the formal sector, defined in Egypt to include all establishments with ten or more full-time employees.

Some Egyptian unions represent wage laborers who are not workers according to my definition. The inclusion of such unions into a singular confederation which also contains the industrially based federations precludes isolating only the latter for study. Therefore, where I talk about the Egyptian *trade union* movement, I am referring to the collectivity of union leaders and their staffs, regardless of the nature of any particular union's membership. The primary focus of inquiry here, however, is not on the trade union movement but rather on a *labor* movement that is at once both more broadly defined to include nonunionized workers, and more narrowly restricted only to those, outside of unionists, who fit the definition of "worker" used here.

Workers are individuals, each with a unique temperament and a different set of personal circumstances which influence their attitude toward their employment and pay, and condition their response to any changes in these. In order to directly influence economic policies at the plant, industrial, or national level, however, individuals must come together, acting as a group with a common goal; i.e. they must undertake collective action. In the literature on collective actions by workers, three alternative approaches—Marxism, moral economy, and rational choice—contend. Although the differences among the three are sometimes exaggerated,[32] I seek to show here that a moral economy perspective provides the best explanation for this phenomenon in Egypt.[33]

In the moral economy view, collective actions are a response to anger generated by violations of norms and standards that the subaltern class has become accustomed to and expects the dominant elites to maintain. Rather than reflecting some emerging new consciousness, then, protests under a moral economy aim at resurrecting the status quo ante.[34] The operation of moral economies among workers has been demonstrated by E. P. Thompson, who showed that eighteenth-century English workers were moved to protest violations of their notions of fair practices in the marketplace for grain, and by Charles Sabel's evidence that third world immigrant workers in Europe would strike when angered by insults to their dignity and/or infractions of the principle of "a fair day's work for a fair day's pay"; Peter Swenson illustrates how even in advanced industrial economies, workers' views of just wages are influenced by distributional norms, the upsetting of which can precipitate strikes. Kraus has suggested that a "moral economy of the poor" was the operative factor behind an upsurge of labor protest in Ghana during 1968–1971.[35]

In the Egyptian moral economy I posit, as developed more fully in chapter 3, workers view themselves in a patron/client relationship with the state. The latter is expected to guarantee workers a living wage through regulation of their paychecks as well as by controlling prices on basic necessities; the government should also ensure equal treatment of workers performing similar jobs. Workers, for their part, provide the state with political support and contribute to the postcolonial national development project through their labor. It is in this sense that they are "with" the state.

This relationship between the state and workers in Egypt was cemented during the Nasir years when the government's role in the economy greatly expanded, and the regime explicitly appealed to workers' nationalist sentiments when encouraging them to increase their productive effort. Nasir's idea of reciprocal rights and responsibilities may have resonated with workers because of preexisting moral economy beliefs shaped during the country's long years as an agrarian society. Arguably it also tapped into Islamic notions of fairness and justice in employment relationships.

In 1973, three years after Nasir's death, jobs in the public sector and civil service accounted for roughly 25.2% of total employment, and 54% of all nonagricultural employment in Egypt. This included workers in the largest manufacturing plants, all utilities, and much of the blue collar service industries such as health and tourism. Although liberalization under Sadat and later Mubarak enlarged the role of the private sector, it did not fundamentally alter this aspect of the Egyptian economy; as of 1986, the state accounted for a larger share—28.6% of total employment, and 51% of nonagricultural employment.[36] Private sector industrial laborers worked largely in the informal sector, but were still dependent on the state for subsidies, price controls, and social services. Although liberalization subsequently decreased the role of the state in these latter realms, recent data indicate that the share of total employment provided by the parastatal sector held roughly steady through the early 1990s, while the civil service sector grew proportionally.[37] Those on the government payroll continued to enjoy benefits in the form of job security, pensions, and various forms of insurance that were not commonly available to private sector workers, causing many of the latter to aspire to public sector or civil service jobs.

The relationship of reciprocal rights and obligations that I am describing differs from the "they pretend to pay us, and we pretend to work" attitude disparagingly ascribed to Egyptian workers by neoliberal reform advocates.[38] Such a situation may in fact have prevailed in the bloated white-collar civil services, where employees were notoriously both low paid and largely without real work to accomplish. Most blue-collar industrial and service employees, in contrast, did exert effort, and apparently believed that they were productive even if their efficiency was below international standards.

The moral economy that I am suggesting here also differs somewhat from the notion of a social contract between the government and the people that is frequently ascribed to Egypt.[39] The social contract argument holds that during the 1960s, *all citizens* surrendered their political freedoms to the regime in exchange for economic entitlements. While this may be true, the reciprocal rights and obligations I posit are specifically between productive sector workers and the government, and the workers' side of the bargain is their labor power itself. The lack of political freedoms per se is not part of this deal, but restrictions on strikes and union activity are concomitant with this understanding. This repression of labor did reduce the incidence of actual work stoppages as a form of protest. But because it implied that the government valued uninterrupted production, it also served to reinforce workers' belief that their toil signified a contribution to the state and entitled them to the provision of a certain package of wages and benefits. This explains why the Mubarak regime encountered difficulties in its efforts to erode the moral economy through propaganda favoring market reforms without simultaneously ending this repression .

The moral economy approach can also shed light on why certain aspects of economic liberalization are more likely to generate rank-and-file protest than others. In particular, anger will be generated by policies that are seen as violations of the implicit agreement between workers and the state, such as wage cuts, subsidy removal, and privatization, to the extent that the latter deprives workers of job security or benefits that are available only to the public sector. However, changing policies that are not associated with this bargain, such as restrictions on foreign trade and investment, should encounter much less resistance. By the same token, the erosion of workers' real earnings through wage reduction will generate more resistance than cutbacks via inflation, which has a less direct as well as, generally, a more incremental effect.

The interpretation of workers' behavior advanced here clashes with theories of workers' behavior associated with neoclassical economics, and in some ways with broader interpretations of rational choice as well.[40] While definitions vary, rational choice theory commonly holds that all social phenomena must be understood as the consequence of individual behavior, with individual decisions based on expected outcomes according to an ordered set of consistent preferences.[41] To use these propositions in order to predict behavior, however, it is necessary for the theorist to make further assumptions about the actual preference structure of individuals. Neoclassical economists have specified that preferences are shaped by selfishness, which in turn is equated with income maximizing behavior.[42]

The classic application of this logic to labor market economics suggests that workers will join unions only when they can provide selective incentives,[43] and will strike only if the expected benefits, in terms of higher pay and benefits, exceed the expected costs, in terms of foregone wages and possible job loss. This should be those times when the labor market is tight, since under these circumstances workers will anticipate fewer risks from striking, and expect to wring more concessions from their employers. Thus, while workers may have grievances, strikes are precipitated by a perception of opportunity, not an increase in discontent. A large body of empirical studies, drawn from advanced industrialized countries, has been generated around this theory.[44]

Written from a Western perspective, the neoclassical literature assumes that workers and union officials have ready access to data about their country's economic performance, employment statistics, the activities of other unions, and so forth. As noted by Nelson, such information is often not available under the strict press controls prevalent in authoritarian systems. This alone is reason to doubt that labor protests in countries like Egypt are the result of careful cost/benefit calculations. Yet Nelson maintains that neoclassical logic should generally hold in lesser developed countries as well, unless workers believe that things are unlikely to improve anytime in the foreseeable future, in which case there is no reason not to strike when bad times begin.[45]

Working within the same framework, Ellis Goldberg has explored how rationality might work differently in the labor market conditions in Egypt. He suggests that in the third world, workers' skills are largely plant-specific, making tenure at their place of employment their only guarantee of a stable income. Achieving job security, presumably through seniority systems, is thus an important motive for collective action, and hence the older workers who would benefit most from seniority protections are the most likely supporters of unions. At the same time, workers will be disinclined to support collective action in industries considered likely to respond with layoffs. Using this to account for interindustry differentials in unionism before the 1952 coup, Goldberg argues that the etatist policies which came after it eliminated the risk of job loss as an impediment to collective activity. However, they also removed the motivation for it, since the state was now providing job security in addition to higher living standards. Hence, workers became disenchanted with unions, and ceased to engage in voluntary collective activity.[46]

However, Goldberg's argument is insufficient to explain the supposed lack of collective activity after 1952, because if workers were opportunity maximizers and in fact risked little by engaging in collective action, there was no reason for them *not* to do so, even for minimal gains.[47] He ignores the outlawing of strikes as a possible reason for a decline in labor protests. Moreover, this study will invalidate his claim that collective labor action ceased in Egypt after 1952. As the vignettes at the outset of the chapter show, workers were moved to collective protest, though often outside of the official union channels and seldom involving actual work stoppages. The pattern of protest, however, was countercyclical; they were scarce when real wages were rising, while occurring most frequently in the opposite situation. This pattern defies neoclassical logic and the arguments based on it.

Many labor protests in Egypt have been precipitated by cutbacks in wages or other benefits. In rallying to defend their accustomed earnings while not seeking opportunistic gains, Egyptian workers exhibit behavior similar to what Tversky and Kahneman describe, as part of their "prospect theory," as "loss averse."[48] In prospect theory individuals do weigh utilities, but according to a situational frame of reference; loss aversion suggests a "frame" in which a good already possessed is deemed more valuable than an equivalent item which is unowned. Situational framing is consistent with the idea that behavior may be shaped by nonmaterial utilities such as norms and values, and in particular with workers' expectations in a moral economy.[49] Tversky and Kahneman maintain, and Jon Elster concurs, that this behavior is irrational, since it means that preferences are not consistent.[50]

Other rational choice theorists would relax the requirement of preference "invariance," allowing for loss-averse behavior to be considered rational.[51] In Egypt's case, however, we must also consider the risks which workers faced when engaging in collective action to prevent losses. Though jobs have been relatively secure for many workers in Egypt beginning in the 1960s, protesting workers risked imprisonment and/or physical abuse, sometimes even death. To believe that workers rationally chose to face these risks, we must accept that they dispassionately contemplated the likelihood of receiving different forms of punishment, and their degree of "disutility," and then concluded that the anticipated gains were more valuable. Ideologically motivated activists who place a high value on building a workers' movement may in fact have done this. But for the hundreds, sometimes thousands, of ordinary workers who participate in such actions, this hardly seems likely.

Alternatively, ordinary workers who participated in protest may have calmly discounted the possibility of punishment. The fact that workers eschewed actual work stoppages in favor of more symbolic protests like paycheck boycotts[52] and factory occupations can certainly be cast as a rational punishment avoidance strategy. Nevertheless, as we will see here, such actions were not spared repression. Thus, I argue here that a more compelling reason for symbolic protests is workers' own commitment to their productivity, as part of the moral economy. In this sense, protesting workers can be simultaneously with and against an authoritarian but developmentalist state.

I hold that workers' participation in labor protests under conditions of repression is not the result of any dispassionate calculation of costs and benefits, but rather reflects a *suspension* of rationality, caused by anger. Anger is a passionate emotion which for some individuals, at high enough levels, demands venting. It enables workers to surmount the fear of punishment by temporarily displacing it. The release of the physical tension caused by anger, and the feeling of satisfaction derived by standing up to the authorities, do give workers "psychic utilities" which, arguably, may balance out the expected disutility of meeting repression. But my point here is that protests sparked by anger are spontaneous, not carefully planned, and that most participants join on the spur of the moment, without stopping to evaluate the potential consequences. In claiming that such protests represent nonrational acts I in no way imply that the workers involved are crazy, only that their behavior in this instance does not conform to the theoretical expectation of carefully calculated activity.

The anger results from the actual or threatened failure of the government to provide the entitlements workers have come to expect under the moral economy. Having met their side of the bargain, workers feel betrayed. My claim here is not that anger will always result in collective action—it is not a *sufficient* condition—nor that moral economies are present wherever spontaneous labor protest does occur. Rather, the broader comparative argument of this study is that anger, however triggered, is a *necessary* condition for collective labor protest under repressive conditions.[53]

Moral economy explanations also conflict with orthodox Marxism. Marxism posits that workers, as they develop class consciousness, will increasingly struggle to renegotiate and redefine the terms of their exploitation. Moral economy, in contrast, sees workers moved to protest in order to restore these arrangements when they have been disturbed. However, moral economy theories can intersect with neo-Marxist views which emphasize the role of cultural and dominant class ideologies in shaping workers' attitudes, delaying or offsetting their development of class consciousness.[54] Beinin and Lockman's study of Egyptian workers in the first part of this century was written in this vein. Questioning why the high level of labor militancy after World War II did not lead to a leftist government, they propose that the nationalist currents running through the anticolonial movement diverted workers from a strictly class-based consciousness. As explained above, such nationalism is a component of the moral economy I posit. Still, from a Marxist perspective one must question why, 50 years after the critical juncture that Beinin and Lockman identify, nationalism was still trumping class consciousness as workers struggled only to maintain a status quo that gave them little power.

While cultural theories point usefully to the influence of elite-promoted philosophies on subaltern behaviors, I reject the extreme form of postmodernism which extends this to the virtual denial of subaltern agency.[55] The essence of Marxist materialism, upheld here, is that visions can only originate, and then find widespread resonance, when they correspond in some manner to underlying material conditions. Thus, on their own, elites do not have the ability to successfully propagate any ideology at all. In the Egyptian context, the government's propaganda campaign extolling laissez-faire capitalism runs up against the reality of workers' negative experiences with the private sector as well as the deleterious consequences for the lower classes of the regime's limited retreat from the economy.

Structuralist variants of neo-Marxism can complement a moral economy perspective. Among other things, structural views emphasize distinctions within the working class based on the varying production processes, and relations to market forces, of different industries. Goldberg's original study of Egyptian labor before the coup took this approach.[56] The moral economy argument made here is at a high level of aggregation, necessitated by research conditions in Egypt (see below), which does not distinguish between sectors of the working class. It may well be that such structural factors help to explain cross-industrial differences in workers' responses to government policies.

The organizing efforts of Marxists themselves, and the political "line" which they apply in this work, also appear relevant to these differences, as well as to intraindustrial variations in workers' behavior. In particular, in Egypt there is evidence that higher levels of militancy obtained in plants where Marxists were active, suggesting that leftist influence somehow rendered workers more willing to confront the state. This does not imply that protests in these plants went beyond the restorative nature described above. In fact, much of the left in the period under study, as we will see, embraced and cultivated the beliefs about reciprocal rights and obligations posited here. Marxist organizing apparently served to reinforce and heighten anger over violated norms, and/or to promote feelings of class solidarity which helped to transform this anger into militancy.[57]

Union Leaders and Economic Issues in State Corporatist Systems

Protests that are generated by anger tend to be spontaneous and informal, i.e., they are not the result of planning and decision-making by official union organizations. Occurring at individual plants, such incidents can nevertheless affect national economic strategy by pushing up wages, thereby disrupting stabilization programs, or by impeding public sector rationalization or privatization. Spontaneous protest can also take on direct national dimensions, as in the case of Egypt's January 1977 riots against subsidy modifications.

Workers' anger (or its absence) can also influence policies indirectly, through the activities of unions. The extent to which this occurs hinges on two variables: the organizational capacity of unions to respond to different types of issues, and the degree to which union leaders respond to pressures from below. Here I argue that under state corporatism, senior unionists will react most strongly to national-level issues which anger broad numbers of workers. In contrast, local-level issues will seldom be championed by leaders of centralized union structures.

Analytically, labor influence on economic policy making can be achieved in two ways, which I label voice and veto. Voice refers to the policy formulation stage, when representatives of workers (who may or may not be union officials) are included in consultations. It can occur at the national level, when workers' representatives sit on government bodies debating such issues as subsidy reform or trade, investment and minimum wage laws. It can also be manifest at the plant or industrial level, when labor leaders sit on bi- or tri-partite bodies to set policy on wages and working conditions.[58] In the context of resistance to reform, the successful use of voice would mean that labor opposition is responsible for the withdrawal or alteration of an elite initiative for change.[59]

If workers can realize their demands through voice, it is presumably because elites fear that labor can disrupt production, or disturb political stability more generally. This potential can become reality when efforts to influence decisions through voice fail, or when labor representatives are denied access to policy makers in the first place. Veto attempts occur when labor actors object to a given policy decision and seek to revise it through protest actions such as strikes, demonstrations, and boycotts.

Both voice and veto can be exercised outside of formal union channels. Where unions are involved in these efforts their success will reflect, in part, the financial and organizational capabilities of the organization. But the successful union exercise of either voice or veto also requires that those empowered to represent workers and/or mobilize union members and resources have the *will* to do so. Where there is will without capacity, the result will be a defeat for labor. However, when capacity exists without will, there is no conflict at the outset. What, then, determines the will of union leaders?

Institutional analysis emphasizes how organizational structures limit the alternatives open to their participants and condition their strategies.[60] Above I concurred that the structure of the union movement, intersected with the economic policy-making structure, shapes the options for action available to unionists in response to various reform initiatives. Given these constraints, their position in the institutional hierarchy is one factor affecting the attitude of unionists toward different issues. However, the latter is also a function of the motivations and identities which unionists bring to the institutional environment.[61]

To the degree that the above discussion of workers' behavior posits nonrationality, the same claim is not made here for union leaders. Workers occupy their positions because they have no choice; they lack the educational background and/or family connections that it would require to obtain more desirable employment.[62] It is the feelings of near-helplessness which this situation engenders that facilitates the development of a shared sense of oppression and a spirit of solidarity among workers. By contrast, unionists at all levels are there because they consciously elected to solicit a leadership position. It is therefore reasonable to assume that they will act in accordance with the same set of preferences which led them to seek union office.

Individuals may solicit union office for a variety of reasons. At the broadest level, we can distinguish between those seeking power or material rewards and those possessing a genuine commitment to mobilizing workers and representing their perceived interests. I refer to the former as opportunists, and the latter as "sincere" unionists. Opportunists thus meet the neoclassical criterion of self-interested behavior, while the altruism of sincere unionists fits only the broader definitions of rationality. My argument here is that the motivations of unionists influence the manner by which they seek to achieve and maintain incumbency and their response to the different issues confronting workers.

Unless they can successfully conceal their motives from the rank the file, opportunists are likely to need the financial backing, and/or electoral interference, of company elites or low-level government officials in order to obtain local office. They will appeal to these forces on the basis of a willingness to support the patrons' goals. With the more intensive government screening of candidates for the upper layers of the union hierarchy, the patronage or at least acquiescence of regime leaders themselves becomes necessary for advancement to, and tenure, there.

The category of sincere unionists includes, but is not limited to, those who approach union work from an ideological perspective. This group consists, first and foremost, of members of the Marxist left. A strong commitment to a moral economy may itself develop into a broader ideology. Egypt's moral economy, as we have seen, became a component of Nasirism. As a version of an "organic statist" philosophy which subordinates the needs of all societal groups to those of the state,[63] Nasirism resembled Latin American populism. It differed in its incorporation of socialist ideas, which were associated with the development of a vast public sector granting workers representation in management and profit sharing. Because of this, and of Nasir's anti-Western rhetoric and foreign policy in the 1960s, Nasirism is commonly considered, and treated here, as a leftist ideology. Unlike Marxism, however, it is linked to nationalism, and does not preach the abolition of either capitalism or the state.

Religious forces may also sometimes be included among ideologically motivated unionists. In Korea, Christian groups took to mobilizing among workers as a vehicle for weakening that country's authoritarian government.[64] Although Islamist forces in Egypt largely ignored labor issues for most of the period under study here, some did begin to contest union elections in the late 1980s, and campaigned in the 1990s against the regime's structural adjustment program.

In contrast to opportunists, sincere unionists are likely to campaign amongst the rank and file on an issue-based platform. If ideological forces opposed to the regime in power succeed in winning union office, it suggests an ability to effectively overcome government intervention in the union election process. This is most likely to occur at the local level, where union leaders have a lower visibility and little or no contact with regime elites, and where workers can more closely supervise the election results. The state-centered literature tends to presume the immobilization of leftist labor activists due to repression. But leftists are generally able to function, albeit at reduced levels, under such conditions; the secretive organization of communist groups, in particular, is designed to enable at least some of their members to evade even the most intense persecution. The dedication of leftists, along with the organizational skills they learn in their groups, contributes to their ability to continue having an effect among workers, even if it is very circumscribed, under repressive conditions. There may also be legal avenues available which the left can use to counter their arrests or exclusions from union elections; Skidmore noted that the courts were a first line of defense for Brazilian workers during the worst periods of repression there,[65] and this has increasingly been true in Egypt as well. In unusual circumstances leftists may also obtain even high union office with the support or toleration of the government, either for the aura of democracy it provides the regime, or as a way of dividing the union hierarchy. In Egypt, Sadat facilitated the entry of Marxists into the confederation in 1971 to use them against their, and his, Nasirist rivals.

Individuals who embrace a regime's own legitimating ideology, such as Egypt's Nasirists in the 1960s, may gain office by either technique. However, if their main support comes from above, their incumbency may be threatened by new rulers after regime change. This was the challenge faced by the Nasirist unionists after Sadat came to power.

More generally, Sadat's ascendancy highlights the dilemma faced by union leaders in a state corporatist system when the government embraces economic liberalization. They will be caught between the anger of the rank and file over policies which adversely affect labor, and pressures from the regime to preempt or suppress outbreaks of militancy. In a system where the government interferes in the leadership selection process, and represses labor protest, the price unionists might pay for supporting their membership's concerns could be loss of their positions and/or jail sentences or physical punishment. At the same time, however, failure to take up rank-and-file demands could cause a loss of legitimacy at the base.

This is a particular dilemma for the opportunist unionists. The insulation from electoral pressures granted them by government interference renders the opportunists more concerned about pleasing their patrons than their membership. Nevertheless, union leaders do face some need to maintain legitimacy in the eyes of the rank and file if they are be successful at mobilizing or demobilizing workers in accordance with the regime's desires. A proven inability to accomplish this will make them less valuable to the elites, and therefore vulnerable to a withdrawal of government support for their incumbency in the future.

How any individual opportunist will resolve this dilemma around a given issue cannot be predicted. Likewise, the position that sincere unionists will take on an issue cannot be predetermined. Ideologically motivated unionists will be guided by the collective decisions of their political groups; independents may disagree over how the interests of their constituents are best served. For these reasons, there must always be some uncertainty over how unions will respond to different policy issues, and any in-depth analysis of a country's labor/state relations must necessarily be historically based and somewhat ideographic. Nevertheless, it is possible to identify a pattern that is likely to obtain, as follows:

The issues around which the legitimacy dilemma will be greatest for senior unionists are those which affect entire industries, or the working class as a whole. Federation and confederation leaders should have voice around these issues, and are in the position to mobilize the full constituency that is affected by them. Industrial and national level policies, and the positions on them taken by senior union personnel, are also more likely to be reported by the press and hence widely known among the base. Sincere lower-level officials may also seek to mobilize the rank and file around industrial or national policy issues. However, as explained above, local leaders in fact have little scope for action around them.

The previous section suggested that rank-and-file pressures on unionists around industrial and national issues will be greatest for those policy changes which generate anger. It is therefore around these issues that opportunistic senior union leaders are most likely to use voice, or the threat of veto, to resist reforms. Sincere unionists would likely weigh in on these issues as well, because of ideology and/or genuine concern to represent the base. In Egypt's moral economy, such issues would include subsidy removal, privatization schemes, and any reduction in the component of wages determined at the national level.

The legitimacy of opportunist senior unionists will be less at stake, however, around reform issues that do not arouse widespread anger from below. On issues such as these, opportunists would have less to fear from supporting the regime. There may be ideologically inspired opposition from sincere unionists at the lower levels, but the capacity of these individuals to affect change at the national level would be limited. In Egypt's case, regulation of foreign trade and investment, in particular, would appear to fall outside the scope of the moral economy. Exchange rate reform could also be exempted from labor opposition, provided that measures are enacted to keep wages apace with inflation.

The pressures on local-level leaders of all types will be greatest around those issues particular to the individual plant. This concerns the plant-based components of wages, and working conditions; privatization of single factories is also included here. For federation and confederation leaders, however, the workers involved in any individual plant protest represent a much smaller fraction of their constituency. This is especially true in cases where, because of press censorship, the plant protests and the position of union leaders towards them are not publicized, so that few workers outside of those immediately involved are even aware of the situation. Upper-level unionists, therefore, are unlikely to view support for plant-level issues as essential to their legitimacy, and will be prone to ignore them or actively seek to suppress them. While Marxists can be expected to support, if not initiate, struggle around such issues, nationalists or opportunists who feel secure in their electoral prospects could be indifferent or hostile. In Egypt's case, it is here that the ideological differences between Marxists and Nasirists became clearest.

The implication of this model for the relationship between labor organization and economic issues should now be obvious. Given a singular, centralized union movement, labor opposition to reform will be strongest around those issues which are formulated at the national level and which affect and anger large segments of the working class at the same time. This is true because such issues combine the maximum organizational capacity of unions to respond with the greatest rank-and-file pressures on opportunistic senior unionists, and also draw on the ideological or other commitments of sincere unionists.

At the same time, local-level issues under state corporatist regimes will only rarely be encouraged or supported by opportunistic federation and confederation leaders, even when they arouse high levels of rank-and-file anger. Under the authoritarian conditions which constrain the locals and make it difficult for individuals committed to worker mobilization to obtain union office, a centralized union structure is at best irrelevant, and at worst an impediment, to these grievances. In fact, as this study will show, virtually all struggles around local issues in Egypt occurred outside of the official union structure. The informal nature of these protests, and the lack of resources available to their leaders, then makes them particularly vulnerable to rapid dispersal in the face of government repression.

Yet, as an economy becomes more decentralized, the outcomes of such struggles at individual plants will have a greater impact on the wages and employment conditions faced by workers. In the Egyptian case, therefore, it may be in the best interest of workers, should the proposed new labor law be enacted and the pace of privatization pick up with no change in the authoritarian nature of the regime, to fight for transforming the existing trade union confederation into an organizational structure which gives greater scope to local-level initiative.

Such changes were being increasingly contemplated by labor activists in the mid-1990s as the impasse over the new labor law wore on.[66] But can they have any impact on the laws governing union behavior? While state-centered theory suggests otherwise, I argue for the contingent ability of societal forces to shape the institutions of labor/state relations.

Union Leaders and Union Law: Institutional Stickiness vs. Change

The preceding sections have highlighted the ways in which the institutional features of union organization and union/state relations determine labor's capacity to respond to different issues and help to shape the incentive structures of union leaders at different levels in the hierarchy. This study also lends support to the argument of institutional theorists that organizational structures, once established, become sticky, shaping actors' preferences and interests in a manner that favors the continuity of these institutions. Nevertheless, I also argue here, more in keeping with what Locke and Thelen have labeled "political constructionism,"[67] that these structures form an arena of political struggle within the labor movement in which forces promoting alternatives to the existing system contend with those seeking to preserve the status quo. Even after defeat, these alternatives remain on the agenda, and are capable of resurfacing and triumphing at certain critical junctures.[68] The transformation of Egypt's economy underway in the 1990s, and especially the tradeoffs incumbent in the proposed new labor law, constitute such an opportunity.

Both the bureaucratic-authoritarian and NIDL theories discussed above posit that union regulations are imposed on labor by the state. In a critique of the Latin American literature, the Colliers have argued convincingly that union legislation in fact results from bargaining between union leaders and the government. It is subject to renegotiation, although the impetus for change, in their analysis, still comes primarily from the state.[69] While not downplaying the salience of government initiatives in this regard, I emphasize here how the basis for the discursive survival and possible triumph of alternatives to corporatism can come from internal contradictions inherent in hierarchical union structures and government interference in union leadership selection, and from the patient organizing of leftists both within and outside of the union hierarchy.[70]

The Colliers propose that corporatist regulations can serve either as inducements for union leaders to cooperate with the government or as constraints on their ability to challenge it. In their view, inducements are those provisions, such as mandatory membership or dues check-off, which strengthen the demand-making capacity of unions. Constraints, such as restrictions on strikes or requirements for government screening of union leaders, limit the legal avenues for union leaders to pressure the state. Labor/state negotiating over union law thus consists of the government bartering various inducements in exchange for union leaders acquiescence to some constraints.

While I concur with this vision of bargaining between union leaders and the regime in power, my claim is that what constitutes an inducement versus a constraint is likely to be a source of controversy within both the union movement and the government. Sincere unionists do not necessarily agree on the best way to strengthen labor. Moreover, opportunists will evaluate union regulations in part by a different and possibly competing yardstick, namely how they affect career advancement and incumbency. Likewise, various forces within the regime may have different perspectives on the best way to control labor. These potential disagreements mean that union law will result from complex political processes whose outcome is contingent.

For example, electoral interference gives the government the right to prevent those whose ideologies make them likely opponents of orthodoxy from obtaining or retaining union office. The Colliers present it as a constraint on demand-making, and certainly the forces within the union movement who seek to sincerely represent workers' interests are likely to resist government screening of union leaders. But government interference can serve to insulate senior unionists from rank-and-file pressures, ensuring them of incumbency even when they face indifference or hostility from below. It can therefore serve as an inducement for opportunists to cooperate with the regime.

Regulations requiring hierarchy and centralization determine how much freedom union leaders at the local and federation level have to respond to issues that affect only their immediate constituencies. These provisions are seen by the Colliers as an inducement for union leaders, and as I argued above, they do enhance the capacity of labor to respond to national and, to a lesser extent, industrial issues. However, it is only senior unionists whose power and prestige are enhanced by hierarchy; their subordinates are in fact hamstrung by such provisions. Ideologically motivated unionists who place a high value on the plant-based struggles which hierarchy restricts may challenge corporatism on this basis. At the same time, opportunist subordinates may come to prefer an expansion of the power of lower-level leaders, especially if their prospects for rising to a higher post appear limited.

Requirements for mandatory membership and a single union in each industry also have a dual character. These provisions do serve as inducements, since they increase the membership base and the financial resources of the labor organization. But they also enhance the perks associated with union leadership which appeal to opportunists at all levels. In Egypt, many ideologically motivated leaders have supported these corporatist provisions because they facilitate the veto threat. However, some leftists have argued that depriving workers of the right *not* to join a union, and disallowing competition between workers' organizations, make co-optation of the leadership and their neglect of rank-and-file concerns more likely.

Under most circumstances, unionists and labor activists seeking to reform the corporatist system are unlikely to win support from the rank and file. Ordinary workers will be primarily concerned with their economic demands, and will not become directly involved in organizational issues.[71] Workers may, if given the opportunity, vote for a dissident candidate in union elections; they may also form or join organizations outside of the official union structure to pursue their demands. But given the enormity of the task, pushing for structural change requires a higher level of both understanding and *ongoing* commitment from workers than one-time, or even occasional, participation in collective protest. Indeed, it is in this realm that dispassionate calculations of (long-term) costs and benefits are more likely to guide workers' behavior than emotions. Workers persuaded that it is in their interests to work for union reform will probably be those who have already joined leftist groups or leftist-sponsored rank-and-file associations.

Thus, unlike the case of opposition to economic reforms which arouse widespread anger from below, unionists pushing to modify corporatism are unlikely to have a mass base of support with which to threaten the regime. But such support may be unnecessary to win change, given disagreements over union legislative issues with the regime itself. Coalitions between leftists and disgruntled opportunists can be sufficient to exploit these divisions, especially when demands for structural change are associated with periods of rank-and-file militancy. In Egypt, such a coalition was able to beat back efforts to tighten hierarchical controls during an upsurge of struggle in the 1970s. In fact, proposals to split the confederation surfaced at that time, and reappeared during a wave of wildcat protest in the late 1980s. The pressures for privatization, and the "right to strike for right to fire" tradeoff being discussed publicly in the 1990s, gave new impetus to forces pushing to restructure the union movement.

Featured in chapters 1 and 2, my analysis of labor/state interactions over union law thus affords a significant role to leftist and especially Marxist forces in the union movement. Precisely because they are usually the primary targets of repression, leftists will tend to struggle hardest against government interference in union elections, and those committed to worker empowerment will also challenge restrictions on local freedoms. To be sure, this commitment to union democracy is not absolute. Ideologically motivated leaders may not object to regime supervision when the government itself meets with their approval, and the Leninist operating principles adopted by many leftist groups may cause them to see hierarchical decision making in a positive light. But leftist perspectives must be investigated, not presumed, and researchers ignore the left's potential to influence developments at their peril. Bianchi's account of the institutional history of Egypt's trade union confederation is both distorted and incomplete because he assumed the irrelevance of the left.[72]

Bringing the State Back In

The discussion thus far has centered on ways in which processes inherent in the labor movement can generate pressures on the regime either in support of, or in opposition to, economic policy initiatives or change in the institutions of union/state relations. My broad intent here, in the face of the recent predominance of state-centered scholarship, has been to reestablish the need for investigation into the role of societal forces. But despite its societal thrust, this study can shed light on matters of concern to state-centered scholarship, namely, both the will and capacity of regimes to initiate and carry through on policies. My argument here is that both state and society must be considered for policy outcomes, with a focus here on economic and labor-organizational ones, to be understood.[73]

This study does highlight problems of the state's capacity to implement policies. In Egypt, a central dilemma appears to be the inability of the president and his loyalists to prevent obstruction of their initiatives by dissenting subordinates, a problem which reinforces the need for disaggregation of the state in scholarly studies. This issue emerges here primarily in the 1950s and 1960s, in the form of resistance by some elites to restrictions on private sector prerogatives, as well as to labor incorporation. The primary consequences of their obstruction were a tradition of unpunished private sector violations of labor laws and the departure of Egyptian trade union practices from some common components of corporatism. Both of these had a significant impact on the nature of labor opposition to economic reforms in subsequent decades.

On the issue of regime will, the central proposition here is that a government's propensity to initiate reforms is shaped in part by perceptions of the likelihood of labor opposition. E. P. Thompson has suggested that the moral economy of the English "crowd" limited the flexibility of that country's rulers.[74] This should be true *a fortiori* when the government itself represents the target of workers' entitlement expectations.

What mediates the government's decision-making is concern for its own legitimacy. Adam Przeworski argues provocatively that regimes do not really need legitimacy to survive.[75] What matters in policy outcomes, however, is not the objective reality but the subjective interpretation of it by decision makers. If ruling elites *perceive* that legitimacy is necessary for their tenure, they will attempt to preserve it.[76] By the same token, rulers may miscalculate the effects of the policies they contemplate. Policy makers concerned about their legitimacy will shy away from reforms which they expect to generate widespread opposition, or initiate policies which they believe will ameliorate it; the actual or potential public response may well differ from their perceptions.[77] This can explain the prolonged reluctance of the Mubarak government to undertake liberalization measures such as exchange rate reform which, when finally implemented, generated virtually no opposition.

In cases where reform attempts have been initiated and do meet labor resistance, ruling elites have numerous options for response: they can simply ignore the opposition, make concessions to it, attempt to repress it, or some combination of the latter two. Which path will be chosen will depend on a variety of factors inherent to the state. Concerns for legitimacy point toward concession, but do not exclude repression.

The model of labor behavior presented here suggests that in Egypt, given its combination of state corporatism and workers' belief in a moral economy, labor opposition would be strongest, and broadest, around perceived violations of the moral economy which occur at the national level. These issues would antagonize broad numbers of ordinary workers, and thereby create the largest legitimacy dilemma for union leaders. Resistance is likely to take the form of senior unionists trying to prevent policy enactment through voice or the threat of veto; if these fail, national protests could erupt. At the same time, reform policies which fall outside the moral economy will generate the least legitimacy problems for the regime. Labor opposition to these policies would emanate mainly from the ranks of ideological union leaders, who cannot make an effective veto threat without the backing of the rank and file.

This study will show variations in the success of government efforts to initiate different reforms at the national level which do appear related to these differences in the degree of labor opposition. As suggested at the outset, through the late 1980s at least, the labor movement was particularly effective in defeating "one fell swoop" privatization schemes which would have shifted millions of workers simultaneously from the public to the private sector. Labor opposition was also instrumental in modifying plans for subsidy reform in the 1970s and, to a lesser extent, the 1980s. Where labor opposition was weakest—in regard to the opening of foreign trade and investment—there was also no evidence concessions from the state.

Conversely, for the reasons explained above, union support for struggles against takeaways[78] at individual plants would be unlikely, especially at the higher levels. The absence of official backing for plant-level concerns does not mean that there would be no resistance around these issues, but does imply that protests, if they erupt, would likely be spontaneous. They would also mostly remain localized; without the organizational assistance of a broad-based organization, there is little potential for solidarity actions to develop at other plants. Although there are obvious differences according to plant size and location, local struggles pose less of a legitimacy dilemma for ruling elites, especially when press censorship can prevent word of the protest from spreading. Individual plant struggles are also easier to suppress than industry-wide or national level protests.

We will see here that Egyptian workers' struggles at individual plants did win concessions, but these were rarely complete. At the same time, virtually all protests at individual plants were quickly repressed. This carrot and stick combination appeared aimed at keeping local struggles under wraps and short-lived, so that the regime could continue to project an image of widespread popularity. It is here that the benefits to the rulers of state corporatism, which contributed to the isolation of local protests, become clear.

The maintenance of state corporatism, however, is part of the authoritarianism of the government. Genuine political liberalization would necessarily entail an end to the screening of union leaders and their co-optation into the ruling party, as well as the lifting of the ban on strikes and other legal constraints on union activities. Because these measures have their corollaries in similar restrictions on the activities of professional associations, opposition parties, and mass organizations, liberalizing union laws threatens to unravel the entire repressive apparatus of the regime. At the same time, as we have seen here, economic liberalization increasingly undermined the philosophical rationale for labor repression. For these reasons, during the 1990s pursuing privatization without democratization seemed to further erode the legitimacy of the Mubarak government.

Researching Egyptian Labor

To gather the information presented in the empirical chapters which follow, I conducted extensive research in the archives of more than ten Egyptian publications. Briefly described at the beginning of the Arabic bibliography, the list included both government-owned and opposition newspapers, the journals of the Egyptian Trade Union Federation (ETUF) and the Ministry of Labor, and several leftist periodicals. In addition, I monitored the major English-language periodicals which focus on Egyptian economic developments over the 1984–1996 period. I supplemented the information gleaned from these sources by interviews with dozens of trade union officials, labor journalists and historians, several officials at the Ministry of labor, left-wing activists in the unions, and ordinary workers. These interviews were conducted initially during 1987–88, with some follow-up at the end of 1994.

The project of researching labor issues in Egypt is surrounded by political sensitivities. One of the major ways these affected my research is reflected in its "macrolevel" approach to the discussion of workers' attitudes. As noted above, some of the more interesting questions about workers' consciousness revolve around how it is affected by structural differences between industries. To explore such questions in the contemporary context would require intensive study of a small number of selected plants, including canvassing their workers. While I had originally hoped to conduct such research, I quickly learned that it was virtually impossible for a foreigner to obtain the necessary government permission to randomly survey workers, or indeed any ordinary Egyptians. In fact, I was advised by numerous Egyptian scholars that merely to request such permission would invite constant government surveillance and possible obstruction of other aspects of my research. Thus, I abandoned this aspect of my agenda.

Even at the macrolevel, political conditions in Egypt affected the testing of theories of workers' and unionists' behavior. For example, labor market economists frequently seek to establish correlations between strike activity and macroeconomic indicators. For Egypt there was no reliable time series on strike frequency available. The statistics published in the annual *Yearbook of Labor Statistics* are supplied by the government and appear to reflect the fact that strikes are illegal and officially frowned upon;[79] no incidents at all were reported from 1960 to 1968, for example, despite documented evidence to the contrary. Furthermore, as already mentioned, one manifestation of the moral economy is that most of the protests by Egyptian workers have not involved actual work stoppages. Hence they would not be reflected even in accurate data on strikes.

In light of these deficiencies in the official statistics, I relied instead on other sources, primarily press accounts, for my information about labor protests. Still, only the largest incidents are ever mentioned—and then only disparagingly—in the government-run newspapers. As discussed above, this censorship appears intended to prevent the spread of wildcat protests and deny any evidence of popular discontent with the government. In addition, reflecting the union hierarchy's neglect of local-level issues discussed above, the ETUF newspaper as well ignored all but the largest labor protests, and I found most senior unionists reluctant to acknowledge or discuss them beyond uttering general condemnations.

For the most part, then, labor protests were covered only in the leftist press; this reflects the Marxists' historic orientation towards organizing among workers and adopting their concerns. I supplemented the available press accounts by interviewing leftists active in organizing among workers, whose acquaintance I made through the official left-wing party known as the Tagammu' (see chapter 2). There were obvious sensitivities involved here, since admitting to playing a role in recent protests could subject an activist to imprisonment and possibly torture. Much information about events in the 1950s through 1970s was provided to me by individuals eager to see labor's untold story brought to light. But to protect myself and my sources, I never asked about their involvement in then-current events, or possible membership in any of the illegal Marxist groups who work with the Tagammu.' For similar reasons, the identity of some of these sources is kept confidential here.

Because these sources are indispensable to the study of workers' role in Egyptian political economy, the neglect of labor in many works on Egypt suggests their authors' distrust of, if not disdain for, the left. The same is true of those studies which limit their treatment of labor to the activities of the government-affiliated union leadership. At the same time, a reliance on leftist sources is not unproblematic. Several limitations to the data I obtained in this manner should be mentioned.

Because worker activism can be seen to vindicate the Marxists' faith in inevitable workers' revolution, the obvious danger is that leftists will exaggerate its prevalence. Thus, leftist accounts may inaccurately portray the number of incidents which occur and/or the number of workers who participate in them. Marxists might also, in an effort to ignite further protests, embellish the demands of workers and/or the degree to which these were realized. Where there is competition amongst Marxist groups, each may seek to glorify events in which their members were involved. Finally, to heighten popular anger at the state, Marxists might magnify the amount of repression that was used against workers.

At the same time, there are ways in which leftist sources may actually understate the level of labor activism. For one thing, during periods where opposition forces are not permitted to publish at all, incidents may occur which go uncovered. In addition, even at its most accurate, leftist reporting is limited by the reach of their organizations. Protests which erupt in areas where there are no leftist forces may thus remain unreported. Finally, given the multiplicity of interpretations of Marxism, whether and how spontaneous and informal protests will be covered in the leftist press hinges upon the particular ideological perspective of the groups in question. In Egypt's case, the forces dominant in the Tagammu' were supportive of the Nasir regime in the 1960s, and became loathe to challenge Mubarak in the 1980s, fearing that destabilizing his government could facilitate a coup by more right-wing forces in the military, or a takeover by Islamic fundamentalists. One manifestation of this was the absence of coverage in the Tagammu's newspaper of certain spontaneous labor and other protests.

I attempted to minimize the possibilities of data distortion by interviewing a wide range of activists representing various and sometimes rival tendencies. I checked their accounts against each other, and against printed stories, where available. I also consulted individuals outside of the left who were knowledgeable about labor affairs. In this manner I could gain from one source information that was concealed by, or unavailable from, another. This cross-checking method was used to construct the accounts presented here about labor protests as well as conflicts over union/state relations.

I also interviewed about thirty ordinary workers. These included employees of engineering and textile factories, public bus and privately owned van drivers, and skilled construction workers. For the reasons specified above, the workers I spoke with were not randomly selected; rather the discussions were arranged through various contacts. This type of arrangement, and Egyptian culture generally, dictated the nature of these interviews: they were generally conducted in groups at the home of one of the workers, and frequently joined by relatives, neighbors, and coworkers. Thus the conversations were necessarily free-flowing and open-ended.[80] For the reasons specified above, I refrained from asking workers about their political views or possible participation in protests. Instead I focused on their attitudes towards their earnings and working conditions, and particularly on their feelings about the differences between public and private sector employment and the effects of economic liberalization on their lives.

Besides the political problems, Egypt's relatively low level of economic and technological development also affected my research. Accurate economic data, needed to correlate protest behavior with economic indicators, is notoriously difficult to come by in developing countries. Here I use the real wage index (RWI) to indicate changes in workers' earning power. This must be derived from nominal wage statistics, and there are a number of different official sources of nominal wage data in Egypt, each reporting different information. In particular, the Labor Force Sample Survey, Population Census, and Survey of Employment, Wages, and Hours of Work all differ in the timing and method of data collection, the age of employees included, and the type of establishment surveyed. Aggregate employment figures also reflect these discrepancies.[81]

Real wage calculations also hinge on the accuracy of the deflator employed. The sources cited here used either the general or the urban consumer price index (CPI), both of which are officially based on a market basket heavily weighted with domestically produced and price-controlled items. Since the mid-1970s, as consumer preferences have turned towards imports and controlled items have become more difficult to find, the CPI has increasingly understated inflation. For these reasons, I sometimes show several different RWI tables, from different sources, and acknowledge conflicting interpretations. Where the information is available, the coverage of the original nominal wage series and any limitations this poses to the analysis are reported in notes to the statistical tables.

Statistics relevant to union activity are similarly problematic. At Egypt's confederation headquarters in 1988, for example, one was hard pressed to find an electric typewriter, much less a Xerox machine, fax, or computer. Membership data was collected irregularly and recorded by pencil in a ledger. Moreover, the membership figures reported by the ETUF are for sessions ranging in length from two to four years; the exact date of data collection is not specified. There are also notable discrepancies in the figures. For example, the membership data for the 1979–83 session published by the ETUF in its 1982 report[82] differ quite substantially from those given me by the organization's membership office (3 million-odd total members versus 2.3 million). The membership secretary himself told me that his office (which was rarely staffed) accepted without question the data supplied to it by the member federations, even though some of these had reason to inflate their figures.[83]

Data on grievances filed, and their outcomes, was not systematically collected and was reported only sporadically in ETUF publications. Finally, I did not inquire into dues receipts because the past-president of the ETUF was being investigated by the government on charges of "financial irregularities" during the initial period of my research (see chapter 2). Information on union finances did not appear central to my research, and I did not wish to invite suspicions by making inquiries about them at such a sensitive time.

Outline of the Study

The empirical support for the arguments made here is organized into five chapters. Covering the period from 1952 through the early 1960s, chapter 1 shows how the formation of the Egyptian Trade Union Federation was shaped by internal contradictions within both the labor movement and the new military regime. It also traces the changes in the regime's economic strategies and their effect on workers, highlighting limitations in the government's capacity to implement its programs.

The internal dynamics of, and conflicts within, the trade union movement after confederation are the focus of chapter 2. It shows how the union movement remained intertwined with, and dependent on, the state, and reveals the multiple forms of government intervention in union affairs. It also demonstrates that in spite of the continuity of corporatism, the latter was a consistent source of internal conflict within the labor movement. Finally, it illustrates how opportunism and ideological identities among unionists come into play in shaping these conflicts and their outcomes.

Chapter 3 centers on rank-and-file struggles at individual plants and their impact on government policies. The timing and nature of these protests, as well as the demands raised during them, support the argument that workers' behavior is best explained by a moral economy perspective. By showing that these plant-level protests occurred largely outside of the union structure, the material also demonstrates how corporatism weakens the capacity of the formal workers' organizations at the local level.

The role played by union leaders in shaping national economic policy is the subject of chapter 4. It shows how the strength of union opposition to economic reform hinged on the degree to which the policies were perceived as violations of the moral economy, the number of workers simultaneously affected by the policy, and the ideological perspectives of the unionists. This combination led to defense of the public sector becoming the *raison d'être* of the union movement throughout the 1970s and 1980s. The chapter also provides evidence that regime concessions on reform initiatives appear linked to the strength and breadth of labor opposition.

Chapter 5 traces the Mubarak regime's heightened efforts to undo *etatism* in the 1990s. It reveals how the government's newfound commitment to a privatization program resulted from increasing pressures on the regime from its multilateral creditors, and in turn led to elevated government pressure on senior union personnel. This conflict between labor and the state crystallized around the proposed new labor law. I explain how and why its removal of most of the protections for workers that were established during the Nasirist era, in exchange for granting a limited right to strike, is both exacerbating the tensions within the corporatist union structure and eroding the legitimacy of the Mubarak regime among workers.

1

Corporatism and Etatism Take Shape, 1952–1964

From the 1970s onward, Egyptian governmental forces pushing a neoliberal economic agenda clashed with various segments of the country's trade union movement opposed to the reforms. Even earlier, ordinary workers began acting on their own, often unsupported if not actively impeded by the centralized and hierarchical union structure, to counter retractions of their perceived entitlements from the state. Both the etatism that the neoliberals would see undone in the name of economic efficiency, and the corporatist union arrangements that impinge on workers' ability to challenge orthodox reform policies, have their origins in the 1950s and early 1960s. The central purpose of this chapter is to trace the development of these phenomena, which form the backdrop for the labor/state and intraunion battles to come.

Comparative scholarship has revealed a frequent association between postcolonial regimes following import substitution industrialization strategies and labor incorporation. Egypt in the period under study here conforms to this pattern, but it also supports the arguments that political policies are not determined by the selection of economic strategies nor by their consequences.[1] In particular, in this chapter I seek to show how the process of labor incorporation in Egypt was gradual and tentative, its success threatened by disagreements within both the regime and the union movement. Corporatism was not the inevitable accompaniment to etatism, but rather the outcome of complex political maneuvering among and between unionists and ruling elites.

I divide the chapter into three historical periods, although the demarcation line between the first two is gradual and fuzzy rather than abrupt and distinct. The first period runs from the July 1952 Free Officers' coup to late

1954/early 1955. During this time, the officers ruled by collective decision making, and their ideological differences lend an air of incoherence to the policy outcomes. United by a desire to expand industry, they were pressured by an international recession into concentrating primarily on trying to hold onto the country's existing capital stock by reassuring investors. Maintaining political stability was necessary for this, and important in its own right to the new regime. The consequences of these goals for labor was the banning of strikes, the suppression of efforts by unionists to establish a confederation, and labor legislation which, while reflecting the ideological clashes within the regime, interacted with the economic environment to result in declining job security for workers. There is therefore little reason to believe that a majority of workers acted to support Nasir when he and other officers favoring continued military rule clashed with those pushing a return to the barracks in March 1954.

The second period, running through the beginning of 1961, is marked by the consolidation of power by Gamal 'Abd al-Nasir and his turn toward populist appeals to workers and a greater state role in the economy. During this period the trade union confederation was formed as a shell—only at the top—by union leaders working closely with the government. Despite new populist rhetoric from the regime, legislation to improve the lot of workers was slow in coming, and also obstructed in implementation by continued disagreements amongst those in lesser positions of authority. Given their preoccupation with, and ongoing competition for, support within the regime, the new confederation's leaders failed to address the economic concerns of the rank and file; dialectically, government supervision of the leadership selection process enabled them to maintain their posts and their new organizational structure while workers' interest in unions declined.

The "Socialist Decrees" of July 1961 ushered in the third and final period. The decrees marked a tremendous expansion in the size of the public sector and in the government's commitments to Egypt's workers. Nasir's calls for workers to reciprocate for these measures with extra effort on the job formally inaugurated the moral economy. The labor unions were finally permitted to consolidate a centralized and hierarchical structure, but its central role was now defined by the government as inculcating workers with the new ideology and ensuring their contributions to production. Even so, the government continued to distrust the unions, maintaining close supervision of their finances, intervening heavily in their elections, denying them the right to publish their own newspaper, and especially prohibiting them from organizing strikes.

The Free Officers and the Unions

The July 23, 1952, coup which brought the military to power in Egypt was engineered by the Free Officers, a movement of about 100 officers representing all branches of the Egyptian armed forces except the navy. Although Gamal 'Abd al-Nasir was unquestionably the founder of the group, decisions were made collectively by a nine-member executive committee which chose Muhammad Nagib, a general from outside their ranks, to be the public face of the movement. In August Nagib and four others were added to this junta, which now became the 14-member Revolutionary Command Council (RCC).

The RCC was united by its desire to oust the British from Egypt and by a vague commitment to modernization and social justice. But its members lacked a clear program for achieving these latter goals, and in fact had sharply differing ideological perspectives.[2] On the left, Khalid Muhyi al-Din and Yusuf Siddiq were affiliated with the Democratic Movement for National Liberation (commonly known by its Arabic acronym, HADITU, for al-Haraka al-Dimuqratiyya lil-Tahrir al-Watani), then the most prominent of various sects of the Egyptian communist movement,[3] while on the right, 'Abd al-Mun'im 'Amin maintained his ties with the old wealthy class and was a strong proponent of private enterprise and close ties with the United States. Kamal al-Din Husayn and Husayn al-Shafi'i were members of the Muslim Brotherhood, and Anwar Sadat maintained close ties with that group. Nasir himself had worked closely with both the Brotherhood and HADITU in building his organization and plotting the takeover.[4]

While the officers lacked a coherent strategy for accelerating the industrialization of Egypt, they were also constrained by economic factors largely beyond their control. The end of the Korean War had produced a recession in the developed countries which caused Egypt's cotton export earnings to fall. In 1952, Egypt experienced a deficit in the current account and a drop in foreign reserves. The flight of foreign capital from the country after the coup would have exacerbated the dampening effects of these phenomenon on the economy. Accordingly, representatives of the RCC visited foreign embassies to reassure them of the regime's commitment to property rights, and its distaste for socialism. They also reversed the "Egyptianization" policies of the late 1940s by raising the ceiling on foreign-held interests in a firm from 49 to 51%, thereby re-permitting foreigners to control jointly held enterprises. To encourage domestic private enterprise, the RCC increased protection for domestic industries, lowered customs duties on raw material and capital goods, and reduced business taxes. They adopted conservative fiscal and monetary policies, including a temporary decline in current expenditures.[5]

Maintaining political stability was also important to preventing capital flight. Moreover, it was a priority for the regime in its own right as well. Uncertain of its support among the masses, fearing competition from the political forces that had been active prior to the coup, and with a typical military predilection for order, the junta moved quickly to demobilize the societal forces who had been active in the nationalist movement.

For labor, a strike that broke out in the village of Kafr al-Dawwar less than a month after the Free Officers' coup set the tone for the period. On August 12, workers at the Misr Spinning and Weaving Company, one of Egypt's largest industrial establishments at the time, staged a sit-in over a series of economic and job-related issues. Troops were called in, and the violence which ensued resulted in the deaths of four workers, two soldiers, and one policeman, with many others wounded on both sides.[6]

The RCC ordered a military tribunal and tried 29 workers on charges including premeditated murder, arson, property destruction, and refusal to obey police. The presiding officer was Amin, who disallowed testimony by witnesses that would contradict the prosecution's case. Aside from his known sympathies toward the West, leftists accuse him of believing strongly that suppressing worker militancy was necessary to attract foreign investment to Egypt. Two workers, Mustafa Khamis and Muhammad al-Baqari, were convicted of being communists who had incited the incident and were sentenced to death; eleven others received penalties of jail or hard labor.[7]

The military regime's harsh response to the Kafr al-Dawwar incident caused it to lose some of its initial civilian support, prompting some of the new leaders to make special efforts to reach out to workers. 'Abd al-Nasir and Muhammad Nagib toured the industrial areas of Cairo in the fall, "stressing the army's concern with labor conditions and emphasizing the importance of sound labor legislation." Nagib implicitly apologized for the hangings in a talk to workers in Imbaba, an industrial suburb of Cairo, on November 20, pleading with them to understand that the new rulers needed time to formulate a coherent labor policy.[8] He also acknowledged his regrets in his memoirs, writing "after the execution of Khamis and al-Baqari at a time when security was in disarray, I strove to win the confidence of the peasants and workers, so that we wouldn't lose the great masses of people."[9] Nevertheless, the public execution of Khamis and al-Baqari sent a clear message to workers that the regime would not tolerate labor unrest, and two subsequent strikes in 1952 were also broken up by police, and their organizers tried in military courts, although the incidents were much less violent and the sentences far more lenient.[10]

The same concern to preserve order and preempt potential political challenges led the new regime to initiate the first in what would become a series of efforts by the military to purge communists from the unions, and to swiftly squash the efforts of some union leaders to establish a confederation. Numerous attempts at confederation had preceded the Free Officers' coup; they were largely at the initiative of communists but drew in many other unionists as well.[11] Before the coup, these efforts had culminated in the formation of the Founding Committee for a General Federation of Egyptian Trade Unions (hereafter FCGF). The FCGF leaders recognized the need to win the backing of the RCC for the formation of a confederation before the founding congress they had scheduled for mid-September. On the day after the coup Fathi Kamil, the most prominent of the non-Communist FCGF leaders, went with about a dozen other union leaders to the headquarters of the RCC to declare their support for the Free Officers. A week later, the full FCGF met and issued a public statement in support of the coup.[12]

Thereafter Ahmad Taha, a communist leader of the FCGF with ties to the leftists in the RCC, and 'Abd al-Mughni Sa'id, an official of the Ministry of Social Affairs who was supportive of the labor movement, held several secret meetings with some RCC leaders. Taha, however, reported that there was fear of the union movement among some of the coup leaders. Meanwhile, Kamil approached the new Minister of Social Affairs, Muhammad Fu'ad Galal. An expanded follow-up meeting ensued, to which Kamil brought several other unionists and Galal brought another ministry official, Amin 'Izz al-Din, whose memoirs inform much of this chapter. The RCC was represented at this meeting by Kafr al-Dawwar's hanging judge, Amin, who brought with him Sayyid Qutb, known for his conservative leanings and ties to the Muslim Brotherhood. During the meeting Amin and Qutb argued strongly that the formation of a confederation should be postponed until the union movement had purged its ranks of communists. Nevertheless, Galal in the end agreed to support the unionists, and promised to look into creating a legal basis for the confederation.[13]

Accordingly, the FCGF's leaders continued to promote the upcoming congress, and drew in new unions to represent almost 200,000 workers. Anxious to prove to the new government that the union movement did not pose a threat to the country, the unionists invited both Galal and Nagib to attend the opening session of the congress, and were careful not to criticize the regime's actions at Kafr al-Dawwar or challenge the ban on strikes. It therefore came as a surprise to the FCGF to read in the newspapers, only two days before the scheduled meeting, that it had been postponed indefinitely by the ministry.

The reason given was that many labor leaders had been elected in the prerevolutionary era when there were foreign pressures and influences on the elections, a veiled accusation that forces hostile to the regime dominated the labor movement. Unions were instructed to purge their ranks of such elements, after which the ministry would reschedule the meeting.[14]

The origins of this decision remain unclear. Although the cancellation was announced by the Ministry of Social Affairs, 'Izz al-Din blames Amin and Qutb, noting that by this time Amin had assumed responsibility for labor affairs for the RCC and taken an office in the ministry next to Galal, who fell under his complete supervision. He further states that Galal was himself known for anti-communism and pro-American views, adding that the selection of Amin and Galal for these posts was indicative of a prevailing attitude of hostility toward labor within the RCC as a whole. However, 'Izz al-Din also acknowledges that 'Abd al-Mughni Sa'id, in his own memoirs, claims that the surprise postponement was actually decided by the security forces, headquartered in the Ministry of the Interior. Sa'id indicates that the security apparatus prepared a series of reports to the RCC, each warning of a different and dangerous influence within the trade union leadership, i.e., Wafdists, communists, and Muslim Brothers. In this manner, proponents of all the different tendencies within the RCC could find some reason to be fearful of the proposed labor organization.[15]

The FCGF leaders were not prepared to challenge the regime's decision, but let their disappointment be known in a public statement that was published in the leading daily newspaper, *al-Ahram*. It read:

> The committee was the first to support the new regime, and was anxious to hold the congress to announce the formation of the confederation thinking that its existence would be one way for workers to cooperate with the regime. And we were careful to inform the relevant authorities about all the preparatory steps we took for holding the congress. Today, although the meeting has been postponed for various reasons, *the committee still believes that the confederation is the effective tool for purging the ranks of the trade union movement of, and protecting it from, those forces foreign to the workers which are contaminating it.* We have no choice but to content ourselves with the postponement, in order to demonstrate that the intentions of the workers are good, that they believe this regime is their regime, and that they share the views of the Free Officers.[16]

After this statement was issued, the Interior Ministry forbid meetings of the FCGF, and it was effectively disbanded. This was followed by further warnings to unionists to purge their ranks of communists.[17]

The regime achieved the desired results. According to Beinin and Lockman, "by the beginning of 1953 many forces favoring a trade union movement independent of the regime had been stripped of their positions in the unions, and by the end of the year virtually all of the most important communist labor leaders were in jail."[18] Although the purge was aided in no small part by heavy police infiltration of the unions where leftist influence was strong, it is important to note here that some unionists did cooperate. Their motivations in this regard cannot be firmly established. Some may have acted out of fear, or the sincere belief that compromise with the regime was the only way to advance workers' interests. However, the FCGF statement itself indicates another basis for compliance by the non-Marxist unionists. For the Marxists in the group, its reference to "forces foreign to the workers" could mean agents of both the former palace and the Wafd Party, which represented segments of Egyptian capital. But the communists themselves were vulnerable to this accusation, because the various Marxist organizations had mostly been founded by foreign workers, and contained a disproportionate number of Egyptian Jews who were increasingly classified as "foreign" as a result of the Zionist movement.[19] This provided regime members concerned about "communist infiltration" of the unions with a convenient wedge to use against the Marxists.

Toward a New Labor Policy

With the threat of an independent labor movement preempted, the regime moved to formulate a proactive labor policy. After the FCGF's congress was postponed, the RCC initiated a tripartite committee to review existing labor legislation and promulgate a new law. However, the ideological differences amongst RCC members plagued this effort. Rather than a coherent and comprehensive approach, the result was a series of contradictory compromises.

Fathi Kamil was appointed by the government to represent workers, and was permitted to bring fellow unionist 'Abd al-'Aziz Mustafa along with him. While this marked the first time that labor representatives were given a voice in policy making, the fragmented state of the unions and the ongoing repression of their most militant leaders meant that Kamil and Mustafa had little capacity to threaten the regime with action if their ideas were ignored. Further, although the project had been proposed by the RCC's leftist generals, Khalid Muhyi al-Din and Yusuf Sadiq, it was Amin who was appointed to represent the RCC on the committee. Minister of Social Affairs Galal was placed at the head of the committee, and two other officials of the Labor Department (a division of the ministry) were also included;[20] there was also one representative from the Federation of Industries, a businessmen's group.[21]

The prevailing opinion in the Labor Department at the time was to replace the multitude of laws which had been passed in the previous decades, which were a source of confusion among both the rank and file and union leaders, with a single piece of legislation. However Amin, reminding others on the committee that he spoke for the RCC, rejected this plan of action. And although Galal was nominally chair of the committee, it was Amin who in fact ran the meetings, and did so in a dictatorial manner which included prohibiting any of the participants from carrying home notes taken during the meeting. Hassan Isma'il, the head of the Labor Department, resigned from the committee because of Amin's conduct, and subsequently withdrew from the ministry.[22]

The result of Amin's leadership was that three separate labor laws, Nos. 317, 318, and 319, were issued in December 1952. These laws are often cited as evidence of the new regime's commitment to workers, or at least its attempts to regain any support lost after Kafr al-Dawwar.[23] In fact, each contained contradictory aspects for labor. I deal here with the effects of the first two laws, which concerned work contracts and labor disputes. The subsequent sections cover the application of Law 319, which involved issues of union structure and union/state relations.

The new legislation regarding employment contracts and conditions has been portrayed by others as consisting largely of concessions to trade unions, resulting in the successful co-optation of the labor movement and widespread support for the military regime among workers.[24] While such claims do describe a process that began in the latter half of the decade, they are for the regime's first two years inaccurate and premature. The legislation was the outcome of clashes and compromises between more conservative forces keen to create their vision of a favorable investment climate and leftists eager to provide workers with higher wages, more job security and better benefits; the center sought to incorporate the rightists' concerns but at the same time avoid alienating the working class. Moreover, whatever their intent, the consequences of the laws in the context of the overall economic situation was a decline in both job security and wages, thus giving workers little reason to embrace the regime.

Law 318 on Arbitration and Conciliation provided for speedier hearings and improved representation for workers in grievance cases, and mandated that employers' pay the workers' court fees in cases where the latter won their complaints. But at the same time, it made arbitration necessary in all labor disputes, thus effectively continuing the ban on strikes. The law also placed restrictions on lockouts: employers were required to notify the Ministry of

Social Affairs (hereafter MSA) in writing of "compelling reasons" for shutting down or reducing operations, and could carry out their plans only if the ministry had not ruled otherwise within 15 days. In practice, however, and perhaps because the ministry was still disorganized and understaffed, such requests were seldom reviewed.[25]

Conditions of employment were covered in the Individual Contract Act, Law 317. In concessions to the labor movement and the left-leaning members of the RCC, the law mandated increases in severance pay and annual vacations, and the provision of food and transportation to workers commuting from distant areas, in all plants with over 50 employees. Owners of plants employing more than 500 workers were further required to provide free medical care to their workers.[26] However, the law also *quadrupled* the probationary period (*fatrat al-ikhtibar*) from one three-month stint to a six-month period that could be renewed once. As probationary workers were not entitled to the same wages and benefits as permanent employees, and were commonly fired when the period ended, this provision was a significant setback to workers.[27]

The law required employers to give notice, and pay severance, to workers upon terminating their contracts. However, at the initiative of Amin, it also specified numerous violations for which workers could be fired summarily. Amin agreed to allow for fired employees to seek compensation through the courts only with the expectation that workers would not take advantage of this right, and the relevant clause (Article 39) contained no provision for the courts to reinstate workers who were fired unjustly. The labor representatives on the committee interpreted this package to be effectively granting employers the right to fire workers without cause, and went along with Amin in the mistaken belief that he represented the consensus of the RCC on this point.[28]

In fact, Khalid Muhyi al-Din was deeply opposed to granting employers this prerogative, and brought his concerns to a joint RCC/cabinet conference in March 1953, convened to discuss a draft law on encouraging foreign investment. Muhyi al-Din sought a complete ban on summary firings, but was outvoted by others present who feared that such a restriction would discourage entrepreneurship. However, when he then threatened resignation, a compromise ensued: Law 317 was modified in April 1953, to prohibit punitive firings for trade union activity, and courts were now authorized to order the rehiring with back pay of workers who proved that they had been unjustly fired for this reason. Law 318 was revised around the same time with the creation of an MSA "Committee of Factory Shutdowns;" employers were now prohibited from undertaking any mass workforce reductions without the expressed approval of the committee.[29]

The modifications to Laws 317 and 318 signify the regime's first attempts to regulate employment termination, but the new government lacked the capacity to accomplish this. When combined with the regime's broader economic policies and the unfavorable international economic environment in which they were implemented, the net effect of the new legislation was a slowdown in growth and a decline in industrial employment (see table 1.1). Arguably, the modest concessions to workers in the law contributed to capitalists' decisions to disinvest, but economists attribute the downturn to the recession in the industrialized countries after the Korean war, and the decision of the regime not to combat its domestic effects with expansionary fiscal or monetary policies.[30]

Moreover, and despite the grumblings of factory owners and managers,[31] the new government apparently also lacked the unity of will to restrict capitalists' prerogatives to shed workers individually or *en masse*. In 1953, after the anti-lockout law took effect, the government processed 131 requests from companies for permission to shut down operations; approvals caused about 13,000 workers to lose their jobs. The textile industry in particular suffered a rash of closings that year. Many employers continued to dismiss individual workers as they pleased, especially those involved in union activities or overtly sympathetic to what socially progressive pronouncements were forthcoming from the regime; the same year saw 38,309 complaints filed over broken labor contracts. The low proportion of decisions in favor of workers at the courts—averaging 35%—shows that they provided only limited relief to workers from summary dismissals.[32]

TABLE 1.1

Industrial Employment, 1952–1960

Year	Industrial Employment	% Change
1952	264,927	-
1954	263,863	-0.4
1956	253,255	-4.0
1957	268,151	5.9
1958	261,287	-2.5
1959	305,659	17.0
1960	325,166	6.4

Source: Abdel Fadil, p. 8, based on the Census of Industrial Production.

Note: Includes only establishments employing 10 or more persons. Data for 1953 and 1955 not available. Mabro and Radwan (p. 260) have a similar table drawn from the same source. Their numbers, inexplicably, vary slightly from those shown here, especially for 1954.

At the same time, the quadrupling of the probationary period exerted a downward pressure on wages and increased the vulnerability of workers to job loss. The largest industrial sector at the time was textiles, and Textile Federation statistics show a dramatic increase in the proportion of their members working under probation, from 10% in 1952 to 50% in 1956. Real wages fell in 1953, and the estimates shown in table 1.2 may in fact be more optimistic than the reality.[33] Sayyid Fa'id, then an activist in the federation, called these years "the worst period in the history of Egyptian labor." The British Labour Attache, writing in the fall of that year, concurs that the new restriction on dismissing workers had failed to reduce unemployment, and concluded that the regime was, "in spite of genuine efforts to improve their lot, generally unpopular with the mass of the workers."[34]

Refuting numerous scholars who have argued that the Free Officers' coup was motivated by their desire to promote capitalist development, Robert Vitalis maintains instead that the insurgent officers were united by antimonopolism, and failed to encourage a private market-based economy. "The post-

TABLE 1.2

Hours of Work and Real Wage Estimates, 1950–1960

Year	Hours Work/Week[a]	Weekly RWI[a]	Weekly RWI[b]
1950	50	100	111
1951	50	100	112
1952	51	120	134
1953	51	116	129
1954	52	132	146
1955	52	138	152
1956	51	137	147
1957	50	139	150
1958	52	145	155
1959	50	143	152
1960	49	141	151

[a]From Abdel Fadil, p. 33, based on the *Survey of Wages and Working Hours* (SWWH) published by CAPMAS. The SWWH includes salaries for managerial, technical, and clerical employees, which were generally higher than those of industrial workers during this period. As well, the survey is based on responses from employers, and there is no attempt to correct for variations in response level from year to year. Finally, the SWWH excludes establishments employing less than 10 persons, where wages were generally lower. Such small-scale firms accounted for at least 33% of industrial employment in the 1950s. Beinin and Lockman, p. 268; al-'Issawi, pp. 2–3; Abdel Fadil, pp. 41–42.

[b]From Mabro, p. 335. The nominal wage figures are from the ILO's *Yearbook of Labour Statistics*, but these are also drawn from the SWWH. The base year, however, is 1937. Mabro notes an upward bias in the figures after 1953.

1952 state was either unable or unwilling to supply the necessary incentives that investors had come to require if they were to act as capitalists: concessions, subsidies, protection, self-regulation, monopoly rents, *a tightly controlled if not completely hostile environment for workers*, and regular access to the top leadership of the state."[35] This contrasts markedly with Beinin's claim that the state in this period gave "preference to the interests of capital" over labor.[36] The material presented here supports the latter view. It may well be the case that Egypt's most successful entrepreneurs, domestic and foreign, were denied access to the regime and hence could not press the new rulers for the restrictions on labor described above; even the nonoligopolistic capitalists who Vitalis shows were treated more favorably by the RCC may not have lobbied for this legislation. Nevertheless, I believe there is sufficient evidence here to show that the *intent* of the government's restrictions on labor was to encourage investment. That the goal was not realized does not negate this intent, especially when economists have attributed much of the industrial disinvestment which did occur to international economic conditions beyond the government's control. The regime's failure to implement those protections for workers it did enact further bolsters Beinin's argument. In spite of the efforts of the RCC's progressive forces, the net content of the regime's policies was to privilege capital, to the detriment of Egypt's workers.[37]

The RCC versus Corporatism

Although government intervention in union affairs increased during the military regime's first years, the initial result of this interference was to promote multiple and decentralized unionism rather than a corporatist structure. At the same time, this phenomenon was as much the result of disagreements within the regime over union policy as of anyone's conscious design.

In some ways, Law 319 represented a retreat from the government's insistence on close supervision of the union movement. It repealed the right of the MSA to disband unions, transferring this to the courts, and although unions were now required to register with the ministry they no longer needed its permission to form. The committee was able to pass these clauses only because Amin was distracted by inner-RCC conflicts, and attended committee meetings only sporadically when the union laws were being discussed.[38] The law also contained some "inducement" provisions which could strengthen unions, such as requiring employers to deduct union dues from paychecks, and a closed-shop mandate, wherever three-fifths of workers voted to unionize. However, other aspects of Law 319 served as constraints. It prohibited the unions from investing any money without the permission of the MSA, and required them to maintain detailed financial records which were subject to inspection by the ministry at any time.

Governmental supervision soon took other forms as well. During 1953 the RCC outlawed political parties, and created an organization known as the Liberation Rally as a vehicle to organize mass support for the regime.[39] The Rally was effectively another arm of the new military government; all its leaders were army officers. Its Labor Bureau, which was charged with building ties with the labor movement, was headed by Major Ahmad 'Abd Allah Tu'ayma. Over time the Labor Bureau supplanted the Labor Department in the MSA as the agency for dealing with workplace and union issues, and unionists were expected to always show their loyalty to the regime by frequenting the Liberation Rally office.[40]

Also because of Amin's sporadic absence, the unionists were able to insert the legal right to confederate into Law 319. Nevertheless, the law objectively served to impede the process of centralization sought by unionists. Whereas union leaders in the previous period had been pushing to establish industrially based unions, the new law encouraged the establishment of separate unions in each individual establishment. It also provided for distinct unions for blue- and white-collar workers within an individual enterprise. It is not clear what positions were taken by the unionists and the government representatives on the committee with regard to these particular provisions. Outside of it, however, at least some union leaders were critical of these aspects of the law, arguing that they served to splinter rather than unite the labor movement. Moreover, in the ensuing years, the security apparatus and some officials within the MSA intentionally promoted multiple unionism and employer-created unions.[41] Statistics on union membership confirm that fragmentation rather than consolidation was the trend of the next few years; the total number of unions more than doubled between 1952 and 1954, but the average membership per union stayed virtually the same (see table 1.3).

The same forces hostile to unions were able to stifle application of other aspects of the law, such as dues deductions and closed shop provisions, when unions independent of the regime seemed the likely outcome. In this climate there were no further moves toward confederation until the end of 1953. Unionists, awaiting a go-ahead from the MSA, did not form any group to replace the FCGF, and the ministry in turn sought a green light from the RCC. It was widely rumored at the time that the RCC did not favor the idea of a confederation despite the new law, and would consider any attempt to start one without prior consultation a hostile act. However, toward the end of the year 'Abd al-Mughni Sa'id endeavored to win the endorsement of the RCC for

TABLE 1.3

Union Membership in the 1950s

Year	# of Unions	# of Members	Avg. Members/Union
1952	568	159,608	281
1953	947	265,192	280
1954	1155	286,671	284
1955	1154	394,245	341
1956	1249	459,029	370
1957	1347	437,751	325
1958	1377	433,000	314
1959	1056	341,169	323

Source: Hasan, p. 106.

what he called a "middle solution": the creation of an unofficial body of union leaders which could "play the role of a confederation" without formal recognition. It would serve "as an experiment for two or three years, until the officials gained confidence in the inclinations of the trade union movement, and their fears of a confederation had dissipated."[42]

Sa'id was able to convince the (unnamed) RCC officials with whom he met, and the group was set up under the auspices of the Liberation Rally's Labor Bureau in the first few months of 1954. It came to be known as the Permanent Congress of Egyptian Trade Unions (hereafter PC), and had its headquarters in the Liberation Rally office. As'ad Ragib, president of one of the petroleum workers' unions, was appointed as its general secretary. During most of 1954 the PC consisted of only a small group of union leaders. The communist labor leaders who were active in prior efforts to establish a confederation were excluded, apparently with the tacit consent of those labor leaders involved who were anxious not to arouse the suspicions of the authorities, and Kamil did not become part of the group until 1955.[43] Nevertheless, the disagreements over labor affairs within the ruling circles continued, and the period from 1953–56 saw a splintering among union leaders as different cliques within the new elite sought clients in the union movement, and as rival union leaders solicited regime patrons.

The March 1954 Crisis: Labor's Role Reconsidered

The divisions within both the union movement and the ruling circles were brought into sharp relief during the crisis of March 1954, when the ranks of the

RCC were split by a dispute over the return to civilian rule. The latter was promoted by Nagib, then still head of the council, while 'Abd al-Nasir favored the continuation of the military government. When Nagib resigned from his positions as prime minister and RCC chairman in late February as a result of the controversy, it prompted an RCC decision in early March to rescind press censorship at once, lift martial law by June, and convene a constituent assembly by July. Nagib resumed his posts as a result. However, on March 25 the RCC declared that it would once again permit the establishment of political parties, and that the council itself would disband before the scheduled constituent assembly. This announcement was apparently intended by Nasir, who had been quietly building his base of support in the military, to spark a popular movement against a return to civilian rule.[44]

A strike by Cairo transport workers two days later played a critical role in bringing about the reversal of these decisions, and the consolidation of power by 'Abd al-Nasir. Now part of the mythology of "the July Revolution," the strike marks the first attempt by forces in the regime to mobilize labor for political purposes; it is often cited as an indication of Nasir's populist proclivities and/or as evidence of strong working class support for continued rule by the military.[45] But the event was more of a harbinger of what was to come than proof of its arrival. Populism is not only a matter of rhetoric but also of concrete program and, as we have seen, the regime had given workers little on which to base their support. What actually transpired indicates a more contradictory and measured response by both unions and ordinary workers, albeit one that had an important impact on the future course of union/state affairs.[46]

A meeting of Cairo union leaders to discuss the turn of events was hastily convened on the morning of March 26.[47] The opinions of the unionists present were divided; both Nasir and Nagib had supporters, while others, such as Kamil himself, favored further study of the situation. In the end, it was decided to issue a statement calling on the RCC to resolve its internal problems and preserve the unity of its ranks. Thus the unionists formally adopted a position of neutrality. However, toward the end of the meeting, several of the union leaders present quietly withdrew, having been summoned to another meeting at the Liberation Rally headquarters. Chief among them was Kamil 'Uqayli, president of the Cairo taxi drivers' local and a member of the PC. In conjunction with Tu'ayma and Ibrahim al-Tahawi, the head of the LR, these unionists decided to call a strike in support of Nasir on the following day. The action was intended to preempt demonstrations in support of Nagib that had been called for March 28. The unionists went to the headquarters of the Cairo Joint Transport Federation (CJTF) in Bulaq, where they announced a hunger strike in support of the following demands: no political parties; no elections; the continuation of the RCC until the British had evacuated their forces from Egypt; and the formation of a National Assembly, including trade union representatives, to serve as a consultative body to the RCC. The statement, broadcast over the radio, called on workers to stay away from their jobs beginning on the morning of March 27.

The events of that morning indicate only tepid support for the strike. Although busses were stopped, the tram and metro ran normally, while taxi drivers vied for the added business of stranded bus riders. In some cases the striking leaders were unable to rally support among their membership; elsewhere union leaders actively worked to defeat the strike. Outside of the transport sector, reaction was also mixed. The Cairo Printers' Union announced opposition to the strike, and a group of other Cairo industrial and service unions jointly issued a "patriotic statement" endorsing the RCC's March 25 decisions. In Alexandria there was a general strike in favor of the return to democracy and in Kafr al-Dawwar, workers at Misr Fine Spinning and Weaving, the largest factory there, also supported the RCC's dissolution. Communists with a strong mass base among the workers played key roles in organizing this countercampaign. They and other Nagib supporters also circulated petitions against the transport strike on several main streets in Cairo, and published them in the Wafd newspaper.

Embarrassed by this turn of events, the Liberation Rally leaders sought to achieve by force what their allies in the unions had failed to accomplish by persuasion. Units drawn from the military police, state security and national guard were sent out to put the strike into effect. They cut the electricity for the tram and blocked the train tracks, in addition to physically assaulting tram and metro drivers who continued to work; one union leader was almost killed. Taxi drivers were likewise beaten and had their licenses torn up. As a result of this campaign of intimidation and sabotage, transport was effectively shut down the next day, and the normal functioning of Cairo was seriously disrupted. Meanwhile, Fathi Kamil and about a dozen other unionists were summoned to the Interior Ministry where they were questioned about the previous day's meeting; they were released only after they were able to produce the original copy of their statement of neutrality, which had not yet been issued. Security personnel also fanned out onto main streets seeking to promote spontaneous demonstrations in support of the strike. Thus a perception of widespread support for the RCC was created, and some other unions began to join the strike. On the evening of March 29, the RCC announced the suspension of its decision to return to democracy.

The Liberation Rally offered compensation to the unionists who spearheaded the strike. Kamil 'Uqayli claims to have refused this; al-Sawi Ahmad al-Sawi, the president of the CJTF, apparently received the greater part of the E£5,000 total outlay. Ironically, while 'Abd al-Nasir himself apparently approved this expense ex ante, he had not supported the strike when it was proposed to him by Tu'ayma on the afternoon of March 26, skeptical that the union leaders had sufficient popularity among their members to pull it off.

Yet it was mainly Nasir's popularity, not theirs, that was really at issue here. And while it is impossible to make precise inferences based solely on the available information, my best judgment is that both Nasir and Nagib had equal amounts of supporters among a minority of unionists and workers. The majority in each category were probably neutral and/or indifferent, with some going along with the strike out of fear, and others eventually joining in the familiar desire to feel like they'd "voted for" a winner. The evidence of the previous section clearly contradicts the notion that large masses of workers turned out in a genuine desire to support the continuation of a military government that was improving their lot.

Indeed it is clear that Nasir recognized the need to improve the image of the RCC, and its ties among the workers, after the March crisis. On the evening of March 29, he and two other RCC members attended a rally at the CJTF headquarters, and the RCC apparently discussed concrete measures to win more support from workers in a meeting in early April. In the subsequent weeks Nasir undertook a series of workplace visits and meetings with unionists which included all the major industrial areas of the country, promising social justice, an industrial revolution to increase unemployment, and a role for workers in policy making.[48] Spending on social services and public investment did increase thereafter, and with the improved international economic situation the economy pulled out of its recession in mid-1954. There were, however, no amendments to the labor legislation until the fall of 1955, suggesting continued resistance from some elements in the regime.[49]

At the same time, the ensuing months saw multiple and competing initiatives by government agencies to foster ties with the union movement, suggesting that the ultimate success of the strike convinced Nasir of the potential benefits of a close relationship with the trade union movement. But given the strike's initial failure and the disunity in the trade union movement that it revealed, the regime did not concentrate its efforts to recruit labor clients in any single organization. More labor inclusion under government supervision was encouraged, but a singular confederation was not yet Nasir's preferred strategy. The result was that the union movement as a whole became even more fragmented.

First, the Labor Bureau of the Liberation Rally sought to extend its tutelage over the Permanent Congress. Although the PC had its headquarters in the LR office and Tu'ayma was kept informed of its activities, it was initially able to function independently and without interference. Tu'ayma now sought to have his own delegate attend the Congress' meetings, which resulted in Ragih's resignation as General Secretary.[50] Until 1955, when the PC was enlarged and expanded, the group apparently remained without an official leader.

A second organizing effort took place among civil service workers, who were still banned by law from forming unions. To get around this, the workers had formed "leagues" (*ruwabit*). The expansion of such leagues, and their affiliation into a semiofficial federation, was now encouraged by Tu'ayma. The result was the establishment on July 9, 1954 of the General Congress of Unions and Federations of Egyptian Government Workers and Employees (al-Mu'tamar al-'Amm li-Niqabat wa Ittihadat 'Ummal wa Mustakhdami al-Hukumah al-Misriyah).[51]

But Tu'ayma was not singly entrusted with the regime's relationships to labor. Rivaling the groups under his influence was, first, the Workers' Club (Nadi al-'Ummal; hereafter WC). Although it also operated from the Liberation Rally's Headquarters, the creation of the WC was a direct reflection of Nasir's dissatisfaction with the Labor Bureau and its union affiliates after the March events; its director, Captain Muhammad Wafa Higazi, was personally chosen by Nasir and in direct contact with him. Higazi had formerly served as Minister of the Interior, and soon became office director for Husayn al-Shafi'i, the Minister of Social Affairs. The locals affiliated with the Club were drawn mainly from the petroleum unions, most of which were based outside of Cairo, and known for their conservative leadership. As'ad Ragih, former head of the PC, joined the club after he left the latter; its members also included Anwar Salama, who was known for his ties to the Muslim Brotherhood. The PC unionists, especially Uqayli and al-Sawi, were bitterly opposed to the formation of the club, and campaigned to have it disbanded. Their demand was eventually realized in 1955, but the unionists in the club continued to function together unofficially, and remained close to Higazi; Salama went on the become the first head of the confederation when it was finally formed in 1957.[52]

The PC's other competitor was the Congress of Egyptian Workers (Mu'tamar 'Ummal Gumhurriyat Misr; hereafter CEW). Although the exact links between this group and the regime remain unclear, it was apparently initiated by some pro-Nasir government officials outside of the Labor Bureau during the March crisis, and supported the transport strike. Several weeks later the CEW issued a statement endorsing the principles of the revolution, and proposing to contact workers' organizations around the world seeking support for a British evacuation. The CEW began to organize conferences in areas outside of Cairo, and was soon able to establish affiliates in several other governates. This expansion enabled it to become the most active wing of the trade union movement until 1956.[53]

At the same time, the government opposed independent attempts to create and/or merge unions. Most notable in this regard was the RCC's refusal, from the summer of 1953 until late in 1954, to permit the establishment of a national textile workers' federation, fearing that communist influence in the textile unions was too strong.[54] Leading in the effort to build the federation were Muhammad Mutawalli al-Sha'rawi from Kafr al-Dawwar, and Ahmad Fahim and Sayyid Fa'id from Shubra al-Khayma. When the federation was finally permitted, Fahim was the members' choice for president. Though actually a foreman, and president of the foremen's union, he was an honest and respected union leader; during the early years of the RCC he used his position to defend workers who had been arrested, pressing Tu'ayma to release some who were being tortured in prison. Tu'ayma, fearing that the federation would become too powerful and independent under Fahim's leadership, tried unsuccessfully to find someone who would challenge his candidacy.[55] As we will see in the next chapter, Fahim went on to become the confederation's second president, reconciling himself along the way to government supervision of the union movement.

On the Road to Etatism

Nasir's triumph in the March crisis paved the way for major political changes. The RCC moved away from collective decision making, toward empowering Nasir as the arbiter of disputes. His position was confirmed and consolidated with the approval of a new constitution in June 1956. Prepared by the RCC in the winter and spring of that year and put to public referendum in June, the constitution created a new presidency, and vested it with considerable powers; Nasir, in a separate referendum, was elected to that post. The RCC was abolished at this time, though most of its members became cabinet ministers.[56]

Associated with these developments came moves toward a greater state role in the economy. As the international recession ended, the government entered some joint ventures to foster capital intensive industries, such as iron and steel, adding to its earlier spending on infrastructural projects. The 1956 constitution promised that the state would undertake economic planning and coordinate its activity with the private sector in the service of the national development goals. These etatist tendencies were accelerated during and after the Suez War in October 1956. All banks, insurance companies, and commercial agencies owned by British and French nationals or Egyptian Jews were sequestered during the war and the owners were required in its aftermath to convert them within five years to domestically held joint-stock companies. With a special plan for promoting industrial development, the government now sought to direct private investments into specific areas through such means as lowering rents on new buildings to shift capital away from real estate and into productive ventures, and controlling the flow of bank credit. The state also increased its own capital investments, especially in electric power, infrastructure, and heavy capital projects.[57]

Despite Nasir's pledges to workers in the wake of the March crisis, there was no new labor legislation until September 1955, when new social insurance legislation was passed. It required employers to contribute to workers' pension and disability funds, but at first only in plants larger than 50 workers located in Cairo and Alexandria. It seems unlikely that the law initially signified much improvement for workers; O'Brien notes that there was little objection to the scheme from the private sector because most firms affected already had similar plans in operation.[58]

However, the 1956 constitution served as a harbinger for more progressive labor legislation. Its preamble called for the achievement of social justice in a "democratic, socialist cooperative" society," and Article 53 specified that the principle of social justice was to be applied to employer/employee relationships. Article 52 stated that work is a right of all Egyptians and pledged the government to seek to ensure adequate employment opportunities. It also committed the government to guaranteeing all workers just treatment in jobs, hours of work, wages, vacations, and insurance against accidents. In Article 21, the government promised to increase social insurance and improve public health, and affirmed the right of citizens to financial assistance in cases of old age, sickness, or disability.[59]

These commitments were translated into concrete legislation, although not until 1958. Most significantly, Law 78 of that year forbid the extension of probationary work, limiting it to one six-month period; there were also improvements made in health insurance and disability compensation requirements in some occupations. The 1959 Unified Labor Code then further reduced the probation period to three months, cut working hours to 8 per day, and doubled the differential for shift work.[60] Additional measures that year extended the health insurance regulations to all industries, and provided for more paid holidays and increases in sickness and severance pay. Subsequent ministerial decrees set a seven-hour day for some 26 dangerous or unhealthy occupations. However, there were no amendments to the regulations concerning dismissals.[61]

Conditions for workers appear to have improved somewhat with these measures, in the context of an overall economic upturn as the international recession of the early 1950s ended. As shown in table 1.2, the real wage index apparently rose during the second half of the 1950s.[62] In addition, industrial employment began to pick up after 1956, with a marked spurt in 1959, possibly due to the mandated reduction in the legal working day (see table 1.1).

However, the regime was once again unable to thoroughly implement its new orientation, suggesting either that Nasir was not completely committed to it, or that he lacked the capacity to impose his new thinking over his subordinates running the government's ministries and agencies. Both employers and conservative forces in the regime resisted the changes, and enforcement of the new laws was hindered by these objections. Official statistics show only a slight decrease in the work week after Law 91 took effect, and many workers' complained about violations of the 8-hour day laws (see table 1.2). Moreover, arbitrary individual firings continued throughout the period and employers continued to defy court rulings that workers illegally dismissed for union activity must be rehired; during the entire period from 1953 to 1961 not a single fired worker was actually returned to work.[63]

Toward a Union Confederation

Accompanying Nasir's consolidation of power and the turn toward etatism was the corporatization, albeit incomplete, of the labor movement. Nasir finally assented to the establishment of a singular and hierarchically ordered trade union confederation with the proviso that it remain under state tutelage. Egypt thus lends support to O'Donnell's theory of an "elective affinity" between inclusionary corporatism and the early stages of industrialization under an import substituting regime.[64] What I want to emphasize here, however, is the contradictory and tentative nature of the process in the Egyptian case. On the one hand, Nasir himself was never fully committed to the idea of singularity, and subordinates with stronger reservations were able to both obstruct the organizational consolidation of the confederation and ensure that rivals continued to exist in some establishments. At the same time, ideological and personal rivalries among union leaders themselves threatened the process and weakened the organizational shell which resulted.

Kamil traces the change in Nasir's approach to the idea of confederation to Egypt's emerging closeness with Yugoslavia, as a result of the April 1955 Bandung conference which was attended by Nasir and Tito. With the working class and workers' organizations given a high status in Yugoslavia's formal ideology, at least, if not in practice, Nasir became embarrassed at the underdevelopment of labor politics in Egypt. Thus, following the visit of a Yugoslavian trade union delegation to Egypt, the Labor Bureau took the initiative to convene a large meeting of trade union leaders where the establishment of an informal organization to function unofficially as a confederation was proposed. This body, then, was an enlarged version of the Permanent Congress; the meeting was attended by Fathi Kamil and numerous others who had not been affiliated with the earlier, smaller group. One of the new group's functions was to represent Egypt in official exchanges with Arab and international trade union bodies.[65]

Like its predecessor, however, this new PC still had the vulnerable status of being an unofficial organization under the auspices of the Labor Bureau, and its leadership clearly needed to have the approval of the latter. Kamil was nominated and unanimously elected to be general secretary of the group. After this, he reports, he was passed a list by Tu'ayma, which he understood to be the Bureau's choices for the remainder of the executive committee. Kamil was able to add some of his own suggestions to this list, in order to make the leadership more representative of the different tendencies in the union movement, but felt he could not oppose any of Tu'ayma's candidates.[66] As the group began to function, Kamil rejected a proposal to maintain the office of the PC in the Liberation Rally building, but he did keep Tu'ayma informed of the organization's activities; there was often an official from the Bureau present in the PC's office. Thus Kamil himself adopted the position that certain compromises with the regime were necessary in order to proceed with the establishment of a confederation.[67]

In the ensuing months, delegations from the PC met regularly with their counterparts from other Arab countries, and these interactions resulted in the formation of the International Confederation of Arab Trade Unions (ICATU). Kamil was elected general secretary of that body as well, and its headquarters were set up in Cairo. The Nasir regime, eager to extend its influence beyond Egypt, obviously attached importance to these relations; the government paid half the costs of the PC's travel for this purpose. The other Arab countries involved, however, did have official trade union confederations, and Kamil used this fact to lobby the authorities to speed the establishment of a confederation in Egypt.[68]

Nasir, however, either did not attempt, or was unable, to impose his thinking on this matter upon others in the ruling circle. Though backed by the MSA,[69] under the leadership of al-Shafi'i and Higazi, the idea of a confederation remained controversial elsewhere in the government. The Ministry of the Interior, representing the security apparatus, continued to oppose it; the security forces were particularly wary of the mass political activity that might be entailed in having union locals throughout the country elect delegates to a founding convention. Such elections were also opposed by the Labor Bureau out of concern that its closest affiliates in the trade union movement might be defeated, and by some unionists who also feared losing their posts. These disagreements were finally resolved in 1956 with a compromise proposal to form the confederation from the top down. No mass elections would be held; rather, a small core of unionists, in conjunction with the authorities, would select a list of the largest unions and federations to be represented at the founding convention, prepare a set of by-laws, and choose the leadership for the group.[70]

The preparatory steps to forming the confederation were therefore taken by an informal group of unionists, in unpublicized meetings, under the watchful eye of the regime. Three different clusters of unionists were represented in this group: 1) those close to the Labor Bureau, including 'Uqayli and al-Sawi; 2) members of the defunct Workers' Club and others close to the MSA, most prominently Anwar Salama and Ahmed Fahim; and 3) independents within the PC, most notably Fathi Kamil. The leadership list drawn up by this group included members of all three tendencies, as well as a number of unionists whose affiliation was not clearly defined. However, when Fathi Kamil declined to become the leader of the confederation-to-be, because of his position with the ICATU, the group could not agree on an alternative candidate. The government officials involved were also divided, and 'Abd al-Nasir himself, who had heretofore delegated the responsibility for overseeing trade union affairs to others, was drawn into the decision making.[71]

Nasir finally chose Anwar Salama on the recommendation of Higazi and al-Shafi'i, who had arranged for a secret meeting between the president and the unionist. Nasir's precise reasoning remains a source of speculation and debate. Salama was affiliated with a pro-Western international petroleum workers' federation,[72] and was both more conservative than 'Uqayli and more experienced in the affairs which more established unions overseas dealt with regularly; of 21 collective agreements concluded between unions and employers in Egypt during the 1950s, 19 were in the oil industry.[73] Thus some claim that, the Suez war notwithstanding, Nasir was looking for a unionist who was both familiar with, and acceptable to, the West. Salama himself hinted at this, suggesting that he was chosen because 'Uqayli was less educated, particularly in foreign languages, though he also noted that 'Uqayli was not an industrial unionist. However, Khalid Hakim, a Syrian unionist who spent considerable time in Egypt during this period, maintains that the operative factor was Nasir's fear of the strength of 'Uqayli's union.[74]

Whatever the case, the selection of Salama shows clearly that Nasir felt no further need to compensate the unionists who had spearheaded the March 1954 strike in his support. They had, in fact, expected one of their own ranks to be rewarded now, and felt betrayed; 'Uqayli was especially vigorous in his opposition to Salama, and tried unsuccessfully to convince the authorities to change their choice. Many other unionists were surprised and angered because of Salama's Western ties. Fathi Kamil was among them, but felt that these objections should not stand in the way of forming the confederation, and the plans for the founding convention were delayed while Kamil worked to convince 'Uqayli of this.[75]

The confederation was finally formed on January 30, 1957, at a founding convention attended by 101 delegates representing 17 large unions and federations with a total of 242,485 members. It initially took the name Egyptian Workers' Federation (al-Ittihad al-'Amm al-Misri lil-'Ummal; hereafter EWF).[76] The convention itself was a brief and perfunctory affair, since all the important decisions had been made prior to the meeting, and in conjunction with the authorities, by the planning group. After a few opening speeches, the delegates approved the by-laws and the leadership list without any changes, and the meeting was adjourned.[77]

Thus a bargain was struck between the unionists and the state: the formation of a singular confederation in exchange for the government's right to choose its leaders. The motivations of the unionists who cut this deal cannot be precisely determined. Bianchi, relying largely on 'Abd al-Mughni Sa'id's memoirs, suggests that the "non-Communist" union leaders uniformly believed that these compromises were in the best interests of workers, and that independence from the regime could be achieved over time.[78] But the history of competition amongst these unionists, and especially their rivalries for patronage from regime elites, suggests that more than a few were acting out of opportunism.

Moreover, the functioning of the organization initially depended on the government as well. When the confederation was formally formed, its dues base was insufficient to cover expenses, and the unionists sought and received financial assistance from the Labor Bureau.[79] The extent of this funding, and its duration, is unclear, and the fact that there are no references to it in the official confederation histories or the memoirs of those involved would seem to suggest that it has become a source of some embarrassment today; it is certainly possible that the confederation had to accede to some regime demands in order to obtain this money. In any case the funding served to amplify the dependence of the trade union movement on the state.

Nevertheless, Nasir and many of his subordinates continued to distrust the confederations' leaders despite their manifest moderation and dependence on the regime. The government would not, for example, allow them to disseminate information and ideas to workers and lower level unionists. The EWF sought permission to publish its own newspaper shortly after its founding convention, claiming that the daily newspapers, then privately owned, ignored news about workers' concerns. It is apparent, although stated only by innuendo in the organization's official histories, that approval was not forthcoming. The confederation was able to produce only an irregular newsletter, whose significance is called into question by the fact that only the subsequent official confederation publications give mention to it.[80]

The confederation's official organizational histories herald its founding as a great triumph for the working class. Governmental interference is completely overlooked, as is the role of the left in all the prior attempts to establish a confederation.[81] But if the account presented above belies this revisionist history, it also discredits the claim by some scholars that corporatism was strictly a project of the state.[82] The sincere forces pushing unification of the labor movement viewed the formation of the EWF as an important step forward in spite of the compromises with the regime that it entailed.[83] Corporatism was a bargain, not an imposition. Nevertheless, the nature of the process could not help but tempt even the honest unionists to turn their backs on the rank and file, since their posts were obtained by support from above, not below.

1957–1961: Consolidating the Confederation

After the event was announced, the Federation of Free Workers voluntarily disbanded itself. In addition, the two years prior to the establishment of the confederation had seen total union membership grow more rapidly than the number of locals, so that individual unions had become larger. Nevertheless, the problem of fragmentation remained serious. The initial membership of the confederation represented only a little more than 50% of the total union membership at the end of 1956 (see table 1.3), and many industries were characterized by having numerous locals that were not united into federations; in some cases, as mentioned above, individual plants had more than one local.

The confederation leadership made its primary task for the coming period expanding its membership base by incorporating the unaffiliated locals, and restructuring the trade union movement along industrial lines. In this endeavor, they continued to be hampered by the persistent opposition of some quarters within the regime. They were plagued as well by resistance from some of the independent unions. While the history of efforts to establish a confederation demonstrates that incorporation was not strictly an imposition by the state, this resistance indicates that the pressures from below for this type of organization were limited, thus reinforcing the turn of its leaders to the state for support.

The military establishment, in particular, remained opposed to the idea of a single, hierarchical confederation. In this they were joined by Muhammad Tawfiq 'Abd al-Fatah, who headed the executive branch of the Ministry of Labor and Social Affairs, and by some elements of the National Union (NU), which replaced the Liberation Rally in May 1957 as the RCC's attempt to organizationally structure its relations with the masses.[84] The LR's Labor Bureau continued to function, now as part of the NU, but Tu'ayma was replaced as its head by Khalid Fawzi. Still active in NU affairs, however, Tu'ayma was one of those who worked to undermine the unity of the labor movement. During this period, the NU encouraged the growth of the leagues, organizing them not only in the civil service but also in industrial establishments where they served as parallel organizations to the unions; these leagues were then united into regional workers' congresses. Aware of ongoing disagreements between Fathi Kamil and Anwar Salama, Fawzi tried to enlist Kamil's aid in this endeavor, but Kamil would not cooperate. Elsewhere within the union movement, leaders of the skilled craft locals opposed the consolidation of the confederation along industrial lines, fearing that their smaller locals would lose power as a result.[85]

The confederation's leadership did still enjoy the support of Husayn al-Shafi'i and the central branch of the Ministry of Labor and Social Affairs; the latter joined the unionists in pushing to establish a legal basis for consolidating the labor movement. They won a victory with the passage of Law 91, the Unified Labor Code, in April 1959. Initially supported by Nasir himself, the law provided for the reorganization of the labor movement into 65 federations along industrial lines, united into the confederation. But as Nasir became increasingly preoccupied with the merger with Syria, which had joined with Egypt to form the United Arab Republic in February 1958, the opponents of consolidation set out to sabotage implementation of the law. When members of the oil and textile workers' federations went to file their papers, they found that employees of the Labor Bureau had been instructed not to accept them. Later they were called in by the security forces, interrogated, and warned not to persevere in their efforts. This situation persisted for a period of months, with each side appealing directly to Nasir. The Ministry of the Interior continued to send him reports warning of the dangers of a strong confederation, which al-Shafi'i tried to refute in personal meetings with the leader. Nasir at one point became so irritated with the controversy that he told al-Shafi'i the whole confederation project should be scrapped.[86]

While this struggle was going on, rivalries among labor leaders also continued, as did the interference of different arms of the government in union affairs. It was in this context that Anwar Salama resigned as president of the confederation in February 1958. 'Amir maintains that this was an act of protest against efforts by Tu'ayma to persuade unionists to organize strikes in support of the candidates he had backed in the elections to the National Assembly, most of whom had lost; when Salama refused to cooperate, Tu'ayma considered him an enemy, and contacted various members of the executive committee to agitate against him.[87] 'Izz al-Din, however, reports the resignation as a protest against the establishment of the regional congresses of the leagues, particularly after the government blocked efforts by the confederation to set up branches in the provinces to rival these congresses.[88] While it may be that both accounts are true, either one serves as further evidence of continued efforts by subordinates of Nasir to establish clientalist relationships with the labor movement.

Salama was supported in his actions by the majority of the confederation's executive committee, who left his seat as president empty for almost two years as a sign of solidarity with him. He nevertheless remained active in confederation affairs during this period, although Ahmad Fahim assumed the day-to-day leadership of the organization. And despite the symbolic protest, both men endeavored to maintain good relations with the regime, reflecting a belief that trade union independence from the government could be won only gradually.[89] However, it is clear from Salama's acceptance of the presidency based on Nasir's decision, and from his actions regarding the Arab confederation, that his opposition to government interference in union affairs was partial at best.

Salama's co-optation by the government can also be seen clearly in the events leading up to the removal of Fathi Kamil as general secretary of the ICATU in 1959. This was in part a consequence of the contradictions between Salama and Kamil, and the unionists grouped around each, which had continued after Salama's election as president of the EWF. Salama urged the Arab confederation to endorse his plan to bring Arab petroleum workers closer to a Western-backed oil workers' federation; when the ICATU rejected this, he appealed to the Egyptian authorities for support, accusing the Arab confederation of being pro-communist. Relations between the Arab and Egyptian confederations soured as a result.[90]

Kamil also irritated the authorities, and furthered the impression that he was a communist sympathizer, by pursuing an independent approach to Arab affairs and refusing to subordinate the ICATU to the foreign policy of Egypt.

In particular, Kamil maintained close ties with the communist-backed trade union confederation in Iraq, at a time when the Nasir regime was supporting the Ba'athists, who backed a rival confederation. Shortly before the 1959 ICATU elections, the patrons of the Egyptian confederation in the Ministry of Labor and Social Affairs initiated a move to have Kamil ousted from its leadership. Husayn al-Shafi'i convened an ad hoc committee of ministry and government officials, and unionists, to decide on Egypt's candidate for the ICATU post; they chose As'ad Ragih. Kamil refused to go along with this, and was supported by 'Uqayli and his other close friends in the EWF leadership, and initially by most of the other Arab delegations. However, the majority of the Egyptian confederation, which was closer to Salama and Fahim, endorsed the decision, and they were able to persuade the Arab delegations that the choice of an Egyptian candidate should be made by the Egyptian confederation. Kamil reports that the Egyptian confederation was given E£150,000 by the government as a reward for its cooperation in his removal.[91]

The events in Iraq, and the position of the Egyptian communists in regard to them, ultimately precipitated a new and more extensive round-up of the Egyptian left. Beginning at the end of 1958 and through 1959, more than a thousand communists were arrested; they were subjected to brutal tortures in prison over the next few years. This campaign removed almost all of the remaining activists from the union movement, as well as the lawyers who had provided legal support to it, such as Yusuf Darwish. There was some limited resistance to this from among the rank and file, but the confederation's leadership voiced no protest at all over the arrests.[92]

The incarceration of the communists weakened the arguments of the security forces against consolidation of the confederation, and al-Shafi'i was finally able to prevail over them by calling on the Interior minister of Syria. The latter convinced Nasir that one hierarchical union structure is less of a threat than many smaller ones, since it is easier for the government to keep undesirable elements out of the leadership of a single confederation, and since a handful of leaders are easier to manipulate than many. The confederation leaders were then able to proceed with the reorganization of the union movement. By January 1961, 59 of the 65 federations designated by Law 91 had been formed. The confederation at this point changed its name to the Egyptian Trade Union Federation (al-Ittihad al-'Amm lil-Niqabat 'Ummal Misr, hereafter ETUF) to highlight the change in its structure.[93]

The unofficial requirement that the regime give prior approval to the confederation's leadership remained intact thereafter. Before the organization held its first convention with the new structure in place in January 1961, 'Abd al-Mughni Sa'id, the Labor Ministry official, met with Khalid Fawzi from the Labor Bureau of the National Union to go over the candidate list. They agreed that Anwar Salama should be returned to the presidency, while Ahmad Fahim would continue as vice-president. Beyond this, the two arms of the government apparently concurred that new faces should be brought into the confederation leadership: of the 19 others elected to the executive committee, only three had served on it previously.[94] As before, the union leaders did not challenge this interference; the list of candidates was approved by default as well as acclamation.[95]

Law 91, the 1959 Unified Labor Code, superseded Laws 317–19 and the subsequent modifications to them. Ironically, in a victory of sorts for the left in absentia, the law repealed the closed shop provisions of the earlier legislation. This stipulation had been opposed by some of the key activists at the time, who argued that it allowed company owners to establish "yellow unions," and then compel the workers to join. However, it was not leftist pressure which accounted for the change. According to Hasan, the law was a cause of embarrassment to Egyptian delegations, both governmental and labor-related, at international gatherings, because the International Labour Organization was opposed to closed shops. The provision was apparently repealed by mutual agreement of the labor leaders and the government.[96]

Otherwise, the new law maintained all the restrictions on unions from the previous legislation. Most significantly, it continued to require conciliation and arbitration in all labor disputes, outlawing strikes while these processes are in effect.[97] As before, there is no record of opposition or complaint from confederation leaders about these restrictions.[98]

There are two ironies in this outcome. The first is that Egypt's leftist labor activists, who had first proposed the idea of confederation in the period before the coup and had devoted the greatest time and energy toward that goal, were virtually completely excluded from the labor movement when the organization was finally formed. Second, the left had championed the cause of union centralization in the belief that it would strengthen the labor movement, enabling it to bargain more effectively not only with employers but also with the state. But their dream became reality only when the rulers of the state were convinced that confederation would facilitate government penetration and subordination of the unions.

And indeed, only a minority of union leaders continued to promote the economic demands of workers. This was especially true in the Textile Workers' Federation, which had been a stronghold of leftists before the coup. Several, such as Fa'id and al-Sha'rawi, managed to escape imprisonment and continued to play prominent roles in the organization. It was a textile federation conference in 1956 which first demanded doubling the minimum wage and the creation of a wage council. Thereafter, the federation held well-attended educational conferences every six months to discuss economic problems and other work-related issues, and agitated for reform of the labor laws, especially with regard to the probationary period.[99]

Most unionists actively involved with the EWF, however, were absorbed in organizational matters and keen to convince the regime that labor would not be disruptive. The few union leaders who were elected to parliament in 1957 with EWF backing did push for improvements in labor legislation, but there is no evidence to show that the EWF as an organization mobilized around these demands. The confederation's founding convention lasted just two hours, with the agenda consisting only of approving the executive committee slate and the by-laws, which make only a vague commitment to defending workers' rights. 'Amir observed that there is "no trace of the words 'hours of work' or 'social insurance' or 'wages,' " in the EWF literature of the period, while 'Izz al-Din writes that the confederation was concerned with consolidation "to such a degree that it seems as if it had completely abandoned the issue of wages, just as it had previously abandoned the problem of unemployment."[100]

Correspondingly, workers' enthusiasm for unionism waned; as shown in table 1.3, after peaking at 459,029 in 1956, union membership fell steadily to 341,169 in 1959, a drop of more than 25%. Another indication of workers' disenchantment with the unions is the fact that between 1955 and the first half of 1956, 24,745 workers had their membership canceled for lack of dues payment. Of these, only 22 individuals even bothered to submit a complaint about the decision.[101] Moreover, there was a marked (40%) decline in the average number of grievances filed by workers from 1955 to 1958, over the previous three years, suggesting that labor discontent was abating.[102] This can support Goldberg's argument that workers no longer saw a rational reason for unions since the state was providing their basic needs.[103] Still, as Beinin notes, the number of grievances remained well above their pre-coup levels, and the evidence presented above indicates that many workers were still vulnerable to summary dismissals and/or subjected to illegal labor practices by their employers. The fact that unions were largely doing nothing to combat these conditions cannot be discounted in explaining why workers were abandoning them.[104]

Nasir Calls for Socialism

In what came to be known as "the socialist decrees" of July 1961 'Abd al-Nasir greatly expanded the sphere of the state in the economy, and the protections it afforded to labor. Explicitly suggesting that these measures obligated workers to dedicate their efforts to expanding the national economy, Nasir formally endorsed the idea of a social contract implicit in a moral economy. With this came a different understand of the purpose of the newly centralized union movement.

A large number of manufacturing and commercial concerns were wholly or partly nationalized by the socialist decrees, greatly expanding the size of the public sector.[105] Concomitantly, a series of laws were issued which aimed at material improvements in workers' living standards.[106] Law 133 of 1961 limited the work week in industrial establishments to 42 hours, six days of seven hours per day. Because some workers actually experienced a salary reduction from this law due to loss of overtime, it was adjusted in 1962 to specify that where overtime pay had been regularly included in a worker's weekly earnings before the law was issued, such additional earnings must be restored to his salary despite the reduction in the working day. A compulsory social insurance scheme was introduced in the 1961 laws and modified in 1962, increasing the employer's contribution from 7 to 17 percent of salary. Further, in an effort to reduce unemployment, the government committed itself to provide administrative jobs to all university graduates, and manual employment to all graduates of secondary schools. Finally, Law 262 of 1962 doubled the minimum wage from 12.5 piasters to 25 piasters per day for many workers.[107]

In addition, the government obligated itself to guarantee the compliance of private capitalists with minimum wage standards and other laws protecting workers. The Nasir regime also took on more of a distributive function by extending the previously existing system of price subsidies to include many essential food items, as well as energy. This was coupled with a wide-ranging system of price controls, also aimed at ensuring the cheap supply of basic necessities to the populace.

There were additional benefits for industrial and service workers in the public sector. Law 114 of 1961 provided for worker representation on the board of management of all public sector firms. One laborer and one clerical employee were to be elected as part of the seven-member boards. Law 111 of 1961 gave public sector workers a share in company profits. It specified that three percent of net profits was to be distributed to workers in cash, according to a formula based on their earnings. An additional 4.5 percent of profits was earmarked for social services like housing and recreation.

As before, our concern here is not primarily with the factors that motivated 'Abd al-Nasir to decree these changes,[108] but rather with their implementation and their effect on the labor movement. Incorporating much of industry under government control did enhance the regime's capacity to enforce its labor laws, and workers clearly benefited from this. The nationalizations and accompanying decrees initiated a period of rapid industrial expansion. During the first four years of the 1960s the economy overall grew at an average annual compound rate of 6.6%, with industrial output growing by an average of 12.3% annually.[109] Employment in manufacturing establishments with 10 or more workers grew from some 325,000 in 1960 to about 509,000 in 1964/5, an average growth rate of 11.8%.[110] And, as shown in table 1.4, workers made substantial gains in the 1961–64 period, both in terms of wage increases and a reduction in the work week, as a result of the new legislation.

Nevertheless, by enlarging the number of government employees and agencies that interacted with labor, 'Abd al-Nasir had also increased the potential for policy disputes and interagency rivalries. The newly appointed managers of the parastatals were in a particular position to impede implementation of the decrees. Many were not committed to the new approach, and were particularly hostile to the provisions for workers' representation in management (hereafter WRM). In the early years, managers used several different means to circumvent the system: holding secret meetings without the workers' representatives, interfering in the WRM elections, and firing the elected representatives.[111]

In this case, 'Abd al-Nasir acted fairly quickly to stem the worst abuses in the public sector. In 1963, the WRM law was amended to give employees four seats on a nine-member board which was now required by law to meet monthly. To

TABLE 1.4

Real Wage Index and Hours of Work, 1959–1964

Year	Hrs. Work/Wk	RWI/Hr.	Weekly RWI
1959	50	na	143
1960	49	na	141
1961	48	147	140
1962	47	151	141
1963	45	177	159
1964	44	190	168

Source: Mahmoud Abdel-Fadil, *The Political Economy of Nasserism*, p. 33. Reprinted with permission.

Note: Base year 1950. Based on the *Survey of Employment, Wages, and Working Hours* (hereafter SEWWH), published by CAPMAS. The SEWWH includes salaries for managerial, technical, and clerical employees, which were generally higher than those of industrial workers during this period. It excludes establishments employing less than 10 persons, where wages were generally lower. Wage figures are obtained from employers and include the basic wage as well as all bonuses and incentives.

restrict the managers' prerogatives to use promotions and demotions as reward and punishment, 12 different job categories, each with a set minimum salary, were defined, with annual raises guaranteed for all workers who perform satisfactorily. Also, Ministerial Decree 96 of 1962 made it more difficult for public sector managers to fire workers by requiring that they first consult a tripartite committee consisting of representatives from the union, the Ministry of Labor, and management. Workers who felt they were unjustly dismissed could appeal their cases to this committee before turning to the court system.[112] Nevertheless, managers objected to these new provisions, and some continued to defy the legal barriers to dismissals by refusing to honor court rulings which called for the reinstatement of fired workers.[113]

In the private sector, which continued to dominate small-scale manufacturing although virtually excluded from the most strategic industries, the scope for evasion of the new laws was greater.[114] With owners and managers not subject to performance reviews by the regime, it was up to the ministries of Labor and Industry to supervise application of the new laws and, in the case of violations, prosecution was the province of the courts. Enforcement posed a challenge to the weak regulatory capacities of the regime.[115] The early years of the socialist transition saw another big increase in the number of grievances raised by individual workers and union locals, and in 1963 the Ministry of Labor had to expand the number of labor offices from 22 to 35 in order to deal with the increased volume of grievances. The majority of these complaints were over violations of the new laws by employers.[116]

Making up for the bulk of employment in establishments of less than 50 workers, private entrepreneurs were in fact discouraged from hiring more full-time employees because the socialist laws also mandated a modified form of WRM in plants larger than 50 workers.[117] Fifty full-time employees was also the minimum required for the establishment of a single union local at a plant; otherwise, one or more additional firms in the same industry had to be incorporated, making unionization more difficult. Private sector employers seeking to expand therefore often did so, without challenge from the government, by keeping regular workers under temporary work contracts and thereby deprived of the benefits legally due to full-time, permanent employees.[118]

These discrepancies notwithstanding, 'Abd al-Nasir now actively promoted his regime as a socialist one, dedicated to improving the lot of the lower classes and elevating their status. An integral part of its new policies was the establishment in 1962 of a new, mass-based party, called the Arab Socialist Union (ASU), to replace the National Union.[119] New laws required that 50% of all elected leaders of the government and the ASU come from the workers and peasantry. In turn the lower classes were expected to be both docile and grateful, the former manifested in an absence of demands for anything more, and the latter to be shown by extra effort on the job. Thus 'Abd al-Nasir promoted the idea of reciprocal rights and responsibilities inherent in moral economies, paving the way for workers' dissatisfaction when the regime reneged on its commitments.

A New Role for Unions

Nasir's expectations of workers became clear in his approach to unions. Having finally been won over to allowing a centralized union movement, Nasir now defined for it a specific role in the new moral economy. The new function for unions was to ensure that workers fulfilled their responsibilities.

This was first indicated in official propaganda which appeared at the end of 1961, in the official magazine *Al-Ahram al-Iqtisadi*:

> [Unions] are no longer groupings for the seizing of rights or defense of interests in opposition to employers, but have become centers for the concentration of workers and parliaments for the expression of their opinions; unions are no longer charitable societies helping the distressed and treating the sick—they must become centers of revolutionary radiation and instruments for pushing forward the wheels of production.[120]

Nasir's view was expounded again in the spring of 1962 when he proposed the National Charter, a constitution-like document. The Charter called on the unions to:

> become the leading vanguard in social and economic development. Labor organizations can exercise their responsibilities through serious contributions to intellectual and scientific efficiency and thus increase productivity among labor. Unions can fulfill their obligations by safeguarding labor's rights and interests and by raising the workmen's material and cultural standards. This includes plans of cooperative housing and cooperative consumption as well as the organization of vacation and other free periods to increase the health, psychological and intellectual welfare of the workers.[121]

The regime organized a nationally televised congress, with elected delegates, to debate and approve the document; it served as part of the groundwork for creating the ASU. During the congress, Nasir clarified his view of the reasons for changing the role of unions:

> Our society is in the process of changing from feudalist/capitalist to socialist. And the methods and tactics which unions followed in the past, during capitalist society, have to change. Otherwise we will find ourselves having

> contradictions and conflicts. This is the responsibility of unions. . . . Unions today have a very big role, a big part of the responsibility for administering work will be transferred to the unions in the future. This must take the place of the labor contradictions in the country.[122]

The National Charter was initially controversial among federation and confederation leaders.[123] Although they did not challenge its sections on unions explicitly, a group of them did organize to do so indirectly. At one point during the televised discussions, when Nasir was in attendance, then-ETUF president Anwar Salama rose to criticize the draft document for its failure to address issues of youth. Salama began to enumerate proposed revisions, presenting them as collective demands of the Egyptian workers. When he exceeded a 15-minute limit that had been imposed on all speakers, Nasir tried to break off his comments. But Salama had prearranged for other unionists at the meeting to chant for him to continue if there was an attempt to cut him off.[124]

Nasir initially backed off under this din, but then gave a clear indication of his paternalistic attitude toward workers and the relationship between them and the government that he anticipated. At the conclusion of Salama's remarks, he responded with demonstrable anger, "The workers don't demand; we give."[125] He then argued that aggression by unions was unnecessary, because his regime had consistently demonstrated its concern for the workers:

> From the beginning of the revolution the laws of the individual work contract and the banning of arbitrary dismissals appeared. And until today the gains which workers have realized are not the result of demands. No workers' union demanded 5% or 10% or 20% or 25% (of profit sharing). Nor did they demand workers' representation in management. It seems to me that they were surprised by the decisions regarding WRM. And by those regarding the minimum wage. . . . This decision also came without any demand.[126]

It is ironic that this challenge to the government's appropriation of the right to define the unions' raison d'être was mounted by unionists who had already conceded the government's right to interfere in union affairs and forsworn militancy in their efforts to win the government's support for union centralization. Although this can only be speculation, I suspect that Salama and those closest to him, given their conservative inclinations, were actually motivated by a fear that the regime was becoming too radical. Others involved, however, may have genuinely believed that the labor movement's previous concessions to the government did not preclude its eventually achieving independence, but now feared that Nasir's new orientation would foreclose that possibility.

After this incident, Nasir appointed Salama to head the Ministry of Labor, which had just been hived off from the Ministry of Labor and Social Affairs.[127] In his new capacity Salama became part of the temporary 18-member supreme executive committee of the ASU, and was chosen by that body to sit on its seven-member permanent executive.[128] Nasir may have felt that choosing the leader of the trade union movement was the best way to enhance government supervision of the latter although Salama, apparently at his own initiative, resigned the presidency of the ETUF to take the new post. It is also possible that Nasir preferred to have someone of Salama's conservative leanings in his cabinet to act as a counterweight to more radical elements.[129]

In any case, the spectacle at the Congress can only have reinforced the regime's historic reservations about unions. Although in its aftermath senior union leaders publicly embraced the new philosophy of unionism, the government would not empower the confederation solely, or independently, to carry out its new assignments. The regime's distrust was manifest in its ongoing supervision of, and interference in, union affairs, as well as in its continued sponsorship of rivals to the trade union organization.

Two attempts by senior union activists to assume some independent responsibility for inculcating workers with the new ideology were initially resisted by the government. First, the ETUF was still denied permission to publish its own newspaper. The Ministry of Labor did begin to publish a monthly magazine, *al-'Amal* (Labor), in 1962. Although the ministry was then headed by a former union leader (see below) and the magazine regularly featured interviews with unionists and workers, the confederation's leaders did not feel that it could serve as their voice. Hence they made arrangements for one of the daily newspapers, all of which had since been nationalized, to publish a workers' newspaper that the ETUF would produce. The ETUF provided the money for the endeavor, but the paper never appeared.[130]

There was also conflict over the Workers' Educational Institute (Mu'assasat al-Thaqafah al-'Ummaliyah, hereafter WEI), charged with training union leaders. The 1960s opened with the authority to choose the director of the institute resting formally with the institute's management committee, and actually with the Minister of Labor. The first person appointed was 'Abd al-Mughni Sa'id. But in November 1961, one month after his appointment, he was arrested on political charges and remained in prison for a full year.[131] In his absence the directorship went to Hilmi Murad, a university professor with no previous ties to the labor movement. The confederation leaders were displeased with Murad and pressed for implementation of a 1960 agreement that would allow them to choose the head of the management committee, and to gradually assume full control over the project. The leading body of the ASU, which assumed responsibility for the management committee after the National Union was disbanded, did accede to this request in 1963, but when the organization's leadership changed two years later it again assumed control over the WEI.[132]

Also, and over the objections of union leaders, a new labor law issued in 1964 maintained the right of the government, through the Ministry of Labor, to interfere in union affairs in several key ways. Permission from the agency was required for the federations to hold their annual conventions. The law also gave the ministry the right to prescribe conditions for membership on the executive committee of the locals and federations. Finally, in addition to setting down percentage limits on the distribution of union funds, the law denied the federations the right to invest their portion of union dues without permission from the ministry, and required them to submit their financial records to the latter on a regular basis. Ministry inspectors were accorded the right to examine the federations' books at any time.[133]

Government interference in the union leadership selection process continued, and in fact took on new and more extensive forms. In July 1963, as preparations were being made for the 1963–64 union elections, a law was issued requiring that all candidates for union office be members of the ASU.[134] This law gave ASU officials the ability to keep individuals they disapproved of out of union office by denying them ASU membership.

The two previous incarnations of a mass party—the Liberation Rally and the National Union—had both intervened in union elections primarily at the national level, because of both their own organizational weaknesses and the uneven levels of functioning at the union base; the chaotic nature of the union movement at that time meant that elections at the local levels were mostly irregular and disorganized.[135] The consolidation of the confederation in the early 1960s meant the systematic organization of elections throughout the union structure, which would facilitate greater supervision by a party apparatus. At the same time, the ASU developed into an organization more capable of such intervention than its predecessors. It had a larger membership base and branches in all the major workplaces in addition to geographically based chapters. Thus the ASU could "reach" into local unions in a way that the Liberation Rally and the National Union could not.

One important implication of this is that ASU interference in union affairs could be implemented by individuals who were much further removed from the inner circles of power in Egypt than were the bureaucrats of the LR and the NU. Also, the fact that the ASU functioned in a decentralized manner meant that such interference could occur without the knowledge or approval of party or government elites. Thus, the irregularities that occurred at individual factories may have reflected more the local power situation than any national policy. There was scope for state managers, wealthy elites, and union incumbents to influence ASU officials to deny membership certificates to certain contestants and thereby have them disqualified from the elections; workers in numerous establishments complained of such violations.[136]

The 1963–64 local election results were also marred by *tazwir* (falsification of the results), either through ballot box stuffing or deliberate miscounting. Sayyid Fa'id, who was elected to leadership of his local, reports being arrested on election night, and beaten for 16 consecutive hours, after exposing such irregularities, thus indicating that government agencies such as the police and the Labor Ministry were also involved in the interference. And at the higher levels of the union structure, intervention by forces closer to the inner elite is indicated. Muhammad Gamal Imam, who was a confederation functionary at the time, maintains that the old ETUF leaders were almost not reelected in 1964 because the rank and file leaned toward choosing young new faces; the existing leaders prevailed in the end only because of outside interference.[137]

In addition to all its restrictions on, and supervision of, the unions' activities, and to its interference in their leadership selection process, the regime sought to ensure that the unions did not undertake economic militancy through prohibitive legislation. Although the emergency laws which banned strikes were lifted in 1964, the labor law effectively outlawed them in the public sector, and in key private sector establishments, by giving the Ministry of Labor the right to seek legal disbandment of a federation or local for "abandonment of work or deliberate refusal to work by persons performing public functions, employed in public services, or working to fulfill a public need." The law also maintained the prohibition on strikes during the conciliation and arbitration process set out in the 1959 Labor Code. Penalties for illegal work stoppages were specified in the Penal Code.[138]

Yet in spite of all these controls, the regime was still not prepared to rely on the union structure as its exclusive channel to the mass of workers; rivals to the ETUF were tolerated, if not encouraged, as a hedge. The workers' leagues, in particular, remained a thorn in the side of the confederation. The 1964 labor law called for abrogating the leagues, and transferring their funds to the local, in all establishments where a trade union local existed. Although some leagues were dissolved on this basis, most continued to function in defiance of the law.

They were particularly strong in the civil service where unions had previously been outlawed—according to a confederation report, 56 existed in the railway authority alone.[139]

The Free Officers' coup brought to power a collective of officers with diverse views on the appropriate relationship between state, private capital, and labor. In the early years the rightist elements within this group had the greatest influence in setting labor policy. The result was a ban on strikes, a prohibition on the formation of a labor confederation, the purge of communists from the ranks of union leaders, and legislation which extended the probationary period that employers routinely used to deny workers benefits and keep wages low.

Nasir's emergence as the undisputed leader of this group coincided with a shift toward more overtly progressive labor policies. But he and others in the regime continued to distrust unions. The establishment of a singular labor confederation was therefore permitted only after its leaders had forsworn militancy and ceded to the government the right to select the organization's leaders. We cannot be certain about the motivations of the unionists who cut this deal with the state. Had they not opted to capitulate, the unionists might well have become victims of the same brutality visited on the communists, and seen the establishment of the confederation delayed indefinitely. Nevertheless, their cooperation with the regime instead established the precedent for the government interference in union affairs which was still operative four decades later.

In economic matters, the 1950s ultimately saw the enactment of legislation which promised workers shorter hours, safer working conditions, and better pensions and insurance. Yet, for reasons of regime insincerity and/or incapability, opponents of these measures were largely able to impede their enforcement, as well as the introduction of more extensive protection from summary dismissals. With most union leaders having abdicated their role in promoting rank and file concerns, and especially given the ban on strikes, workers increasingly took their grievances over these matters to the state rather than the union movement. This paved the way for the relationship of reciprocal rights and responsibilities between labor and the state that Nasir soon introduced.

The union movement was permitted to consolidate a centralized, hierarchical structure only after Nasir's formal embrace of socialism, but with this the unions' role was officially defined to be inculcating workers with the regime's new ideology and ensuring that they repaid the government's generosity with enhanced effort on the job. The expansion of the public sector and progressive labor legislation enacted in the early 1960s did bring quick improvement to the living standards of many workers, especially those in the public sector, and this rendered both workers and unionists susceptible to Nasir's philosophy. At the same time, the government proved unable to enforce its new labor laws in the private sector, and its commitments to the lower classes soon put a strain on the country's finances. The economic burdens of Nasirism underlay the attempts at liberalization which would characterize the next three decades; the contrast between conditions for public and private sector workers explains why privatization has proven to be the most politically difficult reform for the government to enact.

2

The Continuity and Conflicts of Corporatism

From the mid-1960s onward, successive Egyptian leaders have been engaged in efforts, of varied intensity and even more varied success, to reduce Nasir's original commitments to Egypt's workers and to liberalize the economy in other ways as well. Some of these efforts have met with resistance from all strata of the Egyptian labor movement, some from only select groups within it, and some have encountered little or no opposition at all. The structural features of the Egyptian labor movement which took shape alongside the etatist economy have influenced the nature of this reaction, in two ways. First, both the distribution of power within the confederation and the legal opportunities for union demand-making set by the government contribute to the capacity of the unions, at different levels, to respond to different types of policy reform. Second, the government's ability to intervene in the unions' leadership selection process affects the will of union leaders at different levels to challenge the government's economic restructuring initiatives.

This chapter traces the evolution of the confederation's structure, demand-making capabilities, and political relationship with the successive regimes from the mid-1960s through the early 1990s. I show how the essential features of the corporatist system which took shape in Nasir's first decade of rule remained in effect during this period: regime elites continued to select the senior leadership of the ETUF; the latter remained the single official union organization, with a hierarchical structure allowing only one federation to represent workers in any given industry; strikes remained illegal, and unionists sought to influence government policy mainly through membership in the ruling party and whatever subordinate positions in the government itself this might confer.

In addition to setting the legal/institutional stage for the economic battles to come, this chapter seeks to contribute to the theoretical literature on union/state relations in the third world. The "pessimistic" approach which has characterized much of this scholarship would see in the institutional stasis described here confirmation of the state's predominant role in setting the boundaries of union operations. Continuing the argument in the previous chapter, I aim to show here instead that the state's intents were not always singular or even readily identifiable. Rather, as in the 1950s, government policy toward the unions was often the outcome in part of clashes within the regime.

Furthermore, both the final shape of the successive regimes' policies and the degree to which they were implemented were influenced by their interaction with the union movement. The Colliers have demonstrated that the legal environment in which unions operate is an arena of bargaining between labor leaders and the state. I seek to augment that finding by showing here that union/state relations often form a source of conflict within the union movement itself. In these conflicts, the majority of unionists are motivated by concern for their own institutional advancement or tenure. But the structure of the union movement itself dictates that such motivation will not result in a common perspective among unionists. Rather, the interests of those on the lower rungs of the union ladder may clash with the upper level personnel, especially regarding the distribution of power to the different levels of the hierarchy and the consequences of government intervention in leadership selection on the possibilities for individual advancement.

However, opportunism is not the only explanation for intra-union disputes. Some unionists are instead motivated by ideological beliefs and affiliations. This chapter will demonstrate, in particular, that the various forces which comprise Egypt's left have had a significant effect on union/state bargaining over institutional matters. The role of the left underscores the need for blending a purely institutional analysis of union actors' behavior with an interpretive framework emphasizing the salience of ideas and identities. This combination of factors implies that the outcomes of clashes over union/state relations must always be somewhat indeterminate.

The empirical support for these points is presented here in chronologically ordered sections, subdivided into analytical categories. The historical presentation should facilitate easy cross-referencing with the economic issues covered in the next three chapters. In addition, it serves to highlight the critical junctures which union/state relations passed through during the period under study, and the extent to which the outcomes at one juncture shaped subsequent interactions.

Unions, the State, and the ASU in Nasir's Final Years

Established in 1962, the Arab Socialist Union (ASU) was the third and last of Nasir's attempts at creating a single party channel between the masses and the regime. The party became, for union leaders, an important vehicle for communication with policy makers. At the same time, the ASU emerged as a major rival to the confederation, as well as the principal means by which some forces in the inner elite sought to penetrate and subordinate the unions.

Unions Under the ASU?

The key institutional player in the effort to subordinate the unions to the party was an ASU bureau known as the Workers' Secretariat (WS). It was initially created in 1964 by Anwar Salama in his ASU leadership capacity. Salama was joined in this endeavor by 'Ali Sayyid 'Ali, from the Petroleum Workers' Federation, who became vice president of the ETUF in 1964. They invited 30 others to join the body, drawn from among the confederation's executive committee and other noted unionists not then in ETUF leadership positions, and created a bureaucratic structure to administer labor affairs. Ahmad Fahim, the ETUF's president, was among the members. These WS positions were attractive because they were financially rewarded and afforded political prominence, and the bureau at first served mainly to divert the energies of key labor leaders away from the union hierarchy. Salama and Ali's motivations in establishing it are unclear, as is the question of whether their actions were supported by others in the inner elite. Salama, as we have seen, was not a strong supporter of socialism, and 'Ali was apparently initially of the same mind. 'Izz al-Din reports that the two were unpopular among most unionists, and that their actions as WS leaders increased the dislike of their colleagues for them.[1]

The secretariat emerged as an agency for union control the following year, as a result of political changes triggered by the economic difficulties the country was facing. Nasir replaced Prime Minister 'Ali Sabri with Zakariyya Muhyi al-Din, the rightist former Minister of the Interior, in September 1965. Sabri, though he lacked roots in the traditional left, was commonly perceived as left-leaning because he strongly supported Nasir's foreign and etatist economic policies. Thus the switch was widely understood as a signal by Nasir to the West that he was now willing to consider the IMF "bitter pill" which he had earlier refused to swallow (see chapter 4).[2]

However, and again in an apparent effort to maintain a balance of ideological forces in the regime, Nasir put Sabri in charge of the ASU, replacing the moderate Husayn al-Shafi'i. Salama, who was close with the latter, was at this point removed from the General Secretariat of the party. He was replaced as head of the Workers' Secretariat by 'Ali Sayyid 'Ali, and as Minister of Labor by Kamal al-Din Rif'at, another former Free Officer sympathetic with Nasir's socialist rhetoric and policies.[3]

As a means of expanding his own power base, 'Ali Sabri then undertook to subordinate the workers' and peasants' organizations to the ASU. 'Ali Sayyid 'Ali aligned himself with Sabri. He removed the 30 unionists that had been added to the Workers' Secretariat, replacing them with two assistant secretaries both from his own petroleum federation. Together these three, working with several allies on the ETUF's executive committee, including 'Abd al-Latif Bultiya, the general secretary, tried to restrict the activities of the confederation.[4]

These efforts were repeatedly challenged by Fahim and others. However, the dimensions of this power struggle were not, as stated by Bianchi, a battle between "the leaders of the labor confederation" on the one hand and powerful individuals in the regime on the other.[5] Rather, the union leadership was itself divided. The majority of confederation leaders and other unionists supported Fahim, and will henceforth be called the confederation group. However, it must be stressed that the group headquartered in the Workers Secretariat (and to be called hereafter the WS group) itself consisted of several key confederation personnel and other elected union leaders,[6] who evidently saw the WS as a better route to expanding their power than the ETUF.

Nor was the dividing line over support or opposition for government interference in the union movement, as Bianchi claims, since both sides were complicit in this. The WS group was closely tied to the 'Ali Sabri "power center" in the ASU, and 'Ali and his cronies also forged strong links with several key ministers. However, the ETUF group had solid ties to certain powerful government officials as well, most notably Kamal Rif'at. The main confederation leaders were also involved in the upper echelons of the ASU, where rival centers of power could be enlisted for support. As in the 1950s then, disagreements within the trade union movement overlapped with, and were influenced by, power struggles within the regime.[7]

As an outcome of the WS/ETUF group clashes, the WEI was put more firmly under the ASU's control.[8] The WS was also somewhat successful in frustrating an ETUF initiative to hold "production conferences" at major parastatals, to further the national development campaign. However, the most enduring battle was over the question of union elections, which were scheduled to begin in the spring of 1967. The WS first tried to impose a system of elections by list. This would have enabled individuals like 'Ali and others, who were distanced from the workplaces because of their ASU responsibilities, to group themselves with popular candidates and thereby retain their union positions, despite having little support from the rank and file. The confederation group, aligned with Rif'at, was able to defeat this proposal. However, shortly after the election process began at the local level, it was suspended because of the June 1967 war. After the war, the WS circle successfully blocked repeated efforts by the confederation group to hold the elections for the rest of the locals as well as for federation and confederation leadership.[9]

The events that followed the war generated an atmosphere in which the confederation group felt it could push its case against the WS. Exposes of corruption in the armed forces led to widespread expectations of political reform. In this climate, the confederation group organized a large dinner party in January 1968, to which Kamal Rif'at, 'Ali Sayyid 'Ali, and all federation presidents were invited to discuss various issues in the relationship between the union movement, the ASU, and the government. 'Ali was clearly put on the defensive, denying any contradiction between the party and the ETUF as well as any personal interference in confederation affairs.[10]

The following month, the lenient sentences handed down for the army officers convicted of corruption sparked mass protests. On February 21, worker and student riots broke out in the industrial area of Helwan as well as in several other cities (see chapter 3). After calm was restored, Fahim wrote an editorial defending the workers for *al-'Ummal* (The Workers), the ETUF newspaper,[11] and in the ensuing months Fahim and those other members of the confederation group who sat on the higher bodies of the ASU raised their concerns in the secret meetings of these bodies. This was during a period of self-criticism within the party encouraged by Nasir, in the context of his promises for greater democracy in a March 30 document.[12] The document called for elections at all levels of the ASU and, in what I read as an implicit acknowledgment of the confederation group's concerns, for "unleashing the creative force of the unions." The ETUF organized a conference in the fall of 1968 to discuss implementation of this clause, which resulted in resolutions reiterating the concerns raised at the January dinner.[13]

However, 'Ali Sabri and the WS were able to stymie the ETUF group. Plans for the resumption of union elections were repeatedly announced in *al-'Amal* and *al-'Ummal* in 1968 and 1969, but they were never held. The WS group apparently feared the loss of their own positions, although the public justifications were the preparation of a new labor law which would change election procedures, and conflicts with the other elections being sponsored by the ASU. There is some suggestion that the WS appealed to powerful elements in the security apparatus, who they convinced that the union elections would be too disruptive in the post-war climate. This may have been the decisive factor in the inability of Rif'at and the confederation group to win Nasir's support for the elections. These intrigues are also the apparent reason for the fact that promulgation of a new labor law was repeatedly delayed.[14]

Throughout 1968 and 1969 the confederation group continued to raise their concerns through conferences and in their independent press. But other than receiving permission to publish *al-'Ummal* weekly rather than monthly, they remained frustrated in their goals. Nasir proved unable or unwilling to take on the 'Ali Sabri clique, apparently because of its potential ramifications for his own political position. Union issues were subordinated to larger, national political considerations, and this gave the upper hand to the WS group. In addition, in the summer of 1969 Ahmad Fahim became seriously ill and unable to lead the confederation group's campaign. His official duties were taken over by 'Abd al-Latif Bultiya, who formally assumed the ETUF presidency after Fahim's death in December of that year. Bultiya, as we have seen, had been close to the WS since its inception, and he immediately put an official end to the conflict between the two organizations by dropping the push for union elections and generally asserting the right of the ASU to supervise union affairs. For unionists who remained loyal to Fahim and the confederation group's principles, however, the contradiction remained.

The infighting between the WS and the ETUF group overshadowed, but did not eliminate, other contradictions over government interference in union affairs. The confederation objected to provisions of the 1964 labor law which required federations to first inform the police and then obtain permission from the Ministry of Labor before holding their annual conventions, and which gave the ministry other rights to interfere in union affairs. These issues were raised repeatedly in the aftermath of the 1967 war, with the ETUF calling for the confederation to take over some of the administrative and technical functions of the Ministry of Labor. It appears, however, that both Kamal Rif'at, the Minister of Labor, and Nasir felt that the preferred path was to put the ministry itself under the direction of leading unionists. Nasir had intended for Salama to maintain his presidency of the confederation when he was appointed as minister, and after Salama's removal the post was apparently offered to Ahmad Fahim several times. Fahim declined, maintaining that such an overlap would infringe on the independence of the union movement. But Bultiya did not share Fahim's reservations about serving simultaneously in government, and shortly after his assumption of the ETUF presidency Kamal Rif'at "joyfully" turned the Ministry of Labor over to him.[15]

These intrigues and their outcomes affected the unions' propensity and capacity to respond to changing economic conditions. Their significance is not limited to this, however, but extends to the broader realm of high politics. The actions of the WS were an important reason why many unionists later supported Sadat in his "corrective revolution" against the power centers, as we will soon see below.[16]

The Issue of Hierarchy: Cracks in the Union Wall

The disputes with the ASU notwithstanding, the confederation was able to proceed with its organizational consolidation. The 1964 labor law revised the structure of the ETUF, consolidating the number of federations from 65 into 27. However, the hierarchical structure which resulted from this soon became controversial with lesser union officials, as well as the rank and file, revealing the inherent contradictions of union centralization. The immersion of the confederation leaders in the intrigues of the ASU served to detach them from other confederation matters, and to divorce them from the lower levels of the union movement where this dissatisfaction developed and grew.[17]

The hierarchical structure of the confederation was never without controversy among middle and local level unionists. As we saw in the previous chapter, only 59 federations had been formed by the ETUF's 1961 convention, and at least 13 of these failed to send representatives to the gathering.[18] Although the reasons for this can only be guessed, the fact that many unionists were not part of the inner circle involved in forming the confederation is certainly suggested. Relatedly, the structure established in that law provided for only 21 individuals on the confederation's executive committee, meaning that the other 44 federations would not be represented there.

The 1964 law was issued without the full approval of the ETUF, and many of its provisions proved controversial with confederation personnel and other union leaders; numerous meetings to discuss revisions were held, of which the largest was an ETUF-sponsored conference in Alexandria in October 1965. One of the key objections raised by mid- and lower-level unionists there was to the manner by which the federations were merged. They complained that the consolidation was haphazard, throwing together industries which had little in common, thereby rendering coordination of union work more difficult. Moreover, six federations still wound up without official representatives on the confederation's executive committee. This spawned calls for further reducing the number of federations to 21 so that each could have a seat on the ETUF board. Although some unionists felt that this would further aggravate the coordination problem, a majority was won to the idea, and at the end of 1966 the

ETUF submitted a proposal to this effect to the Ministry of Labor. A few months later the Ministry issued a law approving the consolidation, but differing from the confederation's proposal in several aspects of federation composition. This further angered some second- and third-tier unionists.[19]

While federation leaders' concerns focused on this issue of representation, many local leaders had complaints about their limited maneuverability. The hierarchical structure of the ETUF made the locals strictly subordinate to the federations. In the 1959 law, the locals were actually denied legal personality. This was granted to them in 1964, but they still could not negotiate a collective agreement without the approval of the federation, and the latter was also empowered to withdraw its confidence from the locals. In addition, the law placed limits on their access to membership dues. The 1964 law required the locals to turn 10% of these over to the confederation, and 25% to the federation. An additional 25% was earmarked for administrative expenses at the local and federation level, and 5% was to be held in reserve. This left only 30% for the local to use for services to its members—items such as housing projects, day care centers, trips, and social gatherings—which, as defined by the regime, were to be the stuff of local activity at the time. While this figure represented an improvement over the 20% allotted to them in the 1959 law, local officials repeatedly complained that the funds available to them were insufficient and called for a change in these stipulations.[20] This subordination of the locals to the higher bodies of the union movement, which continued thereafter, served to stifle the initiative of local leaders and restrict their ability to address the day to day issues confronting workers.

Some unionists also objected to aspects of the indirect electoral system specified by the 1964 law, whereby rank-and-file workers chose only the members of their own local board. The local board then appointed representatives to the federation convention, where the federation board would be selected from among the local delegates; the ETUF board was similarly chosen at a convention of delegates appointed by the federation boards. The by-laws of most locals and federations, as well as the ETUF, further specified that the boards at each level would select their own president and other officials.[21] The way this system functioned is portrayed graphically in figure 2.1. Objections to it stemmed from the fact that the number of convention delegates permitted to most locals was larger than the size of the local board. Therefore it was possible for an unpopular individual to run for local office and lose, but nevertheless be chosen to attend the convention where he could then run for federation office. This in fact happened in at least two cases, one of which involved Salah Gharib, who became president of the Textile Workers' Federation that year and head of the ETUF itself in 1971.[22]

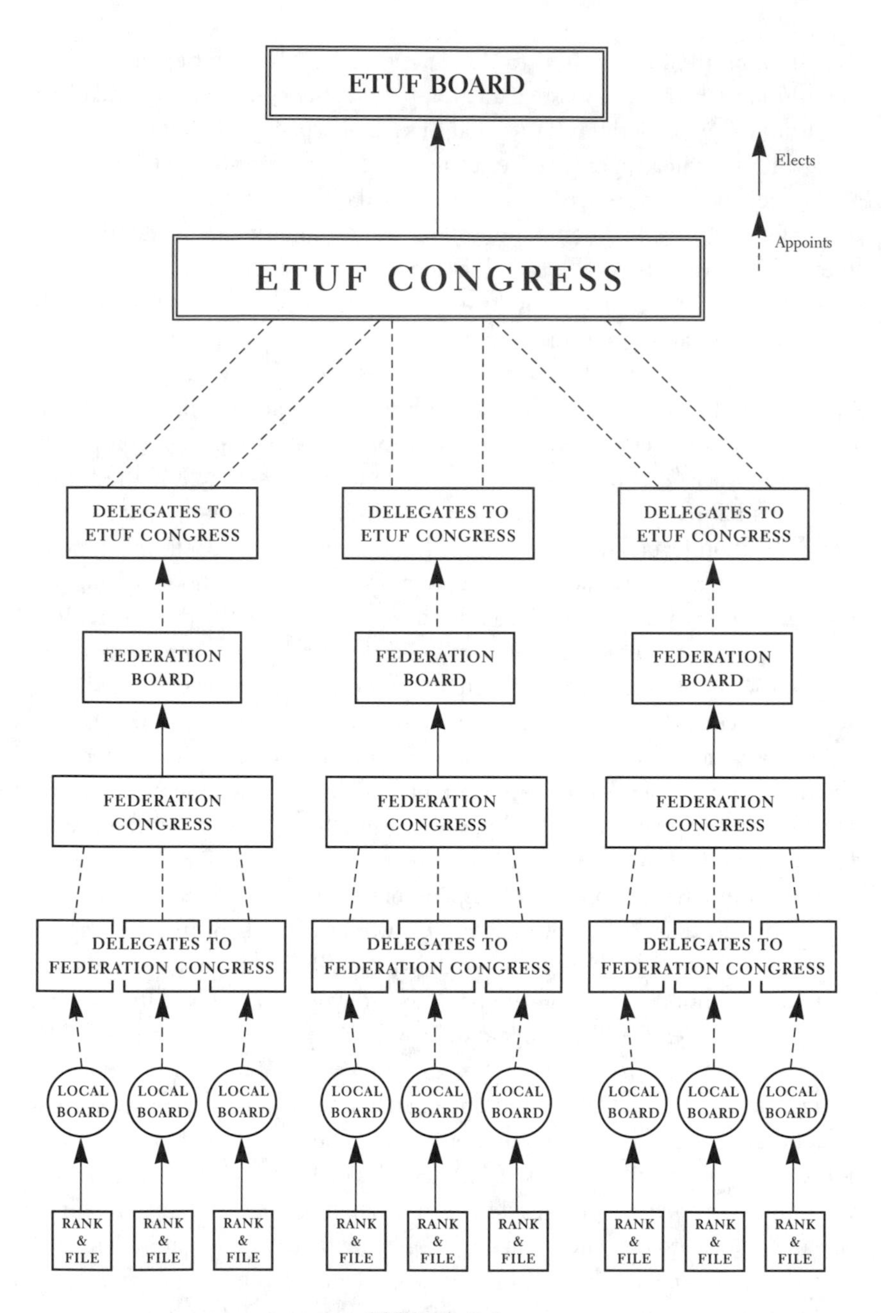

FIGURE 2.1

Structure of the Egyptian Trade Union Federation

From the base, the sharpest and most widespread criticisms were directed at the functioning of the leadership bodies, which were charged with failing to keep the rank and file informed of their activities, neglecting the lower bodies' concerns, and not organizing summations of union work in order to foster improvement. In some cases, these criticisms were encouraged by traditional leftists, some of whom had resumed organizing among workers after their release from prison in 1964;[23] they were prevented from attaining union office, however, by the postponement of the 1967–68 elections. Ahmad Taha is noteworthy in this regard. At the confederation level, Taha notes, there was no discussion or evaluation of the organization's performance at all from 1964 through 1966; in addition, the leadership decided to conceal differences of opinion from the membership and the public by publishing only their final decisions in the minutes of their meetings. Furthermore, the minutes reveal no discussions about the problems confronting the lower bodies, but concern only such issues as who will participate in delegations abroad or be nominated as representatives to various nonunion committees.[24]

The federations also came under fire. A major complaint was their failure to hold their annual conventions as required by law; they were also charged with not paying attention to the problems confronting the locals. These issues were raised repeatedly by workers and local leaders throughout the 1960s, and at the end of the decade *al-'Ummal* found that local leaders, despite varying ideological tendencies, were in agreement that the trade union movement had stagnated and that contact between its different levels had been lost. It is instructive to quote from some workers in this regard:

> We don't feel the existence of the federation; most of us don't even know where it's located.[25]
>
> In some workplaces we feel the existence of a new kind of feudalism called "union feudalism" . . . the links between the different organizational levels are no longer reliable, and an indication is that in some workplaces the management tries to block the unionists and paralyze their activity and the federation does not intervene. And this causes the workers to lose their trust in the local.[26]

Many blamed this loss of contact on the problem of individuals simultaneously holding office at more than one level of the union structure, a practice known in Arabic as *jam'* (combination). The 1959 labor law explicitly permitted this; in 1964, a limit of two positions per individual was imposed. Workers charged that this limit was violated in practice, however; in addition, higher level union officials were often chosen for posts in the ASU and/or the government. *Jam'* enabled some unionists to enhance their livelihoods by drawing on more than one union salary, and workers charged that as this happened they lost empathy with their poorer constituents. It also hurt the locals because the unionists involved tended to devote most of their energies to their higher level positions. Finally, it reduced the total number of individuals involved in union leadership, and in particular inhibited the development of "second-rank" leaders capable of effectively running locals and performing secondary federation tasks.[27]

Lastly, there was the problem that some enterprises had several branches in different locations, all represented by a single local, and in some very large plants there was little contact between the workers and the local leaders. There was a proposal, initiated by 'Abd al-Mughni Sa'id and supported by leftist labor activists, to create a system of elected workers' delegates (*mandubin*; a rough Western equivalent would be shop stewards) to act as intermediaries between workers and local leaders in these cases. This proposal was formally adopted by the ETUF at a conference in January 1968, and it was expected that it would be incorporated into the revised labor law being drafted after the war. As we saw above, the political infighting between the WS and the confederation group blocked the issuance of that law. However, in some workplaces leftists took the initiative to organize the *mandubin* system on an informal basis.[28]

Whether coming from federation or local leaders, traditional leftists, or ordinary rank-and-file workers, none of these criticisms ever challenged the idea of having a singular trade union confederation. Instead, critics sought only a change in the power relations within the confederation, to give more authority to its bottom layers. Confederation leaders were not unaware of these demands, but proved unable or unwilling to make changes. According to 'Izz al-Din, Fahim himself was sympathetic to the criticisms, but believed that the key to transforming the confederation was changing the leadership, hence he devoted his energies to trying to remove the obstacles to holding new elections.[29]

Yet even if as dedicated to pursuing union independence from the regime as 'Izz al-Din would have us believe, Fahim may not have been equally committed to internal union democracy. Sayyid Fa'id, who had worked closely with him in the Textile Workers' Federation throughout the 1950s, charges that Fahim's ascension to the ETUF presidency corrupted him; in particular, he maintains that Fahim was complicit in his arrest and beating during the 1963–64 election campaign. There is also the case of 'Abd al-Ra'uf Abu 'Alam, an ETUF adviser who wrote several articles for *al-Tali'ah* in 1966 in which all of these criticisms about the functioning of the union hierarchy were made. In response, the executive committee of the confederation took punitive action against him by withdrawing his commission; the decision was almost unanimous.[30] Moreover, the ETUF under Fahim's leadership never challenged the continued prohibition against strikes. Instead, as we will see in chapter 4, the confederation in 1965 called on the rank and file to forswear work stoppages out of loyalty to the government, and endorsed punishments for locals that violated this call.

All of this suggests that the creation of a hierarchical union structure, long sought by trade union activists, in practice did little to serve the needs of its base. Moreover, the responsibility for these deficiencies rests only partially on the interference of government agencies and the WS clique. The leaders of the ETUF and many federations, regardless of whether they belonged to the WS or the confederation group, or neither, were at best negligent in their duties and at worst corrupted by the power and material advantages which came with their positions. As Muhammad Gamal Imam sums it up, "the most progressive leaders were too busy to attend to union affairs, those more limited in experience didn't care, and there was a great deal of opportunism."[31]

Competition

Given the complaints at the base over the ETUF's functioning, workers had reason to look for alternative channels of interest articulation. At the same time, Nasir's conversion to corporatism did not extend to giving the confederation a monopoly on workers' representation. Thus the 1960s saw the flourishing of both officially sponsored and unofficially tolerated rivals to the confederation.

With the conscious intent of the WS group, the ASU itself emerged as the main competitor to the ETUF. The confederation was able to defeat the WS clique's first attempt to subsume its role. At the end of 1965, the ETUF executive committee discovered that the WS, in conjunction with Sidqi and the Minister of Justice, had drafted amendments to the new labor laws. The changes would have eliminated union representation on consultative committees with management and the ASU at the plant and industrial level. Upon learning of this, Fahim exposed it in the pages of *al-'Ummal*, while Rif'at protested directly to Nasir. As a result, the proposal was withdrawn.[32]

However, the WS was able to establish ASU factory committees, whose defined role overlapped with that ascribed to the unions. While rank-and-file workers may not have been aware of the power struggles between the two groups, they could see clearly that there were two organizations playing similar roles at the base, and it was often the case that the ASU committees, because of their political prestige and connections to the power centers, were better able to get results in pursuing workers' grievances. This was especially the case when the Sabri clique sought to use these committees to build their case that public sector managers acted as members of a new bureaucratic class. As a result, workers turned increasingly to the ASU with their grievances, contributing to the atrophy of many union locals.[33]

The contradiction with the ASU became so sharp that many in the union movement began to question its own *raison d'être*. Union leaders at all levels repeatedly pushed for clarification as to the functions of the different organizations at the factory level. Given the distrust of most confederation leaders for the Sabri clique, these calls for enlightenment can only have been intended for Nasir himself. But Nasir, according to Baker, was "not displeased" with the status quo. He apparently preferred having more than one watchdog over the managers in the plant, whose rivalry with each other would prevent any one group from becoming too powerful.[34]

Another competitor to the unions emerged in the workers' representation in management (WRM) councils, charged with preparing the production plans for the workplace establishments, upon which workers' bonuses were based. Although some unionists are said to have seen the WRM as a threat to their own positions, the confederation officially welcomed the concept as a "socialist gain." There was, however, a concern that there was no formal system for the elected delegates to report back directly to the workers, and some felt that this should be done through the locals. In some cases local leaders took the initiative to solve this problem by running for election as a WRM, since the law did not prevent this, but the ETUF's official reservations persisted.[35]

Meanwhile, the union locals' traditional rival, the workers' leagues, remained a particular thorn in their side. The 1964 labor law had called for abrogating the leagues, and transferring their funds to the relevant local, in all establishments where a trade union existed. Although some leagues were dissolved on this basis, most continued to function in defiance of the law. They were particularly strong in the civil service where unions had previously been outlawed—according to one confederation report, 56 existed in the railway authority alone. Local leaders complained that because the leagues were not part of any hierarchy, they were able to keep the full amount of dues they collected, and could therefore provide more services to their members than the locals which were obliged to turn over the majority of their dues to the higher bodies. Confederation leaders lobbied strongly with the Ministry of Labor for enforcement of the 1964 law, and abolition of the leagues was one of the resolutions of the 1965 labor law conference. In 1966, the ETUF prepared a draft law calling for the abrogation of the leagues. The bill was passed by the labor committee to the parliament, but died in the economics committee where it was sent, under pressure from the leagues, because of the financial issues involved. The issue surfaced again after the 1967 war; now the ETUF abandoned the call for elimination of the leagues, but did seek the right to supervise their work. Once again, the unions took the demand to the parliament, but no action was taken there.[36] The inability of Rif'at, a strong supporter of the confederation, to enforce the laws against the leagues suggests that the latter had powerful allies elsewhere in the ruling elite.

Not surprisingly, all this competition had an impact on membership in the ETUF.[37] The 1964 labor law paved the way for substantial growth of the confederation by allowing for the incorporation of agricultural and civil service workers. According to the official statistics, the ranks of the ETUF more than tripled as a result, from 346,491 in 1961 to 1,290,811 in 1964 (see table 2.1). Nevertheless, unionization was especially weak in those industries dominated by the private sector; in leatherworking, for example, only 15% of the labor force was organized. There was also a particular problem in the civil services where the leagues remained strong. Yet little was done to change this situation. Except in food processing, there is no evidence of any coordinated organizing drive on the part of the confederation or the federations. As a result, membership virtually stagnated, growing by little more than 1% annually after the 1964 jump. The commitment of those workers who did belong to the unions was also questionable: Ahmad Fahim himself reportedly stated that if public sector workers had understood that it was within their rights to withdraw from the locals, many would have done so.[38]

The survival, if not vibrancy, of the organizational rivals to the official trade union movement highlights the weakness of the argument that state corporatism provides an effective vehicle for a handful of regime elites to control masses of workers. Workers will not be enthusiastic members of organizations which misrepresent them, and will experiment with more participatory alter-

TABLE 2.1

ETUF Membership in the 1960s

Year	# of Members	Avg. Annual Grth Rate[a]
1961	346,491	
1964	1,290,811	55.0%
1971	1,431,160	1.4%

Source: ETUF 1982, p. 24.

[a] My calculation.

natives. Likewise, as the denouement of the WS described below reveals, when given the opportunity workers will reject leaders who fail to fight for the workers' interests.

Sadat and the ETUF: The Early Years

Anwar Sadat assumed the presidency of Egypt when Nasir died in September 1970. Sadat's rule saw a shift in Egypt's political alliances from east to west, and correspondingly an embrace of economic liberalization and a brief experiment in political pluralism. Inside the ETUF, the most dramatic effect of his tenure was the return of the traditional left to positions of influence in the union hierarchy, along with the entry of new leftist elements. These forces united with unionists still committed to Nasirism to challenge Sadat's economic reforms, and as his allies in the unions maneuvered to contain them, a struggle for union democracy and against government interference in the ETUF became intertwined with these economic battles.

Sadat "Corrects" the Unions

Although Sadat was Nasir's legal successor, his ascension to power did not go unchallenged. Over the next eight months Ali Sabri and his cronies resisted Sadat's authority and plotted against him. Sadat finally moved against the group in May 1971, launching a "corrective revolution" to sweep Sabri supporters from positions of power and supposedly end the worst vestiges of political repression under Nasir.[39]

For the unions, the spring and summer of 1971 marked a decisive moment in intraconfederation relations and in the struggle against government interference in union affairs. The corrective revolution presented an opportunity for the ETUF to make a clean break with the past pattern of permitting government intervention in leadership selection. This opportunity was forgone because of a strategic decision made by a leftist grouping in a position to influence the outcome of the union elections. The result was that the presidency of the ETUF went to Sadat's chosen henchman, Salah Gharib; in the aftermath, the issues of hierarchy and government intervention became inextricably intertwined as Gharib used the power of his post to try to remove all opponents of Sadat's political and economic policies from the ranks of union leadership.

Because of the presence of some allies of Sabri in the confederation hierarchy, and of a strike of metal workers in Helwan initiated by pro-Sabri forces (see chapter 3), it is sometimes claimed that the union movement supported the Sabri clique.[40] However, as we just saw, most of the senior union leaders were opposed to the WS group. At the lower levels of the confederation, there were some pockets of support for Ali Sabri, but there was also a great deal of anger and frustration over the obstacles the WS had placed on union democracy, as well as distaste for the condescending attitude of the ASU elite. As a result, the majority of union leaders and activist workers at all levels welcomed the corrective revolution. As Sayyid Fa'id, the long time leftist activist in a Helwan textile plant, put it, "The workers didn't like Sadat but we *hated* Shar'awi Guma', Ali Sabri, etc., who acted like they and only they understood socialism and spoke for all the people."[41]

When the news of Sabri's arrest got out, the confederation quickly issued a statement praising Sadat, and called on the federations to hold emergency meetings to organize public displays of support. The board of the Petroleum Workers' Federation voted to remove 'Ali Sayyid 'Ali from office and deprive him of federation membership. His position as head of the WS went to Salah Gharib from the Textile Workers' Federation. Gharib was a close associate of Mamduh Salem, the mayor from Alexandria, and together they had helped Sadat time his move against Sabri; Salem was rewarded with the post of Minister of the Interior. 'Abd al-Latif Bultiya, the ETUF president who had been close to the Sabri clique, sensed which way the wind was blowing and quickly switched loyalties to lead the pro-Sadat marches sponsored by the unions in industrial areas such as Helwan and Shubra.[42] Thus he was able to retain his post as Minister of Labor during the initial cabinet changes.

Then, in June 1971, Sadat decreed the disbandment of all mass organizations subordinate to the ASU, including the labor unions, ordering new elections to be held at all levels in July. 'Aziz Sidqi, the former Minister of Industry, was put in charge of forming committees—one for each federation—to supervise the new elections. Sidqi worked closely with Gharib and Salem in forming and supervising these bodies.[43] Although this constituted a new form of government interference in union affairs, the move was generally welcomed by labor activists, and it appears that the elections at the local level were mostly free and fair.[44] In most workplaces the number of candidates vastly exceeded the number of slots available, and the campaigning was enthusiastic. In a reflection of the rank-and-file frustrations discussed above, many former union leaders lost their positions in these elections.[45]

Another significant result of the local elections, and one which bears on the developments at the federation and confederation level, was the reentry of the traditional left into the ranks of union leadership. As we just saw, the suspension of the 1966–67 elections had prevented the communists from reclaiming union posts during the final years of Nasir's rule. Now numerous communists and other leftists ran for and won local office, thus placing them in a position to obtain federation posts or at least influence the direction of the federation elections. Among those who did so were Ahmad Taha in the Printing, Publishing, and Media Workers' Federation, 'Abd al-'Azim al-Maghrabi in the Commerce Workers' Federation, Ahmad al-Rifa'i in the Agricultural Workers' Federation, and Ibrahim Khalifi in the Social Services Workers' Federation.[46]

At the same time, Sadat himself was generally unfamiliar with the trade union movement and its past leaders. As a result, he did not have a clear picture of exactly who he wanted to see in leadership posts, and who he did not. His only clear decision was the choice of Gharib for ETUF president, and he entrusted Gharib with preventing undesirable elements from obtaining positions at the upper levels of the hierarchy. The mechanism used to remove these individuals was to deny them the requisite certificate of membership in the ASU (see chapter 1). Communists who had achieved leading positions in the ASU were reportedly complicit in the act in some cases, apparently part of a tacit alliance between Sadat's inner circle and certain wings of the communist movement who supported his campaign against the Sabri "power center."[47]

Although numerous experienced unionists were prevented from running in this manner, there was no clear pattern to the exclusions. It was not, for example, a question only of association with the Sabri group, since many of those who had been close to Ahmad Fahim were likewise barred. Some feel that membership in the ASU's secret "vanguard" may have been a criteria, but it was not the sole or the definitive one;[48] Gharib also moved to eliminate potential challenges to his election by having certain popular individuals barred from running regardless of their ideology and political affiliation. Among those excluded were Kamil 'Uqayli from the Land Transport Workers' Federation, Sa'd Muhammad Ahmad from the Food Processing Workers' Federation and 'Abbas Mahmud from the Mining and Mineral Workers' Federation. All three had been members of the previous ETUF executive committee, and at least the latter two were free of any association with Ali Sabri. In addition, Gharib sought to diminish the power of rival unionists by splitting their federations; prior to the elections he used his ASU post to order another change in the number of federations, this time increasing it to 26.[49]

As a consequence both of rank-and-file choices and these machinations, there was an overall turnover of about 80% in membership of the federation boards, with many of the new faces representing a younger generation. Nevertheless, Bianchi's characterization of these elections as "the only real purge" the ETUF has experienced is a misrepresentation on two different levels.[50] While negating the purges of leftists which both preceded and succeeded this event, it exaggerates both Sadat's knowledge of union affairs and his ability to control these elections. Actually, despite the intervention, some unionists who were supporters of Ali Sabri and other strong proponents of Nasirism were able to avoid exclusion; in addition some of the communists involved were hostile to Gharib.[51] In fact, in many federations there were two lists of contenders—one grouped around support for Sadat and Gharib and another antagonistic to them. The contests were heated and the results close, with opposition forces able to win the leadership of some federations. In the Land Transport Workers' Federation, for example, despite 'Uqayli's absence the presidency was won by his close friend and ally Isma'il al-Dimilawi.[52]

Furthermore, at the confederation level, the elections were free and fair, and the opposition forces mounted a strong challenge to Gharib. One leading unionist active in the labor movement since the 1960s told me that at this stage, these were the most democratic elections the ETUF had ever seen.[53] Gharib almost failed to win, and his ultimate triumph was the result not of interference or manipulation by Sadat but rather of a political decision taken by certain leftist forces.

As explained above, the procedures for forming the ETUF executive committee (EC) called for the convention delegates to choose only the board's general membership, who would then select their own president and vice president, etc. There were 21 seats on the EC, and when the voting was over Salah Gharib was ranked only sixteenth in this poll. The most votes were received by al-Rifa'i, the communist lawyer from the agriculture federation, then the largest of the ETUF's member federations. This gave al-Rifa'i a strong claim to the presidency.

Gharib's poor showing was the result of several organized efforts to defeat him entirely. One was spearheaded by 'Uqayli, in conjunction with Dimilawi and 'Abd al-Mun'im al-Ghazzali, a leftist adviser to their federation; they were pushing Dimilawi for president. The second effort revolved around al-Maghrabi, Taha, Khalifa, and Taha 'Abd al-Qadr from the Administrative Services Workers' Federation; they agreed to back Taha. And independently of both groups, 'Aisha 'Abd al-Hadi was encouraging her fellow delegates from the Chemical Workers' Federation and others with whom she was close not to support Gharib.

The various groups gathered the night before the elections to plan strategy and come up with candidate lists. According to Taha, the meeting of his group was attended by al-Rifa'i, who promised to support him. Meanwhile, according to al-Ghazzali, the Land Transport group decided that it lacked sufficient votes to elect Dimilawi. Feeling that it was nevertheless important to defeat

Gharib, to "break the wall of fear" around challenging government meddling in union affairs, al-Ghazzali approached al-Rifa'i and offered to back him or another independent candidate of his choosing. To his surprise, al-Rifa'i refused, and when it came time for the EC to choose its president, al-Rifa'i declined to run and threw his support to Gharib.

Al-Maghrabi and Khalifa did make it onto the EC, but Taha came in twenty-second, just missing a seat; he blames al-Rifa'i's "treachery" for his defeat. Likewise, although Dimilawi did make it onto the board, al-Ghazzali holds al-Rifa'i and his associates responsible for Gharib's victory, as well as for 'Uqayli's earlier exclusion from the elections. Both Taha and al-Ghazzali charge that al-Rifa'i had secretly met during the night with Gharib and his associates and agreed to back a different list of candidates, with Gharib at its head. He then went about encouraging others, including 'Abd al-Hadi, to do the same.

Although al-Rifa'i himself would not admit it, other members of the communist cluster he was affiliated with at the time, HADITU (see chapter 1), acknowledge that his behavior reflected their policy, which was not to openly defy Sadat's wishes at this early stage. They also did not want to be responsible for blocking the presence of the textile federation on the ETUF board; since Gharib was the only candidate from textiles, his defeat would have meant that the largest industrial union had no representative on the EC.[54] Finally, they questioned the wisdom of having al-Rifa'i, a lawyer and a member of a nonindustrial union, serving as president of the ETUF. Some friends of al-Rifa'i further claim that Gharib had threatened to cancel the elections to the EC unless al-Rifa'i agreed to back him. Hence they defend al-Rifa'i's action on the grounds that he was seeking to preserve democracy in the union movement to some degree, although the results were ultimately the opposite.

It was therefore not a foregone conclusion that Gharib would win the ETUF presidency simply because he was Sadat's known choice. While they disagree on what *should* have happened, everyone I spoke with concurred that the opposition forces *could* have defeated Gharib if al-Rifa'i had acted differently. This underscores the fact that even in highly authoritarian countries, the results of labor/state interactions cannot be traced solely to the will of regime elites. The 1971 elections highlight the important role that senior unionists can play, and the significance of their individual motivations, whether ideological or opportunistic. These unpredictable individual calculations necessarily impose a degree of indeterminacy on state/labor relations.

Likewise, we cannot be certain of what might have happened if Rifa'i had challenged and defeated Gharib. Sadat obviously could have disbanded the unions by decree again or imposed Gharib's presidency by force, and/or ordered the arrest of the dissidents. But any such repressive moves would have had costs; in particular they would have eroded the credibility of Sadat's pledge to liberalize at a time when he was lacking a popular base and seeking to build legitimacy. And Gharib's rule, if imposed, would also have lacked any moral or legal claim. Such an outcome may ultimately have proved less detrimental to the independence of the labor movement than electing Gharib. However benign their intent, the actions of al-Rifa'i and his allies effectively reaffirmed, at a crucial turning point in union/state relations, the right of the regime to interfere in the ETUF's leadership selection process. Moreover, as Sadat began engineering a fundamental shift in Egypt's political alliances and economic development strategy, these actions bestowed legitimacy on an ETUF president determined to use his position to squelch opposition to Sadat's policies.

Gharib vs. the Left

At the helm of the ETUF, Gharib moved to consolidate a core of allies on the ETUF's board, and within the federations.[55] Sadat gave Gharib enhanced abilities to manipulate union affairs by appointing him Minister of Labor in January 1972.[56] In this capacity Gharib sought to promulgate laws to tighten the control of the confederation hierarchy over the base and make it easier for government-backed candidates to win the federation and confederation elections. Communists from various sects, Nasirists, and independent leftist forces formed a common front to challenge Gharib, so that two opposing camps developed within the upper echelons of the union movement. Over the next several years, these two groups clashed over numerous aspects of union affairs and union/state relations.[57]

The conflict surfaced in February 1972, at a conference on labor law sponsored by the ETUF in Alexandria. Of the six panels at the conference, the most popular and lively was the one on union law, chaired by the leftist al-Maghrabi. Heated discussion broke out over a proposal, apparently backed if not initiated by Gharib, to stagger the terms of leadership at the different union levels. Thus locals would hold elections every two years, federations every three, and the confederation every four, with the higher bodies supervising the elections at the lower levels. This plan would have enabled incumbents at the top of the hierarchy to intervene in the lower levels and possibly prevent potential challengers to their own reelection from gaining office. It was opposed by the majority in attendance. At the same time most participants called for an end to the ASU membership requirement for union leaders, a stipulation that

Gharib defended. In addition, in what could now be interpreted as a direct affront to Gharib, the panel endorsed a resolution resurrecting the ETUF's earlier insistence that all aspects of supervising union work be transferred from the Ministry of Labor to the confederation itself.[58]

The move against the communists came in March 1973. Gharib, having mobilized his allies in advance, simply called a meeting of the executive committee and had them voted off the board. Al-Rifa'i, al-Maghrabi, and Khalifa were thus removed from office. The three tried to mobilize support from among the other opposition forces, but antagonisms which had developed around the 1971 elections now came back to haunt them. Those they had deserted in 1971 refused to support them, and without such help they could not mount an effective campaign versus their expulsion. Thus the communists, having served their purpose to the regime in the summer of 1971, were again purged from the upper ranks of union leadership again, less than two years later.[59]

Several months later, Gharib moved to give the expulsion of the communists a legal basis, and to prevent their reentry in the upcoming elections, by banning *all* professionals from holding union leadership positions, whereas they had previously been permitted to hold up to 20% of the leadership posts iin a local. Using his position as Minister of Labor, he decreed in June 1973 (rule #48) that all candidates for union office must fit the official definition of a worker; the latter had been specified and redefined by the ASU in the 1960s to clarify the requirement of 50% worker and peasant representation on governing bodies. As it stood at the time, this stipulation precluded all those entitled to join professional syndicates, other graduates of the universities or military academies, and all holders of advanced technical degrees unless these were obtained after the individual had started out as an ordinary worker and had not changed his place of work after obtaining the degree.[60]

Gharib also acted to prevent any repeat of his embarrassment in the 1971 elections by arbitrarily reducing the number of federations to 16, with a requirement that each federation have one representative on the confederation EC. He then canceled the 1973 ETUF convention so that his actions could not be discussed, and no elections could be held. Instead, each federation appointed its representative to the EC, and the remaining five seats required by law were left empty. Having thus consolidated his position of power, Gharib no longer felt threatened by the other union leaders from the 1960s who had been barred from the 1971 elections. They were permitted to run again in 1973, and both Ahmad and 'Uqayli regained their federation and confederation positions. At the same time, at the lower levels, denial of ASU membership certificates was again used against rank-and-file militants.[61]

The decree about professionals caused a great deal of controversy within the ranks of the confederation. Because its effect went far beyond the al-Rifa'i group to include many intellectuals, it was opposed by activists at all levels of the trade union movement.[62] Debate raged through 1974 and most of 1975, and spilled over into the pages of the daily press, particularly the labor column in *Al-Ahram*. The decree faced several legal challenges in 1974, with the courts upholding the right of professionals to run for union office. But when the confederation was preparing a new draft union law to submit to the parliament in 1975, the restrictive provision was part of it; Gharib also included it in a second draft law which was proposed by the Ministry of Labor.[63]

These conflicts were exacerbated by Sadat's embrace of economic liberalism, and his increasing enchantment with the West after the 1973 Arab–Israeli war. Gharib's strict support of Sadat's policies enraged the leftist elements in the union movement and exacerbated the divisions within it. With the traditional communists absent from the higher ranks of the union structure, the struggle against Gharib was now led by other individuals, mostly of Nasirist leanings. Foremost among them was Sa'id Gum'a, president of the Electrical, Engineering, and Metal Workers' Federation (hereafter EEMWF), Fathi Mahmud Mustafa (hereafter Fathi Mahmud), president of the Commerce Workers' Federation (hereafter CWF), both of whom sat on the ETUF's board, and 'Abd al-Rahman Khayr, an official of the Military Production Workers' Federation.

As detailed in chapter 4, workers in the commerce sector were the first to be threatened by the calls for privatization unleashed by Sadat's economic opening. Mahmud spearheaded the campaign to defend them, and Gharib alienated many unionists by seeking to obstruct Mahmud's efforts, including ultimately engineering his removal as federation leader. Gum'a was allied with Mahmud in these battles. Khayr played a key role in organizing a January 1975 demonstration of workers at the Bab al-Luq train station, one of the first public protests against the government's new economic turn (see chapter 3). Gharib and his allies supported the government's accusations against participants who were arrested and resisted the efforts of progressive unionists to win legal and financial aid for them from the confederation. Khayr also charges that Gharib sent a close associate down to Helwan in an effort to ascertain the identity of the demonstration's organizers.

The early years of the *infitah* also saw an upsurge in wildcat protests at individual plants (see chapter 3). In a move apparently aimed at calming this atmosphere of rank-and-file restiveness and sharp divisions among the confederation's leadership, Sadat removed Gharib as Minister of Labor in June 1975, reinstating 'Abd al-Latif Bultiya. The latter had taken up work in the International Confederation of Arab Trade Unions after his replacement in 1972, and thus continued to have ties with the union movement. He had also befriended Sayyid Mar'ai, then speaker of the parliament, and through that association he had gained the trust of Sadat. But Gharib continued to claw for power over the next year, and was aided by the fact that his close friend Mamduh Salam became Prime Minister in 1975.

There thus emerged two pro-regime clusters in the upper echelons of the union structure, each with access to close associates of Sadat. In this regard the situation does not differ substantially from what we have seen in the previous decades. The new element, however, was the continued activism of the third, leftist pole that was openly hostile to the regime's new economic policies. These forces were generally a minority, but did have supporters in the press and the parliament, as well as some ties to the rank and file.

The category of "leftist" is defined here in a specific historical and regional context, by the then-standard dichotomies on economic and foreign policy issues. "Leftists" refers broadly to those who favored government intervention in the economy to support subaltern strata and promote national development intended to narrow the gap between Egypt and the advanced industrial economies, combined with opposition to Western imperialism and Zionism. By this definition, the leftist camp comprised both traditional Marxists, Nasirists, and new Marxist groups which began to emerge in the late 1960s. Their programmatic unity around these issues should not mask important underlying differences, however. Nasirism is not an atheistic ideology and is thus not, like Marxism, incompatible with Egypt's dominant religion, Islam. Nasirists also lack the philosophical commitment Marxists have to promoting labor militancy although, as we will see below, some of the Marxist sects themselves abandoned this creed. Finally, whereas Marxism advocates the abolition of capitalism and, ultimately, of the state itself, Nasirism promotes the state as the unifying core and driving engine of the nation. As Beinin has noted, to the degree that Egyptian Marxists embraced nationalism and elevated it over class issues, this represented a capitulation to the popularity of Nasir and a departure from Marx's internationalist philosophy.[64]

For this loose leftist alliance, because of Gharib's and now Bultiya's efforts to manipulate union affairs in the service of the regime, opposition to economic liberalization became inextricably linked to the push for union democracy and independence from the state. However, these latter issues are not taken here to define leftism.[65] As the 1971 elections' story underscores, not all leftists embraced democratic operating principles; the affinity of many

Egyptian Marxists and Nasirists for the Soviet model of government is also grounds for questioning their commitment to democracy. On the other hand, it is perfectly plausible that sincere unionists who did not share leftists' ideological beliefs were actively opposed to the government's interventions and Gharib's manipulations.

Immediately after his appointment Bultiya met with the ETUF board. He promised that the union elections would be held as scheduled, and also pledged himself to seek speedy passage of the new labor law while acknowledging the need to resolve the controversies surrounding it. Shortly thereafter, however, he issued an order postponing the elections for three months, until October. The official justification was to avoid overlap with the ASU elections scheduled for that summer, but it was understood that the real motivation behind the decision was to calm the situation in the confederation.[66]

However, the antagonisms intensified rather than abated. In July Gum'a uncovered an attempt by Gharib to undermine the confederation's refusal to relate to Israel's labor confederation, the Histadrut. Gharib had secretly invited a delegation from the International Confederation of Free Trade Unions (ICFTU) to visit Egypt. Because the ICFTU included the Histadrut, Arab unionists had traditionally shunned it; during the 1960s many of Egypt's federations had affiliated instead with the World Federation of Trade Unions, which was associated with the Eastern bloc. Gum'a and Mahmud exposed and denounced the invitation as a violation of the ETUF's international policy as well as its decision-making procedures, leading a victorious campaign against it in the press and among other federation officials. The denunciation was signed by ten federation presidents, representing more than half the total number of federations, and published in both *al-Ahram* and *Rose al-Youssef.* The EEMWF also insisted that Gum'a withdraw from the confederation board in protest. Gharib, writing in *al-'Ummal,* claimed to have Sadat's support, but the invitation was opposed by Bultiya, and the unionists were told by Sayyid Mar'ai that the president had had no prior knowledge of the affair. Gharib also denied that any serious disagreement existed, and accused *Rose al-Youssef* of lying about Gum'a's resignation. *Rose al-Youssef* then printed the minutes of the board meeting where the resignation had occurred, and sued Gharib for his accusations against the magazine.[67]

Ironically, these battles served for the opposition elements to revitalize the union movement and endow it with a new raison d'être. While unanimous in their condemnation of Gharib as the worst president of the ETUF, most union activists I interviewed considered his tenure the liveliest period for the union movement as a whole. The victories they scored against Gharib highlight the critical role that internal union conflict in general, and beneath-the-surface organizing by leftists in particular, play in shaping legal structure outcomes.

Hierarchy and Singularity Revisited

Gharib further antagonized his opponents, and generated new concerns among lower-level unionists and the rank and file, by imposing new restrictions on candidates for union office intended to assure the reelection of himself and his close associates. He entered provisions into the draft labor laws before parliament that only those who had held local office for two sessions could run for the federation boards, and only those who had held federation office for two terms could seek positions on the EC of the ETUF. He initiated changes in the federation leadership selection process by altering the model by-laws for federations to preclude elections being held at the congress; instead, delegates would cast their votes prior to the meeting at the federation's headquarters. This stipulation, which was adopted by most federations, meant that candidates for federation office could not do their campaigning over the course of the congress, but rather had to get to know the local delegates by traveling to their workplace sometime prior to the vote. It prejudiced new candidates and ordinary workers who lacked the resources for such travel, while favoring incumbents and those having the financial backing of a governmental agency or party. A similar situation obtained, and likewise privileged government-backed candidates, in the local elections of factories having more than one branch, since the law permitted only one local per factory. There was also a renewed debate around restructuring the number of federations again, with Gharib now seeking an increase to 18; Bultiya and others were advocating a return to 21.[68]

At the local level Gharib did legalize the system of workers' delegates in large plants. However, his 1973 decree to that effect stipulated that they be appointed by the local leadership, rather than elected by workers. This was opposed by the leftists, and the issue was hotly debated in the press during the 1974–76 period .[69]

In September 1975, Bultiya decreed another three-month postponement of the elections, until January 1976, to give the parliament more time to consider the labor law. Then Sadat lifted the ASU membership requirement by decree in October, so that for the first time the confederation elections would be supervised by the courts rather than a political agency. Both moves were again understood as attempts to restore calm in the union movement. But Gharib continued to use his powers as ETUF president against the progressive forces. In November 1975, he engineered the removal of Mahmud and his allies from leadership of the CWF and then launched an investigation into its financial affairs.[70]

In this atmosphere Mahmud and Gum'a considered pulling their federations out of the ETUF and forming a rival confederation. In fact, both the Printing, Publishing, and Telegraph Workers' Federation and the EEMWF briefly withdrew from the ETUF in protest of Mahmud's removal. This marks the first time the idea of a singular trade union confederation was questioned from within. These individuals ultimately rejected this plan, not wishing to be blamed for dividing the union movement, especially after the long years of struggle to establish the ETUF.[71] But the idea, once proposed as a serious option, would remain alive among those discontent with the existing state of union/government relations.

Gum'a did make it known, when he rejected separation, that he planned to challenge Gharib for the ETUF presidency in the upcoming elections. In the midst of all this Gharib announced another postponement of the poll, this time for a full year. Once again, the opposition forces took their case to the press and the parliament, and once again Bultiya intervened in an attempt to placate them, promising a full investigation into the CWF affair. Numerous committees were formed to try to restore unity in the confederation.[72]

However, when the labor law had not been passed by January 1976, and it became clear that the elections would be postponed again, Mahmud and others filed a suit claiming that the ETUF's leadership body was illegal since its term of office had expired. At this point Bultiya met with the opposition group and several tacit compromises were worked out. The unionists agreed to accept a six-month postponement of the elections, until June, and to withdraw their suit. In exchange, the offensive provisions of the draft labor bill were removed or modified. The final bill was issued by the Parliament in March. The leftists had lost on the delegates' system issue: the law made no mention of the delegates, and the system continued only informally thereafter, where leftists were able to keep it alive. However, the law did stipulate only that no more than 20% of union leadership at any level could be members of professional or technical syndicates, as opposed to their complete exclusion as sought by Gharib, and the restrictions he had sought on candidates for federation and confederation office were deleted.[73] Bultiya also communicated to the unionists at this meeting that the government was now contemplating alternatives to Gharib's continuation as ETUF president.

The 1976 law generally sanctified the hierarchical structure of the union movement, specifying explicitly that it have a pyramid form where the lower bodies were subordinate to the higher ones. It also formalized the system of having officers at all levels elected indirectly by the boards. On this basis, Bianchi has argued that the law served to buttress the authority of the ETUF hierarchy, increasing the administrative and financial controls of the senior unionists over the federation leaders, and of the latter over the locals; this was portrayed as part of an intentional plan by Sadat to tighten his own control over labor.[74] A close reading of the text suggests, however, that this conclusion is overstated. In two key areas, and apparently as a result of the long debates which preceded its promulgation, the law actually gave locals more independence than they had before. First, it made the basis of union membership the local rather than the federation. This meant that dues collected at the plant should go to the locals first. In addition, the stipulation that locals could not sign collective agreements without the approval of the federation was replaced by a clause stating only that the two had to participate jointly in the negotiation process.[75]

The outcome of these clashes over legal issues represented a clear defeat for Gharib and his allies in the government. Though divisions within the regime played a role, it demonstrates clearly that opposition elements within the union movement can influence the legislative environment in which unions operate. Along with a more unified labor movement's subsequent successes in resisting certain economic reforms (see chapter 4), it may vindicate the decision by Mahmud, Gum'a, and their supporters not to splinter the ETUF at this time. Until his actual overthrow in July 1976, however, Gharib continued to use his position to obstruct opposition to Sadat's policies, turning the centralized and hierarchical union structure into a weapon to weaken labor. And some workers complained that with all the intrigues at the top, the work of the locals was still being neglected. In interviews focused on the performance of the unions, many workers maintained that their locals and/or federations did not hold the annual congresses required by the law. Workers also continued to charge that the widespread practice of *jam'* divorced senior leaders from the base.[76]

Some leftist forces tried to fill in these gaps by establishing informal workers' organizations. At the Delta Ironworks, Sabir Barakat and a small group of others had begun a wall magazine in their shop after returning from the front. It exposed and ridiculed management corruption and agitated around working conditions. Its popularity grew, and by the end of 1975 the magazine had branches in all the shops of the factory. The publishers took on the name "The New Dawn Family," and rented office space in the company club.[77] In Alexandria, textile workers formed delegates' committees and held weekly meetings at the local federation offices to discuss wage- and union-related issues. The meetings grew to include hundreds of workers, and were attended by MP's and, at one point, the Prime Minister himself. The Textile Workers' Federation (TWF), which Salah Gharib headed, was particularly antagonistic to the delegates' movement, and its leaders boycotted the Alexandria meetings.[78] The success of these local initiatives, and the wave of wildcat protests to which they were related, are again evidence of rank-and-file rejection of leaders and organizations which fail to represent their interests and encourage their participation.

When the union election campaign finally opened in June 1976, many local level contests were quite heated. In some plants they became a referendum on the government's economic policies, and management efforts to interfere in the voting process resulted in workers smashing the ballot boxes in a few localities. There were pressures applied to prevent certain leftists from entering the campaign, and in some workplaces the election results were falsified to deny rank-and-file militants their victories.[79] Also, in what organizers see as a conscious scheme by authorities to prevent labor militants from winning local office, the most prominent rank-and-file movements were attacked before the elections opened. The New Dawn office was sealed, the magazines confiscated, and Barakat and several others arrested. In Alexandria, state security forces raided one of the delegates' committee meetings and detained 20 core activists.[80] One of those arrested in Alexandria maintained that it was TWF officials who summoned state security on them and that the local leaders, on instructions from the federation, refused to provide assistance to the delegates during their incarceration. Nevertheless, around 4,000 militants captured local office; this represents about 15% of the total number of local positions. There was at least one leftist in office in every local of Shubra al-Khayma, for example, and the entire board at the Misr-Helwan Spinning and Weaving Company, one of the largest public-sector textile concerns, consisted of labor activists.

The new leftist presence at the federation level increased somewhat and, except for the CWF, where government pressures insured that the recently installed pro-regime leadership was returned to office, the former opposition group leaders retained their positions in the federations. Thus the same divisions existed, and many of the same cast was on the scene, when the congress of the ETUF convened at the end of July to elect the new confederation leadership. Salah Gharib, who was not returned to the TWF presidency, opened the congress by calling for immediate elections before any other item on the agenda was discussed. This infuriated the democratic forces, and a violent brawl actually ensued. Bultiya ordered the congress adjourned and called an emergency meeting that night of all federation leaders and candidates for ETUF office. After prolonged discussion, it was finally agreed to hold the elections two days later, and to reconvene the congress to discuss all other issues in February 1977.[81] But not all the parties were content with this arrangement, and the elections were held under tight security, with no one but the official delegates allowed to enter the building that day.[82]

The newly chosen board proceeded to elect as its president Sa'd Muhammad Ahmad, who was known to be Sadat's choice. In relation to the controversies of the previous years, Ahmad was considered a moderate. Although his federation had supported Gharib on the issue of professionals, Ahmad was not considered one of Gharib's inner circle, and he was one of the federation presidents who had signed the denunciation of the ICFTU invitations. While not a leftist, he was also not a strong supporter of Sadat's turn toward free enterprise. Significantly, the food processing industry contained many private sector plants, and under Ahmad's leadership the federation had been more aggressive than any other in organizing private sector workers during the 1960s.

Sa'id Gum'a had decided not to challenge Ahmad when it became clear that Sadat would oppose his candidacy. Some leftists see this, along with his decision to affiliate with the centrist political "platform" Sadat created for the upcoming parliamentary elections (see below), as an indication that Gum'a had turned toward opportunism, or at best resigned himself to government selection of the confederation's president. Yet while Ahmad was clearly the more moderate and safer choice for Sadat, he was decidedly not the close associate and political ally that Gharib had been. Sadat had apparently concluded that he could no longer have someone perceived as his puppet in the leadership of the ETUF, and saw Ahmad as a compromise candidate who could placate the leftist forces while denying them control over the trade union movement. This result again confirms the importance of the struggles internal to trade union movements in shaping union/state relations, even in authoritarian political climates.

A Union/Party Symbiosis Under New Conditions

Shortly before the union elections began, Sadat initiated the dismantling of the ASU by dividing it into different "platforms" which were to compete for parliamentary seats in the fall; the platforms became political parties when the ASU was disbanded the following year.[83] The turn to limited political pluralism generated new questions for, and controversies within, the union movement.

From the outset, there was pressure on labor leaders to affiliate with the centrist plank associated with Sadat, then called the Arab Socialist Organization. Two of the federations hastened to join the ASO en masse. Gharib himself promptly signed up, and numerous other federation and confederation leaders also signed on as individuals, in some cases enticed by positions of authority in the platform. The unionists who had championed the fight against Gharib were initially outspoken against this development. 'Uqayli's Land Transport Workers' Federation, for example, passed a resolution strongly denouncing the mass memberships as a surrender of trade union independence from the government, and had it printed in *al-Tali'ah*.[84]

At the same time, the leftist plank, known as the *Tagammu'* (Arabic short for National Progressive Unionist Party), was able to recruit thousands of labor activists at the local level. An uneasy alliance of Nasirists, traditional communists, members of new Marxist organizations and independent leftists, it attracted a number of the senior unionists in the anti-Gharib camp.[85] Sa'id Gum'a and Ahmad al-'Amawi, president of the Pharmaceutical and Chemical Workers' Federation (PCWF), participated in the formation of the Tagammu's labor program. Niyazi 'Abd al-'Aziz and 'Aisha 'Abd al-Hadi, board members of the EEMWF and the PCWF, respectively, worked with them and went on to join the platform; 'Abd al-'Aziz became the first head of its labor bureau. 'Abd al-Rahman Khayr and Fathi Mahmud, who had taken up work in the Arab Confederation after his ouster from the CWF, also affiliated with the Tagammu.'

The sentiment against union affiliation with the ASO was strengthened in October 1976, when the first parliamentary elections under the new system took place. Some 30 unionists contested the elections, including six federation presidents and three members of the confederation's board. Most of these were from the centrist platform; all, regardless of their affiliations, failed to win seats. This outcome, which unionists blamed on a lack of regime support for the idea of ETUF representation in parliament, angered the pro-government labor leaders and reinforced the arguments of those rejecting ASO affiliation. Gum'a and 'Uqayli now began to agitate for an independent workers' party which could campaign effectively for union representatives. At a meeting in November, the confederation affirmed that the trade union movement was independent and would not obey any organization or front; a committee was established to evaluate the new political situation and prepare recommendations for discussion at the upcoming ETUF congress in February. The idea of a workers' party gained support, with articles advocating it appearing in several federation magazines.[86]

Sadat had reason to feel threatened by the idea of an independent political organization run by labor leaders. Besides eroding the claim of his centrist plank to be the sole legitimate representative of the Egyptian masses, it could have sparked similar moves by other societal groups, pushing his experiment in democracy farther then he evidently intended. His opportunity to preempt its creation came as a result of the January 1977 riots, caused by his IMF-sponsored decision to lower food subsidies (see chapters 3 and 4). In the midst of the riots, union leaders condemned Sadat's decision and angrily demanded a meeting with him. After restoring the price supports, Sadat acceded to the unionists' request for a dialogue, his first formal meeting with ETUF leaders since before the 1973 war. In this meeting the unionists pushed for wages increases and revival of the tripartite production committees that had existed in plants in the 1960s, with representatives of labor, management, and the ASU; Sadat had allowed the committees to disintegrate after coming to power.[87] The unionists also sought implementation and expansion of the provisions of the 1976 labor law giving them more rights to consultation on legislation affecting workers, charging that these provisions had, until that date, been ignored. In addition, they demanded that these rights be extended to the federations, since the 1976 law had limited consultation to the ETUF. The longstanding demand that the confederation assume control over the Workers' Educational Institute, which had been transferred to the Ministry of Labor after the abolition of the ASU, was also reiterated.[88]

Sadat agreed, at least in principle, to all these demands. Unionists later complained that the production committees were not, in fact, revived, but the ETUF's consultative powers were respected, and the confederation was given representation on various government committees. In return, however, Sadat insisted that the labor leaders abandon the new movement for political independence, and instead tie the confederation to the ruling platform. The unionists accepted the trade off; the workers' party was formally forsworn by Ahmad at the meeting, and it was not raised at the February ETUF congress.[89]

There followed a campaign to entice leading unionists into joining the ASO, which soon became the National Democratic Party. Sadat promised the unionists that they would receive the government's support in future political elections if they affiliated with his plank. Unionists were added to the political bureau of the party and the leadership of its Cairo branch, and Ahmad became the Minister of Labor.[90] Sa'id Gum'a and Ahmad al-'Amawi, after having worked closely with the Tagammu', were thus co-opted into the ruling party. And where enticement failed, intimidation was attempted: 'Aisha 'Abd al-Hadi, for example, left the Tagammu' shortly after the January riots reportedly because the affiliation jeopardized the career of her husband, then working in the Labor Ministry.[91] While the confederation officially still asserted the independence of the trade union movement from all three planks, the pages of its newspaper increasingly spotlighted the activities and positions only of the government's party.

Sadat gained two things from this deal. The first was confederation support for his political initiatives.[92] This was not limited to affiliation with his party, but extended to public endorsement of his peace overtures with Israel, which were the source of tremendous controversy in the country at the time. When Sadat traveled to Camp David to sign the peace accords with Israel in 1979, he insisted that Ahmad and Sa'id Gum'a accompany him. The ETUF suffered denunciation by, and isolation from, the other Arab trade union organizations as a result of this support; correspondingly, Western-backed international federations stepped up their efforts to entice the Egyptian unions into joining them.[93]

The second was the senior unionists' acquiescence to a vicious campaign of repression against labor activists associated with the left. Hundreds of leftist local leaders were arrested during the riots or in the following months; Barakat says that virtually every leftist involved in the trade union movement was imprisoned at some point between 1977 and 1981.[94] He himself was arrested nine times between 1979 and 1981, each time charged with membership in a different leftist group. In addition, there was a rash of firings in private sector companies; in the public sector, where workers were protected from dismissals by numerous legal impediments, leftists were harassed with punitive fines and transfers. Niyazi 'Abd al-'Aziz, the EEMWF federation official then affiliated with the Tagammu', also charges that the government instructed public sector managers to resist any demands raised by local leaders affiliated with the party, so that they would lose popularity among the workers.[95]

In 1978, as part of the retreat from his tentative moves toward political liberalization,[96] Sadat created a new vehicle for keeping leftists out of union office by reviving the Office of the Socialist Prosecutor (OSP). Originally created to deal with the "power centers" in 1971, the OSP was now charged with implementing Law #33 of 1978 which denied the right to run in elections for any governmental, union, or syndicate office to anyone affiliated with a sect whose principles contradicted "the divine laws" of the state. Thus the OSP was to review the candidates for all of the roughly 2,000 locals in the ETUF, adding a new dimension to government interference in union affairs. In the 1979 union elections the OSP objected to 56 candidates, charging them with membership in various communist groups. Thirty-two of these individuals challenged the accusations in court, and of these 27 won the right to contest the elections. All 27 were elected, and both they and outside analysts believe that the government's opposition to them contributed to their victories.[97]

The 1979 elections were also characterized by a high level of management interference; there were 691 complaints of procedural violations filed with the confederation. Along with the other elements of repression, this contributed to a general decline in electoral competition: in almost half of the workplaces, the candidates faced no opposition. The proportion of default victories was highest (about 90%) in the agricultural federation, but even in the industrial cities some 25% to 30% of the locals won by default. An absence of competition was also evident at the federation and confederation levels, where there was virtually no turnover in leadership.[98]

The results of this campaign on the leftist presence in the trade union movement were devastating. Hundreds resigned from the Tagammu' to avoid arrest or harassment; this was true, for example, of all the leftist local officers in Shubra al-Khayma.[99] From 4,000 local officers elected in 1976, after the 1979 elections the left could claim only 120. The repression got worse in the final two years of Sadat's rule, and by 1981 most of the leftists who had obtained local office in 1979 were languishing in prison.[100] The co-opted senior unionists did not challenge the attacks on the left and, when Sadat decreed more severe punishment for convicted strikers, namely hard labor, after the 1977 riots, there was no objection from the ETUF leadership.[101]

Leftists charge that the NDP-affiliated ETUF leaders were either supportive of the government's electoral interventions or at best only weakly opposed to them. When the OSP law was first issued in 1978, none of the unionists who sat in parliament opposed it. Sa'd Muhammad Ahmad did sign a letter in support of the court appeal when some of the accused workers met with confederation leaders at the ETUF headquarters. But the next day, the same workers encountered Ahmad at the Ministry of Labor, where he refused to endorse their appeal. Furthermore, the confederation took no action when the CWF voted to rescind the membership of Mahmud and al-Maghrabi because of their Tagammu' membership. The ETUF also accepted only 150 of the 691 complaints about the 1979 election results.

Moreover, the confederation leaders cooperated with the regime in a tightening of control over the locals that was clearly aimed at reducing the displays of rank-and-file discontent which had characterized the last years of Gharib's presidency. This was done by restoring to the labor laws some of the modifications to hierarchy that had been passed in 1976. In 1979, when the left was ill-positioned to challenge it, the Labor Ministry and the ETUF leadership jointly drafted a new law in which these "loopholes" were corrected. The new draft law, which was never circulated among the base for discussion, was presented to a parliament seated in 1979 by what were widely understood to be "cooked" elections, after Sadat had arbitrarily dismissed the body elected in 1976. The law was finally promulgated in 1981.[102] In addition to reasserting the power of the federations over the locals, the new law also extended the leadership's term of office at all levels to four years. This extension had previously been opposed by rank-and-file activists because it made it more difficult for workers to remove opportunists at the local level.[103]

These reversals stand as a tribute to the degree to which even senior unionists who had earlier espoused and fought for union democracy had been co-opted by the regime. And this case illustrates my earlier argument that for opportunist unionists, government intervention in union elections can serve as an inducement as well as a constraint, since it protects them from being thrown out by an angry rank and file. Nevertheless, it would be wrong to conclude from these reversals that internal union struggles are ultimately insignificant, with authoritarian regimes always prevailing in the end. Sadat's heightened repression succeeded in part because, as the next chapter will show, an economic upturn eroded the impetus for the widespread workers' protests that had earlier strengthened the hand of the reformers fighting inside the ETUF structure. At the same time, the reformers' ideas did not die, but only fell dormant, to be revived with the changing economic circumstances to come.

Union, Party, and Government Under Mubarak

Husni Mubarak became president after Sadat was assassinated by Islamic militants in October 1981. Unlike Sadat, Mubarak initially made no attempt to alter the personnel at the helm of the ETUF, nor to change the structures of union/state relations. However, as an economic downturn combined with Mubarak's cautious embrace of economic liberalism brought a new upsurge of wildcat protest, Mubarak too moved to change the confederation's leadership, and appears to have permanently abandoned the practice of appointing the organization's president to his Cabinet.

In his early years, Mubarak eased up on the elevated levels of repression that had characterized his predecessor's final years. Political prisoners were gradually released, the censorship of the opposition press was substantially lifted, and more parties were permitted to contest elections. However, throughout the 1980s the emergency laws remained in effect, parliamentary elections were never freed of fraud and manipulation to favor the ruling party, and the audio-visual media were under strict government control. In the 1990s political reforms have essentially been put on hold, and repression intensified, in response to a growing threat from Islamic groups. Nevertheless, an important consequence of Mubarak's limited political liberalization was giving the left more room to maneuver in its campaign for union reform.[104]

Sa'd Muhammad Ahmad's Second Semester

Sa'd Muhammad Ahmad's reign emerged as the heyday of union work for middle- and upper-level personnel. Federation and confederation officials continued to enjoy easy access to economic policy makers as well as the national media, and Ahmad strove to ensure that, unlike under Gharib, the ETUF presented a unified stance. He also worked to instill a friendly and cooperative atmosphere in the organization's headquarters, instituting perks such as free beverages for all building employees.

Ahmad's largesse was facilitated in part by an increase in the resources available to the confederation as membership grew (see tables 2.2 and 2.3).[105] Opportunities for union-related travel overseas increased as Egypt's relations with the West blossomed. At the same time, the 1976 law gave ETUF and federation officials the right to invest union funds in a variety of projects (of which vacation lodging for members became a rapid favorite), creating numerous opportunities for kickbacks and nepotism.[106]

Accordingly, unionists seeking greater influence in policy making, as well as those motivated by material self-interest, had reason to be content during these years. At the same time, the sponsorship of the government, as well as their ability to make deals with local officials that would assure their reelection, enabled federation and confederation officials to enjoy prolonged tenure.

TABLE 2.2

Growth of Unions and Unionization

	1973	1986	Avg. Annual Grth Rate[a]
Nonagric.union memb.(millions)[b]	1,454	2.523	4.3%
Nonagric. workforce[c]	4.453	7.046	3.6%
Union memb. as % of workforce	32.6	35.8	0.7%

[a]My calculation.

[b]1973 numbers are from ETUF 1982, p. 143; numbers for 1986 are those obtained from the ETUF membership office for the 1983–87 session; the exact date of compilation is not specified.

[c]Assaad and Commander, p. 11 (based on CAPMAS, Labor Force Sample Survey and Population Census). In the original, the numbers are rounded to one decimal point; I obtained the more accurate figures from Ragui Assaad. Workforce numbers include individuals aged 12 to 15 who are not eligible for union membership.

The 1983 union elections saw sparse competition and little turnover in the upper ranks of union leadership. Two federations had been split just before the elections, bringing the total number to 23; of the 21 which had existed in 1979, the presidency changed hands in only four.[107] At the confederation level, the entire executive committee of the ETUF, looking much like the previous board, again won by default.[108]

With virtually all ETUF and federation leaders thus co-opted by the regime, it fell to the left to resume the fight for union reform. Most of the leftists were released from prison in Mubarak's first year, and the Tagammu' began to publish *al-Ahali* again in the latter half of 1982. The left was in a much weaker position in the unions than it had enjoyed in 1976, but used its newfound freedoms to champion four causes: the curtailment of government interference in union elections, legalization of strikes, loosening the hierarchical controls on the locals, and an end to the practice of *jam'*.

On the first of these issues, some headway was made. Those leftists who did hold local office brought resolutions calling for an end to the OSP system to

TABLE 2.3

Growth of Union Membership by Sector

	1971–73[a]	1976–79[a]	1983–87[b]	Avg. Gr.%[c]
Industry and blue-col. serv.[d]	738,485	913,412	1,151,966	3.2
Other services[e]	298,752	421,346	698,551	6.2
Construction	85,577	146,957	249,971	8.0
Land Transp.	130,875	242,980	423,232	8.7
Total Nonagric.	1,253,689	1,724,695	2,523,720	5.1
Agric.	172,471	146,262	196,860	1.0
Total ETUF	1,431,160	1,870,957	2,720,580	4.7

[a]Data from ETUF, 1982, pp. 143–44; date of compilation not specified.

[b]Data from ETUF membership office; date of compilation not specified.

[c]My calculation; assumes 14-year time span.

[d]Frequent changes in the composition of the industrial federations renders a more specific breakdown impossible. Included here are all locals in the following sectors: Engineering, Electrical, Metalworking, Textiles, Food Processing, Leatherworking, Furniture, Printing, Publishing and Broadcast Media, Oil, Chemicals, Mining and Minerals, Military Hardware, Postal, Telephone and Telegraph, Utilities, Dockworkers, and Railway.

[e]Includes locals in Educational, Health, Social, Administrative and Personal Services, Banks and Insurance, Tourism, Commerce, and Air Transport.

their federation congresses prior to the 1983 elections, and the OSP issue was also featured prominently in the pages of *al-Ahali*.[109] Whatever their private beliefs, senior union leaders were hard-pressed not to publicly support the case against the OSP, since it represented such a blatant form of government interference in union affairs. The Tagammu' articles pointed out that the law violated International Labour Organization (ILO) regulations, and the party did send a complaint to the ILO. Thus many ETUF officials and federation leaders now spoke out against the OSP. Sa'id Gum'a introduced a bill in parliament to revise the 1978 law, and the confederation passed a resolution at its congress in March calling for the repeal of all the "exceptional" laws issued by Sadat in his final years.[110]

Although the campaign to abolish the OSP was not successful, there was a reduction in the scope of the exclusions. When the election campaign opened in the fall of 1983, only 12 candidates for union office were opposed, and a generally more open atmosphere prevailed. The campaign at the local levels were described as vibrant, especially in the industrial locals where many new candidates entered and the turnover in leadership exceeded 40% in most workplaces.[111]

Still, leftists charged that the support of the ETUF hierarchy for their campaign was at best lukewarm. Gum'a's draft bill failed in the parliament, an indication that the NDP unionists had not won their party's support for it. Furthermore, leftists pointed out, it was the senior unionists themselves who submitted the candidates' names to the OSP. The union leaders also executed the exclusions, in contrast to the leadership of the professional syndicates who refused to implement the OSP's rulings. Leftists charge that the unionists in some cases feared electoral challenges at the base from popular militants.[112] Relatedly, despite a court ruling in their favor, the expulsions of al-Maghrabi and Mahmud from the CWF were affirmed by a congress of that body two months prior to the vote.[113]

Also, although the candidates excluded by the OSP successfully challenged the ruling in court, changes in the law after 1979 meant that the case now had to go through a variety of legal channels, and the final verdict was not issued until 1986. Management interfered in the elections as well. These factors contributed to another poor showing by the left in the 1983 local elections. While there were some significant victories, particularly in their traditional strongholds like Helwan and Kafr al-Dawwar, the leftists did not increase their numerical presence in the leadership significantly over 1979; some say their numbers actually went down. Activists believe, however, that repression played less of a part in these local results than it did in 1979. Other factors which contributed to the left's weakness were the economic upturn Egypt was still enjoying at the time, and a feeling among the workers that Mubarak might follow a different course than Sadat.[114]

The strike ban was a major issue for the left in both parliamentary and union election campaigns. They maintained that Sadat had effectively legalized strikes in 1981, when he signed the International Agreement on Economic, Political, and Cultural Rights five days before his assassination. Mubarak signed the accord again on December 18, 1981, and the decision (#537 of 1981) was published in the *Official Gazette* of April 8, 1982. Arguing that according to the constitution, the provisions of the accord, including the legality of strikes, now superseded local laws, the Tagammu' sent complaints about the ongoing prohibition to the ILO and other international organizations, including human rights agencies.[115] Although these actions bore no fruit in Mubarak's initial years, the left's exposure of the contradiction between the image the regime was projecting overseas and its actual domestic policies did pave the way for eventual modification of the strike ban (see below).

Finally, the left continued to challenge the hierarchical functioning of the ETUF and its incumbent restrictions on local-level initiative. They linked this with their critiques of *jam'*, now extended to unionists holding simultaneous positions in government or the NDP, not merely multiple union posts. Ahmad's simultaneous occupancy of the ETUF presidency and the Ministry of Labor was particularly targeted. Leftists charged that senior union personnel had become part of a new elite class, since the *infitah* gave them new ways to invest and spend the multiple salaries they earned by combining leadership positions, as well as numerous opportunities for illicit uses of union funds.[116]

On this front, the left could claim no victories; senior unionists stonewalled all critiques of hierarchy and *jam'*. Ironically, however, Mubarak himself appears to have concluded that the overlap of union and government positions was too extensive, especially at the top. In a move that caught the union movement as well as the left by surprise, he ousted Sa'd Muhammad Ahmad from the Ministry of Labor in November 1986.

Deja Vu: The Ouster of Ahmad and the 1987 Union Elections

While it later became clear that Mubarak had broken with the practice of appointing the ETUF's president as Minister of Labor for good, it was immediately evident only that Ahmad had fallen out of favor with the regime's top decision makers. The reasons remain a source of speculation; in union and leftist circles, rumors were rife. The most common explanation was that Ahmad had not tried hard enough to contain a wave of wildcat strikes which had begun to plague the country in 1984 (see chapter 3). The ETUF's resistance to privatization (see chapter 4) was also suspected. A few individuals also suggested that Mubarak had been pressured by the Americans and Israelis, who were angered over the ETUF's continued refusal to deal with the Histadrut.[117] While we cannot know Mubarak's motives with certainty, it is likely that all of these factors were involved to different degrees.[118]

Ahmad's closest associates were targeted as well, as the regime moved quickly to deflate their positions in the party and the government. Mahmud Dabur, president of the of Bank and Insurance Workers' Federation, was removed as an appointed delegate to the parliament, and replaced with a nonunionist. Simultaneously, the NDP appointed Ahmad al-'Amawi, president of the PCWF but not a member of the ETUF board, to serve as the union representative to the Consultative Council; this was done without the approval of the ETUF.[119] Then Yusef Wali, head of the NDP, told the ETUF board in a meeting called to discuss these events that he intended to supervise the 1987 union elections personally, and wanted to see new elements enter the leadership of the confederation.[120]

Mubarak's moves were reminiscent of Sadat's turn against Salah Gharib a decade earlier and, like Gharib, Ahmad resisted the attempts to oust him from the union presidency, mobilizing allies in the ETUF for support. The confederation's executive board called an emergency meeting when the political changes were announced, and passed a resolution by consensus calling for Ahmad to continue as president. They also condemned Dabur's removal and al-'Amawi's appointment. Over the ensuing months the ETUF's newspaper reported on all the rumors and intrigues surrounding Ahmad's removal, while praising his record as confederation president and Minister of Labor.

The conflict spilled over into the parliamentary elections in the spring of 1987. In March, Ahmad, Dabur, and Kamil 'Uqayli, all of whom had served in the previous parliament, were removed by the NDP leadership from the new nomination lists.[121] The confederation then presented the NDP with a list of five different board members, none of whom had previously held political positions, to run for seats. The party opposed all five of these, choosing two others seen as outside of Ahmad's close circle.[122] It became clear at the time that the government had identified a new group of unionists who it wished to promote in the party and to ETUF leadership in the next elections. Its members were 'Asim 'Abd al-Haqq (Textile Workers' Federation), who replaced Ahmad as Minister of Labor, Mukhtar 'Abd al-Hamid (Agricultural Workers' Federation)[123], Muhammad Khayri Hashim (Communications Workers' Federation), Mustafa Mungi (Military Production Workers' Federation), Ahmad al-'Amawi and Ahmad Disuqi (both Pharmaceutical and Chemical Workers' Federation), and 'Abd al-Rahman Khidr (Administrative Services Workers' Federation); the latter were the two the NDP did back for parliament. The "Israeli factor" theory of Ahmad's removal is based in part on the fact that this group included presidents of five out of the six federations that had affiliated with the ICFTU.[124]

In April, the government made it known that it had evidence of misuse of union funds by Ahmad and was pushing for his resignation from the ETUF. The new in-group of unionists had a private meeting and issued a call for the confederation to withdraw its confidence from Ahmad. Ahmad still had enough support on the ETUF board to withstand this pressure, but the threats of financial investigation could not be ignored for long. In the middle of June, he formally resigned the presidency of the ETUF, and Mukhtar 'Abd al-Hamid assumed his duties. It was understood that Ahmad had received a promise from Mubarak that in exchange for his withdrawal, there would be no further publicity about, or legal charges pressed for, his alleged embezzlement of funds. The government now moved to rid the ETUF of his close associates: threats of financial investigations resulted in the resignations of Gadd al-Muwali and Ahmad 'Umar, who was president of the Building and Constructions Workers' Federation.[125]

These events indicate the extent to which the ETUF had come to view itself as a branch of the party, as well as the extent to which the party itself was subordinate to the will of the regime's inner elite. They also suggest the regime's ready ability to impose its wishes on the trade union movement. Yet the rapid emergence of this new group of favored unionists makes it apparent that there were submerged disagreements and rivalries behind the facade of unity that the ETUF presented during Ahmad's tenure, differences which prevented Ahmad from rallying more support from his colleagues.

In addition, the leftists' efforts to discredit the entrenched union leadership with charges of opportunism and embezzlement, combined with the confederation leadership's inattention to local affairs, contributed to a climate in which there was little concern over these intrigues among the rank and file. Indeed, Mubarak may well have known when he targeted Ahmad that the accusations of financial mismanagement would not be widely questioned at the base. There is also little doubt that the charges, which mirror the accusations frequently made by the left, were legitimate. As Mahmud Dabur told me, "everyone was doing it and everyone knew about it; that's how we can be so sure that they really removed Ahmad for other reasons."[126]

As the 1987 elections approached, the government took additional steps to bring about a turnaround in senior ETUF personnel. The new Minister of

Labor, 'Asim 'Abd al-Haqq, promulgated a new restriction on eligibility for office in the 1987 election campaign—that no candidate could be past the age of mandatory retirement in the public sector (60 years), even if still working. This provision was enforced by the special committees to oversee the elections that Wali had hinted at back in January; these were set up and supervised by Wali and 'Abd al-Haqq personally. The age restriction prevented 'Uqayli and Shalabi of the ETUF board, and numerous others in lesser leadership positions, from contesting the elections. Dabur was eliminated because he had been promoted to general director of the bank where he worked; he had been trying for some time prior to the campaign to have the provision preventing directors from holding union office removed from the labor law, but once he was marked by his friendship to Ahmad, he found little government support for his efforts.

Mostly as a result of these measures, the 1987 elections saw an increase in the number of candidates for federation office and a decline in the proportion of default victories. There were changes in the presidency of 9 out of 23 federations,[127] more than double the proportion of leadership turnover that took place in 1983. These changes at the federation level resulted in a dramatic change in the leadership of the ETUF as well. Of the top eight positions on the board, six out of eight turned over, as compared to only one out of eight in 1983. And of the top 20 senior unionists, eight had never served on the ETUF board before, whereas in 1983 none of the top 20 officers were new.[128]

The ETUF elections were preceded by a great deal of rumor-mongering and maneuvering among the presidential aspirants. Initial expectations were that the regime would favor the continuation of Mukhtar 'Abd al-Hamid in the post, but he was rejected, apparently because it was felt that the position should be held by a representative of an industrial union. 'Abd al-Haqq was also keen to obtain the presidency, and had his supporters in the confederation lobbying Wali on his behalf up to two weeks before the vote. But Wali told them that the decision-making elite no longer felt it wise to have one individual head both the Ministry of Labor and the ETUF. The final two contestants were Ahmad al-'Amawi and, once again, Sa'id Gum'a. Both had camps of supporters in the union movement, and arguments between them in the halls of the ETUF headquarters were sometimes intense. But shortly before the vote, Wali met with the newly constituted group of senior unionists and announced his support for al-'Amawi, and once again Gum'a did not contest the regime's decision. The board of the ETUF again selected its office committee by default.[129]

It is ironic that the final choice for ETUF president came down to the two federation presidents who had initially flirted with the Tagammu.' Both were now members in good and long standing of the NDP, and perhaps the best indication of their loyalties is the fact that neither refused to show the list of candidates for office in their federations to the OSP; coincidentally, it was *only* in Chemicals and the EEMWF that candidates were excluded by the OSP. Although many leftists condemned Gum'a as an opportunist for these reasons, his family connections to Egypt's traditional left and his history of opposition to ties between the ETUF and the Histadrut may well have been the reasons for his rejection by Mubarak;[130] the fact that outside of textiles, most of the wildcat strikes plaguing Egypt at the time appeared to occur in the EEMWF could also have been a consideration. More significant than the individual chosen, however, is the manner by which the choice was made. Thirty years after the establishment of the confederation, the government was still selecting the organization's senior leadership, with their acquiescence.

The situation that then emerged is reminiscent in many ways of union/state affairs during the brief period when Gharib and Bultiya, respectively, occupied these positions: as both 'Abd al-Haqq and al-'Amawi maneuvered to find clients within the ETUF and patrons among other regime and party elites, two different clusters of regime loyalists formed within the ETUF elite. Thus, while some hailed the separation as a step toward making the ETUF more independent from the regime, it soon became clear that governmental interference in union affairs was merely, once again, changing in form.

'Abd al-Haqq, who had initially solicited the joint appointment for himself, sought to build a base of support within the federations and the ETUF board. His leverage was the threat of financial investigations against those who didn't cooperate with him, and electoral intervention to favor those who did. In particular, the decree that unionists who reached retirement age must vacate their positions threatened the incumbency of many senior federation and confederation officials.[131] Some Nasirists claim that the ultimate goal of 'Abd al-Haqq's project was to secure ETUF acquiescence to relations with the Histadrut, as a means of advancing his own personal career.[132]

Whatever its ulterior motives, this intervention spawned a reaction from the unionists who chose to remain outside 'Abd al-Haqq's camp. Contradictions soon developed between 'Abd al-Haqq and al-'Amawi, the new ETUF president, who was himself approaching retirement age. He and other senior unionists hostile to al-Haqq renewed the call for denying the Ministry to right to supervise union finances, and agitated against the mandatory retirement decree.[133]

Ironically, although leftist opponents of economic liberalization had vilified Ahmad's joint appointment as the epitome of government/union symbiosis, the net effect of the separation was to weaken the confederation's stance vis-à-vis the economic initiatives of the regime. Al-'Amawi softened the confederation's previous antagonistic posture toward the country's private sector capitalists, and it was during his tenure that the ETUF qualified its opposition to privatization (see chapter 5). Although it does not prove that this was the real, or only, motive behind the separation, the increased vulnerability felt by senior unionists in its wake, and the contradictions it spawned among them, appear linked to these changes.

The Left and the Struggle for Union Reform After Ahmad

While claiming the separation as a partial victory in their campaign against *jam'*, the left continued to push for other union reforms. They were able to claim some success in the battle for the right to strike as a result of a 1986 train drivers' strike (see chapter 3). When that case went to trial in the spring of 1987, lawyers associated with the Tagammu' made the case that Egypt's participation in the international human rights agreements overrode the local laws against striking. The courts upheld the defense and dismissed all charges against the workers. Mubarak was able to undermine the decision by implementing a curious provision in Egyptian law which allows him to reject a court ruling and send the case to a court of the same level in a different locality.[134] But although the ruling of this other court was still pending at the end of 1995, within a few years after the initial decision the government invited the ILO to assist in preparation of a new labor law, knowing that the organization would insist on the legality of strikes (see chapter 5).

The campaign against the OSP also had some results. In the 1987 elections, the OSP objected to 13 candidates, among them several who had won court cases against their exclusions from the previous race. Five of the OSP's victims, including Barakat, 'Abd al-Magid, and 'Abd al-Hamid al-Shaykh, head of the Tagammu's labor bureau, again took their cases to court and won. This time the rulings came more speedily and required that new polls be held with the leftists on the lists, and most of them were elected.[135] However, other forms of government intervention were still operative: The Tagammu' complained in particular about the fact that the election supervisory committees were allowed to reject candidates up until twenty-four hours prior to the poll; Fathi Mahmud was vetoed in this last-minute manner. Some company managers refused to issue Tagammu' candidates the requisite certificates of good work standing, and some pro-government local officials would not issue them the necessary certificates of union membership. There was also an increase in arbitrary arrests of leftist workers just before and during the campaign.[136]

Nevertheless, the left performed better in these elections than it had in the previous two. At the local level the 1987 elections attracted greater interest than previous union campaigns, a phenomenon generally attributed to the deteriorating economic conditions and heightened level of militancy at the time. There were more candidates and fewer default victories than previously, with the turnover in leadership rising to about 50%. The Tagammu' party and independent labor activists were able to field about 400 candidates, and roughly half of these won, resulting in a near doubling of the leftist presence on local boards.[137]

In the ensuing years, the left took advantage of the heightened attention to union/state affairs that first Ahmad's removal and then the disputes between 'Abd al-Haqq and the ETUF had attracted, and intensified their campaign against governmental intervention in union affairs. A committee of lawyers associated with the Tagammu' prepared a draft of new union legislation which would abolish the use of the OSP and eliminate the numerous ways in which the Minister of Labor was empowered to interfere with the confederation; it was submitted to parliament in 1989.[138] The draft also called for legalizing strikes and restoring legal personality to the locals and making them, rather than federations, the basis of union membership. In addition, the annual general meeting of the local would be empowered to withdraw confidence from leaders found to be negligent in their responsibilities to the base. The NDP-dominated parliament was easily able to neglect the proposed legislation, but public debate around these issues would grow wider in the 1990s.

Leftist activists at the base increasingly linked issues of union reform to the economic struggles of workers. Beginning in the mid-1980s, petitions for withdrawing confidence from unresponsive union locals were circulated at numerous large public sector plants. At times, these campaigns were linked with protests over economic issues. At the Helwan Iron and Steel plant, a 1989 factory occupation involving thousands of workers, and ultimately violently smashed by the state security forces, was precipitated when NDP-affiliated local leaders strongly disliked by workers tried to claim credit for gains negotiated by leftist elected WRM's and endorsed the government's ouster of the latter from the management council.[139]

However, the new political conditions also saw historic divisions over labor strategy within the left surface, and new ones develop. Ahmad's removal itself helped expose these, as in the period after his dismissal from the ministry the dominant forces within the Tagammu' rallied to his support. The party's strategy, both for expanding union democracy and fighting economic liberalization, was to seek out sympathetic NDP unionists and encourage them to press these issues with the regime; splits within the NDP camp provided new opportunities for this. *Al-Ahali* became the main source of reports on the unionists' conflicts with the NDP, attacks on the new pro-regime group, and rumors about Mubarak's motivations and forthcoming actions toward the ETUF, while criticisms of the confederation's leadership almost disappeared from the newspaper's pages. Ahmad, who had previously been loathe to talk to *al-Ahali* reporters, now granted them a lengthy interview, which the paper featured prominently.[140] After 'Abd al-Haqq assumed leadership of the Labor Ministry, leaders of the Tagammu's labor bureau sought to widen the disputes between him and the NDP senior unionists.

The Tagammu's implicit backing of Ahmad reflected the position of the Nasirist and formerly HADITU forces who comprised the party's leadership, and was criticized by some independent leftists as well as Marxists from other sects working within the Tagammu.' More generally, these individuals accused the party of abandoning efforts to build labor and other struggle against the regime, in the fear that this might produce a worse alternative.[141] They complained in particular that the Tagammu' failed to encourage or in some cases even publicize the more militant rank-and-file protests, which began to erupt again in the mid-1980s (see chapter 3).[142]

The Tagammu's accommodationism with the government at this juncture demonstrates that for some leftists, the fight for union democracy was not a matter of firm principle, but rather a flexible tactic contingent on their judgment of the particular regime in power. Moreover, internal democracy in the party itself had become increasingly restricted and as a result, some leftists began to leave the Tagammu' in the mid-80s.[143] Others, while continuing to work within it, simultaneously established independent organizations to enable them to voice dissident opinions and pursue alternative strategies. In relation to labor organizing, the most prominent of these groups became Sawt al-'Amil (Voice of the Worker, hereafter VOW), which began publishing and distributing its own newsletter (*Sawt al-'Amil*) in the mid-1980s.[144] Members of this group, and other independent leftists, promoted rank-and-file struggles sometimes shunned by the Tagammu', and in some plants ran slates against more moderate, Tagammu'-backed candidates.

Another difference which emerged between the VOW group and the Tagammu' was the former's willingness to question the longstanding leftist support for a single, centralized union confederation. The Tagammu's draft legislation was silent on this issue, and both the Tagammu' and Nasirist parties remained committed to maintaining the singularity of the ETUF. In contrast, some opposition leftists argued that the ETUF hierarchy would continue to neglect the demands of workers so long as it faced no competition from legitimate rival organizations, and pointed out that each federation's monopoly on representation of workers in that industry violated ILO prescriptions for competitive unionism.[145] Opposition leftists worked concretely to promote competition at the base by establishing plant-based newsletters and rank-and-file groups to fight around shop floor issues, similar to the efforts that blossomed, and were smashed, in the mid-1970s. VOW members also pointed to the continued vitality of the workers' leagues as evidence that frustrated rank-and-file workers were seeking alternatives to the established union movement.[146]

VOW was also more forceful than the Tagammu' in criticizing what they claimed amounted to mandatory union membership in the public sector. Although the laws since 1959 made union membership voluntary, VOW charged that union dues were automatically deducted from public sector workers' pay even if they hadn't signed the requisite forms, and that many workers did not understand their rights under the law. Thus they maintained that the high levels of membership in the public sector were not an accurate reflection of rank-and-file commitment to the union movement, and believed that reform of these practices would compel unionists to be more responsive to rank-and-file concerns.

These disagreements among the left weakened the movement for union reform. Greater unity would have enabled the opposition to the NDP to take better advantage of the rifts within the regime that Ahmad's ouster opened up. Nevertheless, the occurrences described here do illustrate the salience of leftist politics to the ongoing remolding of the institutional structures of union/state relations.

Thirty-odd years after its organizational consolidation, the ETUF remained the singular, hierarchical organization officially representing workers, its leadership still subject to the approval of the regime's inner elite. Some institutional aspects of union/state relations changed in form, but their essentially corporatist content remained the same. The changes are nevertheless as significant as the stasis, because they show the potential for more substantial departures from the current pattern in the future.

The continuity of corporatism cannot be attributed simply to the will of Egypt's successive rulers. We have shown here that Nasir, Sadat and Mubarak were at times only arbiters of disputes over union affairs which arose, without their knowledge or apparent desire, among their subordinates. Moreover, the early years of Sadat's rule, especially, indicate that even initiatives emanating from the presidential palace have encountered resistance and have been modified. The disagreements within the regime at times were reflected in, and at times reflected, contradictions amongst union leaders themselves. These provide evidence that hierarchical structures generate power conflicts between unionists at different levels of the organization.

The third set of players in the story of institutional stasis and change are labor activists associated with the left, including Nasirist and especially Marxist groupings. The heyday of leftist influence was the first six years of Sadat's rule; during this period, arguably, different choices by individual leftists could have put union/state relations on a very different path. Although the left's successes in undermining the hierarchical system were undone during Sadat's final years, and leftists were purged from the union movement during that time, they gradually regained a presence in the lower ranks of union leadership and rebuilt independent organizations at the base.

The response of the rank and file to the struggles over union/state relations is also telling. Workers for the most part did not engage directly in these battles. But many voted with their feet by remaining largely unenthusiastic about the ETUF, while significant numbers joined or supported alternative organizations that were more representative of their interests.

3

Workers at the Point of Production: Moral Economy and Labor Protest

Economic liberalization measures affect workers in a variety of ways and can elicit potentially contradictory reactions. For example, as consumers, workers may welcome the influx of new products into the market fostered by trade liberalization, but they may also fear the job loss that might result from foreign competition with domestically produced goods. Workers as consumers will face inflation from the lifting of price controls and/or the removal of subsidies, but they may not object if these are counterbalanced by wage increases. Privatization threatens the privileges of public sector employees, but workers may welcome the prospect of participation in ownership or expect to receive higher wages and/or have greater advancement possibilities in the private sector. On an individual plant level, efforts to reduce the fiscal deficit can result in layoffs, pay cuts, or forced overtime at different parastatals. Nevertheless, public sector rationalization, like privatization, may be seen by some workers as an opportunity rather than a takeaway.

This list is not meant to be exhaustive, but rather to capture the potential for varied workers' responses to different liberalization measures. This chapter seeks to infer ordinary workers' sentiments on such issues by examining collective action at the point of production, from the early 1950s through the late 1980s. Such protests often constitute, in and of themselves, a reaction to economic reforms, and hence a statement of their participants' perspectives on the particular issue. Collective struggles also provide important clues to the more general economic philosophy of workers, especially in light of the inavailability of other types of data. In Western democracies, researchers seeking to ascertain workers' opinions of different policy issues could conduct random surveys, or rely on disaggregated voting data or the pronunciations of union leaders chosen in fair and freely contested elections. Students of workers in authoritarian countries lack access to such sources. Citizens who can be fined, imprisoned, tortured, or murdered for speaking out against the government cannot be presumed to answer surveys honestly or vote the way they truly feel; the information available to them is in any case filtered through media carefully controlled by the government. Union leaders co-opted into the ruling party and subject to government vetting may or may not represent their members.

This inquiry is guided by three competing perspectives of workers' attitudes—moral economy, Marxism, and rational choice—laid out in the Introduction and reviewed here. Against them I explore the frequency of plant-level struggles and their timing in relation to economic conditions and policies, as well as the nature of the protests and the specific causes and demands raised. My claim is that these factors combined lend strong support to a moral economy interpretation of Egyptian workers' behavior, in which workers objected to reforms that constituted a perceived violation of their belief in reciprocal rights and responsibilities between themselves and the state.

In the moral economy I propose, workers viewed themselves in a patron/client relationship with the state. The latter was expected to guarantee workers a living wage through regulation of their paychecks as well as by controlling prices on basic necessities, and to ensure equal treatment of workers performing similar jobs. Workers, for their part, provided the state with political support and contributed to the postcolonial national development project through their labor.

In support of this interpretation, I will demonstrate here that labor protests occurred more frequently when the economy was deteriorating, suggesting restorative protest aimed at preventing erosion in the workers' standard of living. The immediate causes of these actions also give strong indications of feelings of entitlement. In many cases, workers were seeking to regain earnings that had been taken away. This contrasts with "new" or "aggressive " demands, by which I mean increments to real wages, or improvements in working conditions, that do not grow out of comparisons with the past. Moreover, where workers were seeking actual raises, it was most often in the context of seeking parity with others who had just been granted the increase. Thus it was a preexisting egalitarianism that workers were seeking to restore. Feelings of entitlement were also evident in protests which erupted as a result of unmet promises by management or the government.

Support for this particular type of moral economy also comes from the fact that in almost all protest incidents, the workers' demands were directed against the state. Moreover, workers revealed their view of their own obligations by eschewing actual work stoppages in favor of symbolic protests which signaled that they remained loyal to the cause of production, even while feeling aggrieved.

Another important feature of these plant-level protests is the fact that they occurred almost strictly outside of the formal union structure. In addition to casting further doubt on standard neoclassical assumptions about workers, this phenomenon helps to shed light on how union leaders balance pressures from below with those from above. Finally, the reaction of the government to informal labor struggle is also examined here. The combination of repression and concession employed by successive Egyptian administrations appears aimed at isolating and quickly ending all manifestations of discontent, to preserve the regime's persistent claim to widespread legitimacy among workers.

To support these arguments, the empirical material in this chapter has been organized by analytical categories. This is a departure from the historical presentation that has obtained thus far, and necessitates that various aspects of the numerous collective actions discussed below are parceled out into different sections. While the reader may therefore feel that the full flavor of these individual events has been sacrificed, I believe that the overall picture of local-level labor protest in Egypt is ultimately clearer for this approach.[1]

Macroeconomic Conditions and the Pattern of Labor Protests

Neoclassical arguments assume that workers make decisions about whether or not to halt production based on opportunistic cost-benefit analyses of the likelihood of winning strikes. Downplaying any sense of developing class consciousness, labor market theories hold that strikes are most likely to occur when expected monetary and/or job gains exceed expected losses, conditions usually associated with a tight labor market.[2] In contrast, orthodox Marxism, while positing periodic crises which may contribute to workers' recognition of their common class interest, generally anticipates a fairly linear increase in labor protest, both in frequency and in the numbers of workers involved, as class consciousness develops and solidarity grows.[3] Finally, a moral economy perspective would envision a greater number of protests—but not necessarily strikes—when economic conditions are deteriorating, causing workers to feel that elite commitments to them have been violated.

As explained in the Introduction, the sophisticated correlations of macroeconomic indicators with strike behavior performed by scholars of industrialized and democratic countries are not possible in a developing and authoritarian country like Egypt. Consequently, this section makes no attempt at precise correlation between the two. Nevertheless, I believe that the available evidence does suggest that Egyptian workers are most likely to protest when the economy is deteriorating, thus contradicting rational choice predictions and lending support to the moral economy view. There is also some indication that the overall level of protests which occur when real wages are falling is increasing, so that Marxist arguments cannot be ruled out on this basis alone.

The Nasir Years

As we saw in chapter 1, the two years immediately following the Free Officers' coup were characterized by an economic downturn in which employment declined and, at least in 1953, real wages fell; growing numbers of workers were subjected to an extended probationary period upon hiring. The British Labour Attaché, writing in the fall of 1953, considered the military government to be unpopular among workers for these reasons, and reported frequent instances of labor unrest that year.[4] Workers also sought to resolve their problems through other channels. Beinin has documented a dramatic increase in the number of "industrial disputes" (as measured by formal grievances raised by workers) recorded by the Egyptian Federation of Industries.[5] The number of disputes peaked at 71,841 in 1953, more than five times the 1951 level of 13,658; the three-year average for 1953–55 was 44,453 grievances per year, more than 2.5 times the 1949–51 average of 16,197. Noting that the pre-coup figures were themselves high by historical standards, he suggests plausibly that the post-1952 increase stems from a hope among workers that the new government would be more sympathetic to their complaints, as well as the fact that workers could no longer seek legal recourse through striking. However, the figures may also reflect the heightened problem with summary dismissals, since a growing proportion of these cases involved individual complaints.[6]

Conditions for workers appear to have improved somewhat when the international recession of the early 1950s ended. As we saw in chapter 1, the real wage index rose during the second half of the 1950s.[7] In addition, industrial employment began to pick up after 1956, with a noticeable spurt in 1959, possibly due to the mandated reduction in the legal working day that year (see table 1.1). In response, there was a marked (40%) decline in the average number of grievances filed by workers between from 1955–58, over the previous three years, suggesting that labor discontent was abating.[8] It also appears that labor protest declined dramatically during this period, except for the year 1956. Mutawalli al-Sha'rawi claims that there were 120 sit-ins, work stoppages, and other job actions in the Kafr al-Dawwar area that year; Tomiche documented a strike by dock workers in Alexandria.[9]

The socialist decrees of July 1961 brought immediate material improvements in workers' living standards along with an elevation of their status in society.[10] The minimum wage was doubled for many workers while the industrial work week was reduced, and pensions, injury compensation, and health insurance were enhanced. Additional prohibitions against firing public sector workers were enacted in 1962. The effects of all this legislation on real wages is shown in table 3.1.

Politically, the new legislation had the effect of further tying workers' interests to the state. The Nasir government took on a distributive function by subsidizing the cost of many essential food items, as well as energy, and controlling the prices of many other goods; the regime also obligated itself with ensuring the compliance of private capitalists with minimum wage standards and other laws protecting workers. While workers in the expanded public sector now depended on the state for their very livelihood, these other measures served to institutionalize the economic dependence of private sector workers on the state in other ways.

Nasir explicitly promoted the idea that the state's largesse constituted privileges which workers had to earn by increasing productivity. This was sanctified in the National Charter of 1962: "Every citizen should be aware of his defined responsibility in the whole plan, and should be fully conscious of the definite rights he will enjoy in the event of success."[11] However, Egypt began to retreat from socialism soon after the initial experiment began. After mid-1965, no new social legislation was introduced. Instead, investment expenditures were cut and a forced savings plan was implemented, where one-half day's pay per month was deducted from all public employees' salaries and put into a special account; there were also some price and tax increases and several factory closures. A 1965 revision in the labor code re-allowed paid over-

TABLE 3.1

Real Wage Index and Hours of Work, 1961–1967

Year	Hrs. Work/Wk[a]	RWI/Hr.[a]	Weekly RWI
1961	48	147	140
1962	47	151	141
1963	45	177	159
1964	44	190	168
1965	53	157	166
1966	52	161	167
1967	49	162	158

Source: Mahmoud Abdel-Fadil, *The Political Economy of Nasserism*, p. 33. Reprinted with permission.

Note: Base year 1950. Based on the SEWWH; see explanatory notes to table 1.4.

time, although it did not make it mandatory.[12] As a consequence of these measures, hours of work increased considerably that year, while hourly real wages fell (see table 3.1).

After Egypt's defeat in the 1967 war the government initiated another round of price and tax increases, and there were renewed calls for workers to sacrifice for the "battle" (*al-ma'raka*). The work week was increased from 42 to 48 hours without compensation to workers, forced savings were increased from one half to three-fourths of a day's pay per month, and additional measures, such as cancellation of paid holidays and/or "donation" of compensatory pay for meals, uniforms, shift work, or dangerous jobs, were attempted in some plants over the next year.[13] In manufacturing real wages fell again in 1967 and remained below their prewar levels the following year (see tables 3.1 and 3.2).

Protest activity remained quite low during the early 1960s, with only a few documented incidents through 1964. However, an escalation accompanied the policies of retreat. Despite the support of union leaders for these policies, workers objected; Hussein reports a number of strikes breaking out at the end of 1966 in response to the renewal of overtime, as well as cases of workers evading disciplinary measures, slowing down work, and even paralyzing or breaking machines to express their anger at the deteriorating economic and political situation.[14] There are also indications that workers in some plants successfully resisted forced salary reductions in 1965, and there were numerous protests against them again in 1968 (see chapter 4, and below).

Wages recovered in 1969 and 1970, and there is no record of labor incidents during those years. Helwan, the most industrialized city in Egypt, did become

TABLE 3.2

Real Wage Index in Manufacturing, 1966–1973

Year	RWI
1966	100
1967	95
1968	98
1969	106
1970	103
1971	99
1972	108
1973	107

Source: Gerald Starr, "Wages in the Egyptian Formal Sector," p. 17.
Calculated from the SEWWH.

a center for incipient labor activism at the time. Mass meetings organized by leftists for workers to discuss political and economic issues grew to include 4,000–5,000 workers. Harsh criticisms were leveled at both Arab Socialist Union (ASU) officials and company managers, and in 1969 the Minister of the Interior ordered the Socialist Institute where the meetings were held closed. Labor activists then organized smaller meetings of workers around different plant or occupational issues.[15] However, it was not until 1971, when wages were eroding again, that this groundwork materialized into actual protest activity which challenged Sadat after he assumed power. That year saw at least five incidents, the largest involving 30,000 workers at the Helwan ironworks, and the following spring workers struck several textile factories in Shubra al-Khayma. These actions correspond roughly to another period of declining real wages; in the latter half of 1972 and throughout 1973, when wages were improving, there is again no record of protest activity (see table 3.2).[16]

Three Perspectives on the Data

Taken by themselves, the 1950s data can support a variety of interpretations about workers' behavior, but standard neoclassical approaches are clearly counterindicated. If expected net gains, a function of the likelihood of workers quickly winning their demands, were the main motive behind collective action, why were there more incidents in the 1953 recession, and in 1956 when industrial employment was at its nadir? Neoclassical theory is equally unable to account for the 1960s information. Purely opportunistic logic would dictate that protests increase in the early 1960s, when macroeconomic conditions were most favorable, and decline thereafter, but the opposite occurred. Even if repression and the lack of union support were acknowledged to be the operative factors in preventing collective action in the 1950s, and sharply limiting it in the early 1960s, we would need to show some change in these to account for the resurgence of protest in the middle of the decade. But there was no apparent change in the propensity of the state to punish protesters, the nature of the official penalties for work stoppage, or the loyalty of union leaders to the regime.

Goldberg's application of rational choice argues that collective activity ceased after 1952–53 because the state was providing for workers basic needs, especially job security. Thus any expected gains from struggle would be minimal, and likely outweighed by potential costs. This is somewhat more plausible, especially if we assume some lag between the enactment of the tighter protections against firing in the spring of 1953 and their actual implementation. However, it leaves the 1956 events unaccounted for and overlooks the evidence, cited in chapter 1, that job security was still largely lacking in the 1954–55 period when no job actions were reported.

Goldberg's argument about job security could explain the paucity of protests between 1961 and 1964, but the problem then is to account for the increase in collective action thereafter, since the new job security provisions remained in effect. Moreover, collective struggle can provide workers with material gains other than job security, namely higher wages and other benefits, and Goldberg himself implies that workers with relatively secure jobs engaged in collective action for such gains before 1952. Why then would they not do so even more in the 1960s, when there was supposedly such little fear of job loss?

Without drawing on political and institutional factors, Goldberg can answer this only with the added assumption that workers now saw no reason to pursue further gains even though the risks were minimal. This implies a certain critical package of real wages and benefits to which workers assigned a high utility, whereas additions to it were less valuable. In this case, the expected gains of collective action would generally exceed their costs when the goal is to achieve or restore this package, but costs would exceed gains for increments to the package. Formally, this implies a critical point at which the slope of the utility function changes. However, this exercise in deductive logic runs into a problem because of the wage differentials which existed in the public sector according to skill and experience, as Goldberg himself notes. Because of this, to explain the absence of protest before 1964, we must assume that the critical level of wages was that earned by the lowest-paid workers, since otherwise they would still be motivated to collective struggle to achieve that package. In that case, it should be these workers who resorted to protest against takeaways thereafter; for the higher-paid workers whose incomes remained above the critical point even after takeaways, the costs of collective action would still outweigh the benefits. In other words, if it was older, more experienced workers who were most prone to *aggressive* collective action in the 1940s, as Goldberg suggests, it should be only the youngest and least skilled workers who saw the need for *defensive* struggle two decades later. The available evidence thus far is insufficiently detailed to test this hypothesis, but it is contradicted by the data on subsequent years, presented below.[17]

The 1950s evidence is consistent with the neo-Marxist approach of Beinin and Lockman, and Beinin,[18] who hold that the revolutionary potential of workers in the pre-coup period did not materialize because their class consciousness was supplanted by nationalism, especially after Nasir's triumph over Nagib in 1954. In explaining the decline of labor militancy after 1952, they also emphasize the military regime's repression of labor protest and its purge of the left from the unions, phenomena which Goldberg treats as an *effect* of workers' basic contentment rather than a *cause* of their failure to manifest discontent. In this vein, the reported upsurge in 1956 can be explained not in economic terms, but as a manifestation of nationalism surrounding the 1956 Suez War. Presumably, in that context, the regime would have been less likely to repress labor actions that were directed against foreign-owned plants.

Neo-Marxists might then see in the 1960s' resurgence of collective protest evidence that workers had begun to regain class consciousness. Michael Buroway has argued that under systems of "state capitalism," or "state socialism," workers will recognize that they have a collective interest in opposition, not to private capitalists, but to "the authorities."[19] In this regard, it is noteworthy that few Western scholars have considered Nasirism to constitute any genuine kind of socialism. It is commonly held to be a form of state capitalism, where the latter is defined by a large public sector in an environment where private enterprise continues to exist, the market remains the principal means of distribution, and state ownership signifies little in the way of actual workers' *control* over the means of production.[20] Even so, the specific timing and underlying causes of the protests described here are problematic for this explanation, since they suggest only restorative goals. Why would increasingly class conscious workers not use collective action to seek more fundamental change in the system?

Finally, this data supports the moral economy argument that labor protest is precipitated by workers' anger over deteriorating conditions and/or unmet expectations. The protests in 1952–53, when the new regime failed to improve wages and working conditions, clearly conform to this explanation, and the upsurge in 1956 can be seen as a reflection of disappointed hopes aroused by the new constitution and the increasingly populist rhetoric of the regime that year. There are other reasons to suspect that Egyptian workers were prone to expectations of reciprocal rights and responsibilities between themselves, their employers, and/or the state. Urbanization brought to the working class peasants who may themselves have been reared in an agrarian moral economy, i.e., one in which peasant communities share norms about mutual obligations between themselves and landowning or governing elites. After careful consideration of alternative explanations for peasant protest in Egypt prior to 1952, Nathan Brown found that moral economy offers the best analysis of that phenomenon.[21] These beliefs may then be transferred to employers in an industrial context, as Sabel noted of immigrant "peasant workers" in Europe.[22] There is also a basis in Islam for the belief that workers and employers have mutual obligations to one another; indeed, Nasir's pronouncements to this effect may have resonated among workers partly because they tapped into preexisting Islamic notions of fairness and justice in employment relationships.[23]

The moral economy perspective can account for the defensive nature of collective action in the 1960s and early 1970s with the simple assertion that workers came to accept what the state had given them as an entitlement, and were therefore angered when they began receiving less remuneration than before with no reduction in their responsibilities. While a moral economy approach does not require, as Goldberg's logic does, that the lowest paid workers would be most likely to protest, it is reasonable to speculate that those whose very subsistence is threatened by takeaways would experience the greatest anger. Indeed, Goldberg's risk averse workers who would struggle only to protect a minimum level of income and benefits resemble peasants-cum-workers living in a moral economy.

The 1970s and 1980s

In the immediate aftermath of the 1973 Arab-Israeli war, Sadat's government again increased forced savings. This time deductions rose from three-fourths of a day's pay to one full day's pay per month. Then in the spring of 1974, Parliament passed the *infitah* (economic opening) law, confirming Egypt's new openness to Western trade and investment. Inflation, spawned by demobilization after the war, was exacerbated by the resultant influx of imports. Real wages fell in 1975, and remained below the pre-war level the following year as well (see table 3.3). Then the beginning of 1977 saw Sadat, under pressure from the IMF, announce the partial removal of subsidies on a wide range of items.

While workers remained quiet in the first year after the war, they responded to the aggravated inflation by renewing, and later intensifying, collective activity. This reaction began in the fall of 1974, which saw four protests. At least as many incidents occurred the following year, with some larger factories involved. There was also a demonstration against the government's new economic policies at a central Cairo train station. The workers' chants at the demonstration reveal their dissatisfaction with their eroding earnings: "Where is our breakfast, Hero of the Crossing?" (the latter phrase is a reference to Sadat), and "In the days of defeat, the people could still eat." The year 1976 saw a slight increase in the number of protests, including a strike by bus drivers, which paralyzed parts of Cairo for several days.[24]

When the subsidy lifting was announced in January 1977, workers walked off their jobs in industrial establishments throughout the country. Egypt erupted in rioting that left 79 dead and 1,000 wounded, with widespread destruction of property and 1,250 to 1,500 arrests. As with the January 1975 demonstration, the workers' chants reflect a feeling that things had been better in the past. There were cries of "Down with Sadat," "Nasir always said, 'Take care of the workers,' " "It's not enough that they dress us in jute, now they've come to take our bread away," and the simple word, "Nasir." The rioting ended only when the government reinstated the subsidies.[25]

After backing out of the IMF's stabilization program, Sadat was afforded greater room to maneuver by an easing of Egypt's external constraints. The moves toward peace with Israel contributed to this in several ways: the reopening of the Suez Canal brought an influx of sea-traffic paying tolls, and there was an increase in Western tourism, especially after joint trips to the two countries could be booked. In addition, Western aid began flowing in. At the same time, Egypt benefited directly from the oil price boom in terms of export earnings, and indirectly because thousands of Egyptians found employment in the burgeoning Arab Gulf economies and sent part of their salaries home as remittances.[26]

Numerous other measures to appease labor followed shortly in this context. There was an immediate 10% raise in pensions and an exceptional raise for public sector workers. In the private sector, new workers received raises of 5%

TABLE 3.3

Sectoral Wage Trends, 1973–1987

Year	RWI Public Sector	RWI Private Sector
1973	100	100
1974	101	92
1975	94	90
1976	97	102
1977	108	115
1978	104	114
1979	114	129
1980	113	123
1981	121	127
1982	127	129
1983	123	134
1984	128	147
1985	121	141
1986	108	126
1987	99	115

Source: Ragui Assaad and Simon Commander, "Egypt: The Labor Market Through Boom and Recession," p. 26. Reprinted with permission from Ragui Assaad.

after the riots, while all those with more than one year's experience were entitled to a 12.5% increase. 'Abd al-Latif Bultiya, then still Minister of Labor, issued a number of new directives aimed at speeding the settlement of individual and collective workers' complaints, and generally improving industrial relations. The following year, salary scales in the public sector were revised, increasing the minimum there to E£16 per month, and resulting in about a 20% increase across the board; private sector workers with one year's experience got a 15% raise. This trend continued into the new decade. In 1980, a new monthly cost-of-living allowance for public sector workers was put into effect. That same year the minimum wage was upped to E£20 per month, and the following year Sadat decreed an additional increase to E£25 per month.[27] These concessions helped to keep parastatal workers' wages ahead of inflation. As demonstrated in table 3.3, real wages in the public sector recovered in 1977 to surpass their prewar levels and, though dipping slightly in 1978 and 1980, rose strongly through 1982 to be almost 30% higher than they were a decade earlier.

Part of the regime's approach aimed at narrowing the wage gap between the public and private sectors, and the ETUF was given a greater role to play here. In the private sector, national collective agreements were negotiated between the confederation and the Federation of Chambers of Industry and the Federation of Chambers of Commerce in 1979 and 1980. A 10% wage increase was granted in both agreements. Sadat ordered another 10% raise in 1980, and the 1980 legislation equalized the minimum wage across sectors. Then a new labor law (No. 137), passed before his assassination in 1981, required annual raises for private sector workers. The size of the increase is decided in tripartite meetings between representatives of government, businessmen's organizations, and confederation officials.[28]

Meanwhile, hundreds of thousands of Egyptians left for jobs in the Gulf oil states. Migration bid up the cost of labor in the formal private sector, causing nominal wages there to surpass those in the public sector for the first time, so that real wages there grew more rapidly (see table 3.3).[29] As before, workers did not take advantage of the new prosperity to push for even more gains. On the contrary, there is written documentation of only one incident between 1977 and 1981.[30] It is possible that more job actions did occur, but went unrecorded because of a clampdown on the leftist press after the riots. Barakat told me there were as many wildcats during these years as in the 1974–76 period. However, he was in prison most of this time and could not provide details, and I was unable to unearth any corroborating evidence.

Husni Mubarak, who came to power in October 1981, was initially not as forthcoming with workers as his predecessor. Mubarak seemed bent on reducing the burden of social welfare programs on the state treasury. There were no new increases in the minimum wage, and in 1983 he announced that there was no money to pay for workers' raises; budget tightening measures were introduced in some individual parastatals as well.[31] Nevertheless, the increase in real wages continued during his early years, as shown in table 3.3. Labor protests did resume in 1982–83. Although there was no repetition of large public demonstrations as in 1975–77, the number of incidents during these years does seem to rival the earlier period.[32] The coincidence of these actions with a time of rising wages makes this the only period when collective action appears aggressive, but the events of the subsequent years confirm the general correspondence between worker protest and deteriorating, rather than improving, economic conditions.

By 1984 oil prices were softening and some of Egypt's expatriate workers in the Gulf were losing their jobs; Mubarak intensified his efforts to reform the economy. There were some price increases in 1984, along with an attempt to increase the workers' contributions to insurance plans. A new round of price hikes, especially on energy products, came in 1985 as Egypt once again entered negotiations with the IMF, and currency reform began. These measures accelerated inflation, and real wages began to fall in 1985. There were also frequent calls in the official press for privatization and public sector reform, threatening the benefits enjoyed by workers in the parastatals.[33]

Workers reacted with a marked upsurge in walkouts, factory occupations, and other actions. The level of struggle overall was considerably higher than in the mid-1970s; in 1986 alone, 50 incidents were reported. Some of these protests were quite protracted, and involved large numbers of workers, prompting opposition members of parliament to demand an inquiry into the causes of labor unrest. With the economy continuing to deteriorate, this strike wave remained in progress at the end of the 1980s.[34]

The Three Perspectives Revisited

Marxism can again account for this pattern with the claim that workers were developing an increasing sense of the state as a class enemy; hence the 1980s' strike wave appears more extensive than its 1970s' precursor. Moreover, if Marxist leadership is required to channel this sentiment into collective action, the persecution of the left in the late 1970s and early 1980s can explain the hiatus of protest then. Yet if class consciousness was growing, we should also expect to see that workers increasingly challenged the private sector, showed solidarity across plants, and struggled for systemic change; the succeeding sections will show that this was not the nature of the protests that occurred.

Simple rational choice models are again contradicted by the upsurge in collective protest at a time when macroeconomic conditions made the possibility of winning demands less likely, and vice versa. For Goldberg the most salient feature of this period is the continued absence of enthusiasm for unions among workers, despite the fact that the *infitah* widened income inequalities; this is because there were now "returns to skill that are *sufficiently high* for workers to be more than willing to do without unions."[35] Thus he again raises the notion of a remuneration package above which the costs of organizing exceed the benefits; however, it is now the earnings of those workers at the upper rung of the ladder, whose new alternatives to unionization are moonlighting or expatriate jobs in the Persian Gulf. This contradicts his earlier analysis of the 1960s, which suggested that the critical minimum was the wage and benefit package given to the *lowest* paid workers. Thus, we are again left with no satisfactory explanation for why workers in the 1960s did not resort to aggressive collective action: if workers were willing to pay the not inconsiderable costs of moonlighting or prolonged separation from families in order to achieve higher wages during *infitah*, why would they not pay the costs of collective activity to win such gains in the 1960s, when the risks of protest were supposedly so low?

Goldberg's logic is stymied by the authoritarian controls on labor that he either ignores or treats only as effect. His argument can be reconciled with repression only if we specify some subsistence package of wages and benefits that the lowest-paid workers were willing to risk punishment in order to achieve or restore. To explain the periods of quiescence we have seen here, we must then add that for any income above this minimum, the risks of punishment surmounted *any* potential gains.

However, adding political/institutional factors to the imputed utility function of workers requires believing that workers regularly made everyday decisions based on highly complex cost/benefit equations. The moral economy explanation requires neither training in advanced mathematics nor the assumption that workers used sophisticated calculators to determine why they protested when they did—only the simple understanding that the deterioration of a living standard they had come to expect provoked anger.

Entitlement Protests

If the relationship between protest frequency and economic conditions seems too murky to discern definitively between the three perspectives, the superiority of the moral economy approach becomes clearer when one investigates the specific causes of workers' protests. Virtually all of the incidents reported above grew out of a sense of injustice among workers, a feeling that they were being denied something to which they felt entitled. Anger over elite failures to live up to workers' expectations precipitated protest.

As a guide to understanding what workers may perceive as entitlements, table 3.4 shows the system of wage determination which prevailed in Egypt from the 1960s through 1980s. Significantly, for public sector workers, the supplements shown to the basic wage often exceeded the latter. Moreover, since many of these supplements were provided at the discretion of individual plant managers, they created an arena for struggle between workers and managers at the local level.

Reaction to Takeaways

The most common form of entitlement protest was a reaction to threats to full or partial disruption of workers' customary income stream. In post-coup Egypt, its earliest manifestation was actions in 1952 and 1953 directed against

TABLE 3.4

Components of Wages in the Public Sector

Item	Determination
BASIC WAGE	
1) Minimum wage, plus	Set by law for each of various job categories
2) annual raise (*'ilawah duriyah*), plus	Guaranteed by law on condition of satisfactory performance based on job category
3) "exceptional raise" (*'ilawah istithna'iyah*)	Granted by government at times, generally in response to labor discontent; a percentage of 1) and 2) above
Additions to BASIC WAGE[a]	
1) Compensations (*badalat*) (for meals, uniforms, shift work, dangerous jobs, etc.)	By management; subject to negotiation
2) Monthly incentives (*huwafiz*)	Based on production; percentage decided by management within range set by law; can be negotiated
3) Annual Production reward (*mukafah*)	Guaranteed by law if production plan is met
4) Periodic bonus/grant (*minhah*) usually at start of school year, Mayday, major religious holidays	Declared by government or at management discretion; a percentage of basic wage

Source: Interviews with union leaders, especially Niyazi 'Abd-al-'Aziz and 'Abd al-Rahman Khayr, and workers.

[a]These additions can be up to several times the workers basic wage, but are not included in it when raises are calculated.

job loss. At the large Shurbagi Spinning Company in Imbaba, workers occupied the factory when an entire shift was threatened with layoff. In another incident, oil refinery employees went on a hunger strike to obtain reinstatement of 20 workers who had been fired. Several other protests in the early 1950s revolved around wages and benefits. At the Eastern Tobacco Company in Giza, there was a flare-up over the loss of a religious holiday, and at the Marconi Wireless Company, workers struck over the firm's delay in paying a semiannual bonus.[36]

One of the few strikes to occur in the early 1960s, at the Tanta Tobacco Company, was prompted by management's withdrawal of bonuses.[37] We saw above that the takebacks which accompanied the crisis of 1965 also engendered numerous protest actions. Workers' discontent was manifested in other ways as well; when the leaders of the ETUF urged members to work harder and sacrifice a portion of their earnings there was resistance, and implementation of the ETUF's specific proposals was spotty at best (see chapter 4).

Egyptian workers initially responded to the calls for sacrifice after the 1967 war, but when corruption and mismanagement in both the military and the public sector were exposed in its wake, some workers began to question the sincerity of the "battle" and to resent its burdens. Sabir Barakat, who was involved in one of the protests at that time, explains how participants were motivated by a sense of injustice: "A law was issued in 1968 to deduct 25% of workers' compensations for what was called at the time the war effort. We were surprised when we discovered that this was being applied to our salaries which were really only pennies that didn't suffice—it fed us only because it had to. We couldn't bear any deductions so when this took us by surprise we started to resist it."[38]

Opposition to takeaways also provided the opening salvo to the period of heightened activism which accompanied the *infitah*. In September 1974, workers occupied the Harir textile factory in Helwan, protesting the forced savings which they declared were no longer necessary; with disengagement talks under way, the rationale for workers' sacrifice in the name of fighting Zionism had disappeared. They won a change in the savings plan for workers throughout the country: the deductions were reduced by almost 60% to one and one-half percent of salary, and now applied only to workers earning more than E£30 per month.[39]

A 1974 strike at the private sector Tanta Tobacco Company occurred when the owner suddenly switched workers from a monthly to a daily pay rate, resulting in about a 30% decline in wages. Workers held a sit-in at a public sector plant that year after being denied their annual production reward because the production plan was not met, despite company acknowledgment that the shortfall was not their fault. Bus drivers who struck in 1976 were incensed over the company's delays in paying their traditional holiday bonus; their walkout ended when the bonus was issued.[40] Riots in January 1977 over the subsidies were an obvious manifestation of labor and other popular outrage.[41] Finally, the mid-1970s also saw the first attempts to forestall privatization, perceived as a form of takeaway since public sector workers enjoy certain benefits not commonly available to private sector employees, mainly job security, health and accident insurance, and in some cases access to subsidized housing. Involving all levels of the labor movement, the battles against privatization are covered in the next chapter.

In 1982, Kafr al-Dawwar textile workers occupied their local union headquarters protesting reductions in their incentive pay. Workers of the No. 36 munitions factory in Hilwan held a sit-in in 1983, demanding a cost-of-living increase and payment of incentives that had been withheld. Four thousand workers occupied the Nasr Pipe Manufacturing Company plant in 1984, demanding payment of production incentives and rewards, and the resignation of new management that had changed overtime rules. In March of that year, workers staged a series of protests related to the refusal of the Daqahaliya textile factory to pay overtime; April saw 4,000 workers occupy the Shubra Company for Engineering Products charging that poor management had led to a decline in their incentive pay.[42]

The major national takeaway of 1984 was a new law doubling workers' contributions to health insurance and pensions, first issued in the early summer. Implementation resulted in workers at the Nasr Car Factory and several other large plants refusing their paychecks, and an in-plant demonstration in Alexandria; at the Transport Authority a strike was threatened. These incidents led the Prime Minister to halt further application of the law and form a committee to reexamine the issue. Parliament reissued the law in late September, but this time implementation was staggered to avoid simultaneous protests. The deductions hit first in Alexandria, where more than 10,000 workers in two large textile factories refused their paychecks, an ironic form of protest analyzed below. They began in Kafr al-Dawwar two weeks later, and coincided with a decision by Mubarak to raise prices on a number of subsidized items. The result was three days of strikes and riots in the city reminiscent of January 1977, as workers and other townspeople cut telephone lines, blocked transportation, destroyed rail cars, and set fires.[43]

Al-Ahali charges that a high government policy committee called on public sector managers to lower supplemental pay in 1985, and an upsurge in anti-takeaway protests that year lends support to this accusation.[44] At the Miratex Company in Suez, workers struck for three days in February over declining incentives; the company closed the plant so the action would not spread to the other factories. A new protest broke out there in September over the company's failure to pay the traditional bonus for the opening of the school year. Night shift workers at a Suez refining company occupied their plant in July over declining incentives, and in that same month there was a two-day sit-in over incentives at the Talkha fertilizer plant. In July, workers at the East Delta Bus Company refused their paychecks, protesting reductions of up to 50% in take-home pay; a similar action later occurred at a flour mill in Alexandria because of declining incentives and overtime pay. A one-day work stoppage at the Beni Suaf Weaving Company brought management promises to restore wages and purchase new equipment to offset declines in production, so that incentive pay and bonuses could increase.

In Mahalla al-Kubra, 3,000 workers struck the Sigad textile plant in May 1985, protesting management's failure to pay the Mayday bonus; the action broke up when the company promised to issue the checks. Then in October, 1,200 workers from one shop at the large Misr Spinning and Weaving (hereafter MS&W) plant refused their paychecks for three days because of declining wages. This protest resumed in February 1986 with a demonstration of 500 workers in front of the local headquarters, calling for the union to adopt their demands.[45]

One of the largest incidents of the 1980s occurred at Mahalla al-Kubra in September 1988, after Mubarak announced that the parastatals would no longer issue workers the customary grant used by workers to purchase new clothing and supplies for their children. In what came to be known as the "uprising (*intifada*) of school grants," women workers were the first to walk out in anger. As the ranks of marchers grew, the Minister of the Interior ordered the plant closed for three days, which swelled the number of protesters to 30,000.[46]

Some of the most prominent and prolonged protests over takeaways in the 1980s were at wholly private sector or joint venture enterprises that had been created under the auspices of the *infitah* laws. When the tax incentives they had been granted expired, various of these companies began to scale back operations and/or cut workers' salaries. Thus some 700 workers held a sit-in for three days at Arabb, an electrical products joint venture, in February 1986, objecting to declining compensation and incentives. At the Arab Wood Furniture Factory (Atico), management stopped paying workers' salaries in May 1986, and announced in July that the factory would close, laying off 920 employees. The parliamentary committee supposed to approve such closures rejected the company's proposal, but the company refused to honor the committee's decision. On August 12 and 13, more than 650 workers occupied the headquarters of the ETUF, demanding back pay from May and the reopening of the plant. Despite a court ruling in their favor, workers had not received any of their back pay by the end of 1987, and were threatening a new round of sit-ins.

A similar situation developed with the American-owned McDermott Company, which had requested permission to suspend operations in July 1985, and began to dismiss workers before receiving an answer. In January 1986, in a measure workers charge was aimed at forcing them to quit, McDermott cut the wages of its remaining workers by 50%, in addition to deducting one-third of their salaries for the previous two months. In the fall, workers began a series of protest actions aimed at company as well as government targets. The McDermott case continued into the spring of 1987, when a court ruled that the company must rehire 300 workers and sell assets to provide them with back pay. When the company refused to implement the ruling, the workers began a new round of sit-ins.[47]

Finally, the occupation of the Helwan Iron and Steel plant in August 1989, also discussed in chapter 2, represented a protest against a takeaway of a different type. This action was precipitated when the Minister of Industry ousted two leftists who had been elected as workers' representatives to the management council. The minister's actions constituted a blatant attack on the workers' established right to choose these representatives.[48]

The theoretical issues surrounding takeaways are similar to those of protest in deteriorating economic conditions, and have already been addressed in depth. For Marxist analysis, the defensive actions described here represent a problem only in the absence of a greater number of aggressive struggles for more far-reaching change. This evidence weakens Goldberg's case for rationality, however, because the sheer numbers involved would seem to belie his necessary prediction that defensive protest is the province of only the lowest-paid workers. The better-compensated employees who participated in such defensive protests were thus either making mistakes in weighing their utilities—where *is* that calculator?—or else they were acting altruistically in support of their less-fortunate colleagues, and hence not selfishly rational. Moral economy avoids this conundrum by assigning the motivating factor behind protest to anger over violated entitlements, rather than cost/benefit calculations.[49]

The Demand for Parity

A second type of entitlement protest revolves around notions of fairness in the wages earned by different types of laborers. Distributional norms reflect workers' evaluations of how much they contribute to production in relation to others. Egyptian workers have demonstrated a belief in parity, i.e., that similar work should yield similar rewards. They have also exhibited opposition to widening disparities between the wages of manual workers and those afforded to civil servants and company managers.

After the 1961 socialist decrees, discrepancies in benefits and protections available to public versus private sector workers (see chapter 4) became an issue for the latter. This was behind several sit-ins and work stoppages at the private sector Tanta Tobacco Company in 1961 and 1962, as the workers sought and ultimately won a new minimum wage that had been declared for the public sector. Then, in the winter of 1972 the Cabinet agreed in principle to entitle workers in large private sector establishments to the same minimum wage, working hours, and holidays that public sector workers received.[50] Delays in implementation of the decision (which became Law 24), and resistance to it from company owners, prompted a large demonstration by private sector textile workers in Shubra al-Khayma, and several smaller strikes there over the next few months.[51]

Expectations of parity with public sector workers were also implicit in the demands of workers in the *infitah* companies mentioned above, since such lay-offs could not have occurred in the parastatals. Some of the incidents also involved explicit demands for other protections afforded to government workers. The workers at Arabb, in addition to seeking restoration of their previous pay levels, also objected to the company's overall labor policies, which included arbitrary firings, false work contracts, oppression of temporary employees, and restrictions on union activity. Some 7,800 workers at the Johns Company, an American concern, struck for one day demanding various compensations and the application of Egyptian labor laws to foreign companies.

Parity protests have also emerged within the public sector, when workers at one plant see their counterparts at another receive an increase in the discretionary component of wages. Thus the most prominent concern behind the 1971 action at the Harir textile factory was the demand for a 5% "exceptional" raise, which had been granted to workers at a similar plant in Imbaba. Later that year, when the government was negotiating with the Helwan steelworkers demanding a "nature of work" compensation, the president of the ETUF went to the nearby Harir plant and promised its workers parity with whatever the steelworkers won, to preempt any new outbreaks of militancy there.[52]

A 1975 protest at the MS&W factory in Mahalla al-Kubra was the culmination of a prolonged effort by returning servicemen to receive full wages for time spent on the battlefront. The employees who worked during the war earned an hour per day overtime, but the returning soldiers had received only a straight seven-hour per day compensation. After months of complaining to union, ASU, and management officials, the workers heard that a decree in their favor had been issued, and that their counterparts in Harir had already begun to receive the overdue money. When management denied these reports, workers staged a sit-in at the local union headquarters demanding an investigation, and shortly thereafter they occupied the factory itself.[53]

That same year workers occupied the Shubra al-Khayma cableworks, demanding parity in incentives and compensation with the nearby Delta Ironworks employees.[54] And while the bus drivers who struck in 1976 were seeking their traditional holiday bonus, they were moved to action only after learning that their counterparts in Heliopolis had recently received two such grants.[55] Spinners in Minya occupied their factory in 1983 seeking pay parity with workers in other shops. In 1986, after workers at the Esco textile factory in Shubra al-Khayma won a prolonged battle for holiday pay (see below), the protesting workers at the MS&W plant added this to their own list of demands. Their victory in turn sparked a series of similar actions in Mahalla al-Kubra.

While the socialist ideology of the 1960s praised the contributions to society made by manual workers, managers were rewarded more. Workers accepted this, but did object to government measures which would have widened these discrepancies, or elevated the status and privileges of lower-level white-collar employees. Sabir Barakat's account of the 1968 protest against takeaways at Delta Ironworks reveals how such concerns contributed to the workers' actions:

> We had senior civil servants and workers who used to get something called gas compensation—if someone took their own car to work he got a compensation of 25 pounds per month. In those days a skilled worker earned only 12 pounds a month. So we made a campaign around the issue, saying "if there is a serious need to expand the Treasury, in front of you is an amount that can be borne by the people who can afford it. The worst that will happen is that they won't come to work in a private car and will have to come in company cars like the rest of us—which is really no sacrifice—yet you demand from us a very hard life." . . . They gathered us in the cafeteria of the company and gave rhetorical socialist speeches to us, saying "If you have in your hand a piece of bread, don't you give your brother a bite? Even if all you have in your hand is one piece and it isn't enough to satisfy you, you have to sacrifice a bite to the army for the war." The response of the workers was: "When we have in our hand a piece of bread and our brother is hungry, but we have a third brother who has an expensive cake, it's very natural that he who has the cake gives up half of it, then I'll give up a bite."

The New Year's Day 1975 demonstration in Cairo mentioned earlier was precipitated by the prospect of increasing disparities between blue- and white-collar public servants. Toward the end of 1974, the parliament had begun discussing an employment reform bill which proposed long-delayed promotions for thousands of civil service workers, and revised the job classification scheme to eliminate the lowest-paid categories. Union leaders serving in parliament pushed to have the reforms extended to the public sector as well, but 'Abd al-'Aziz Higazi, then Prime Minister, spoke against the union's proposals. 'Abd al-Rahman Khayr circulated copies of Higazi's remarks among Helwan workers, charging that they reflected a hostility toward manual laborers characteristic of the new government. His coworkers not only agreed, he said, but were also incensed by newspaper advertisements for expensive and luxurious New Year's Eve celebrations at Cairo's fancy hotels, one of the first manifestations of the new flamboyance of the upper class under the *infitah*. Khayr and others therefore planned their protest for the morning of January 1, at the time when the party-goers would be heading home from one of the nearby exclusive hotels. Demanding extension of the employment reforms to blue-collar workers, the workers also chanted against the ostentatiousness of the wealthy.[56]

The reform issue also had a role in the 1975 incident at MS&W in Mahalla al-Kubra, mentioned above. After the workers' initial action, the company president promised to release the soldiers' checks. But before the disbursements were made, the news broke that parliament had decided on employment reform only for civil servants; workers took over the factory the following day, demanding the overtime pay, extension of the employment reforms to industrial workers, and improved health conditions. When security forces stormed the plant three days later, their families and other townspeople raided the homes of the company's managers and put their luxury goods out for public display, thus also reflecting a sense of injustice at the perception of widening income disparities between manager and employee.[57]

In the moral economy view, all of these incidents involve perceptions of fairness and justice, and while workers were seeking real gains in income, the protests were also restorative in demanding a return to previously established patterns of wage differentials. However, for rational choice and Marxist explanations, the aggressive aspect of these incidents would seem to be most important. Marxist explanations can now show that workers will try to renegotiate the terms of their exploitation, and, especially in the latter incidents, there is evidence of developing class hatred toward state managers. The question for Marxists, though, is why didn't this class consciousness go further, to a broader movement that advanced a different program for society altogether?

The neoclassical approach must explain why workers engaged in protest for incremental gains that were not worth the risks before someone else obtained them. There is no logical reason why the utility of these gains would suddenly change, so we must assume that the answer lies in the expectation of achieving them, i.e., that some workers' getting a wage increase made it seem a winnable demand to others for whom it previously appeared unobtainable. This, however, means discussing how workers' *expectations from the state* are derived, and that is the province of norm-based explanations. Even more importantly, this also requires believing that workers actually did compute their hypothetical cost/benefit equations, which means they assigned money equivalents and expected likelihoods to each and every possible form of punishment. This assumption was not necessary when protest was purely defensive, since we could say that workers would risk *any* punishment to protect their subsistence, but it is needed to explain collective action for incremental gains beyond that point. Thus, the project to impute rationality to workers must now propose that they made complex and dispassionate calculations of the expected disutility of a beating, bullet, or whiplash. And, by trivializing these measures, it also has the perhaps unintended normative consequence of legitimizing them. Perhaps Goldberg's denial that collective labor protest indeed occurred in Egypt after 1952 was an attempt to avoid this dilemma.

Entitlement Protests: Unmet Promises

A third type of entitlement protest is one which does involve new demands. These are, however, demands to which workers feel entitled because of promises made by company management, the government, or the courts. Thus, anger over unmet expectations is the impetus to workers' actions.

The incident involving most demonstrators in the early 1960s, at the MS&W plant in Kafr al-Dawwar, was such a protest. Prior to nationalization, the company's owners had routinely deducted a portion of workers' salaries for an insurance fund. After the state assumed this responsibility, workers expected that these back deductions would be returned to them. When they were unable to win this demand through months of negotiation with the plant's new managers, they occupied the factory; the three-day protest ended with an agreement by the company for immediate partial repayment of the deductions.[58]

Unmet promises were also behind the 1972 Shubra al-Khayma incidents that followed the Cabinet's decision to equalize benefits and protections between the two sectors. The Ministry of Industry delayed in issuing the law limiting private sector working hours and, believing that the government was reneging on that issue, hundreds of workers walked out of their plants and marched through the town. A second march ensued the next day, when the papers published the minister's decree and the workers returned, only to find themselves locked out of the plants by employers who were hostile to the new legislation.

The February 1983 sit-in at the Nasr Company for Chemicals and Pharmaceuticals occurred after the company had reneged on verbal promises to meet the workers' demands for increased incentive pay, and compensations for lunch and back-shift employment. March 1984 saw 400 workers do a sit-in for several hours at a military plant in Heliopolis, demanding payment of a previously announced raise. The above-mentioned protest at the Johns Company was primarily in response to the company's failure to fulfill a two-year-old promise to provide workers with health insurance.

At the Esco textile factory, 1985 saw workers awaiting a decision on a court case arguing that a 1981 law entitled them to pay for one day off per week. An initial ruling issued in their favor in October 1984 had been rejected by the management, which appealed the case. The situation boiled over in January 1986 after an appeals court again found in the workers' favor. When the company failed to recognize the ruling, workers refused their paychecks. That action brought no result, and on January 30 about 10,000 workers—more than half the company's total employment—took over the factory. The incident ended on February 2 with the establishment of a ministerial "Committee of Five," including representatives of the local union and textile federation officials, to examine the issue; however, workers left their blankets in the plant to facilitate resumption of the occupation if deemed necessary.

The Committee of Five announced a compromise decision to pay Esco workers for two days off per month. Workers rejected this, refusing their February paychecks, and won a new decision that two additional days off would be paid beginning the following year. The holiday pay would not be retroactive to the issuance of the 1981 law, however. Many workers accepted the deal only reluctantly, and several months later, in the context of continued tense relations between employees and management, there was a new occupation demanding the retroactive pay. This sit-in was smaller than the previous one, but nevertheless it involved thousands of workers. After it was smashed by police, workers began to collect donations to take the back-pay case to court.[59]

A strike by train drivers (as they are called in Egypt) in the summer of 1986 is especially illustrative of how protests can erupt from prolonged official frustration of workers' expectations. The wildcat strike grew out of demands originally raised by the workers in a 1982 slowdown. The drivers waited for three years and got no response from either the government or their federation officials. Finally, in December 1985, the Minister of Transportation and the head of the Railway Authority met with them. These officials promised on the spot to resolve some of the issues, and pledged to investigate the others and meet with the workers again.

Four months later nothing had changed, and the workers began a renewed campaign of sending telegrams to the authorities. After ten more weeks elapsed with no official response, the drivers and conductors announced a sit-in at the headquarters of their league on July 2. The occupation ended later that night when the Deputy Minister of Transportation came, and promised that his superior would meet with the workers again on July 7. That day about 1,000 workers gathered for the meeting, only to learn that it had been put off because the minister was busy again. The council of the league sent an urgent telegram to Mubarak, the Prime Minister, and the Minister of the Interior expressing their anger and frustration. Only when there was no reply did the trains stop running.[60]

The 1989 factory occupation in Helwan can also be understood partly in terms of unmet promises. The two leftist workers' representatives to management ousted by the Minister of Industry had succeeded, after many frustrating months of effort, in winning a promise from management to increase workers' incentive pay. When they were removed shortly after the increase was announced, workers had reason to suspect that the promise would be retracted.

Rational choice explanations require that these outbreaks of protest must be attributable to some change in workers' cost/benefit calculations. Here, presumably, the promises made by elites led workers to revise their expectation of gain and perhaps also to lower their expectation of punishment. But this cannot explain the long gap evident in many of these cases between the first sign of renege and the actual protest—why didn't collective action occur right away? Perhaps time was needed for workers to organize and mobilize, but the more time elapsed, the more expectations of gains should have lowered again, while expectations of risks increased. Moral economy has an answer for this: the steadily mounting anger of workers which ultimately reached the point of making them willing to risk repression in order to restore justice.

Spontaneity and Informality: The Bonds of Solidarity

A striking feature of the incidents described here is their lack of union leadership. Unless there was representation by leftists on the board, local leadership was at best not involved and at worst actively hostile to these protests. Workers, in turn, increasingly accompanied their protests with attempts to invalidate the unions.

For example, at one of the public sector protests in 1975, local leaders told workers they had no hopes of receiving the customary annual production reward. When the workers held a sit-in anyway, the company yielded. After this, the workers withdrew their confidence from the local and formed a committee of delegates to meet regularly with management. The federation involved, however, issued orders canceling the activities of the delegates' committee.[61]

Similarly, the leadership of the MS&W local in Kafr al-Dawwar condemned the 1984 uprising there, and assisted the government in trying to identify leftists to be prosecuted as "agitators." Workers in this plant had begun a petition campaign to withdraw confidence from the local because of its unresponsiveness to the base even before the outbreak of events. At Esco, members of the union local supported the workers' original action, although they did not initiate it. After accepting the government's compromise on behalf of the workers, however, they condemned the second sit-in, which rejected this offer, and declared themselves "not responsible" for the workers' behavior. Finally, the 1989 occupation of the iron and steel plant at Helwan targeted the union local, which had supported the government's removal of the two leftist representatives to the management council and tried to claim credit for the management's decision to increase incentives. After the occupation began, the union leaders fled the factory and urged the authorities to storm the plant.

Federation and confederation leaders often ignored these protests; if called upon by the government to represent the workers, senior unionists routinely condemned their tactics first. After the bus drivers' strike in 1976, the head of the Land Transport Workers' Federation publicly apologized to Sadat, and Salah Gharib, then president of the ETUF and the Textile Workers' Federation, issued a statement of "deep regret" after the 1975 protest at Mahalla.[62] And in response to the 1986 train drivers' strike, the president of the Railway Workers' Federation issued an official apology to the Prime Minister, and asked that the workers' league involved in the protest be dissolved.[63]

In numerous cases, workers donated part of their salaries to support the families or pay the legal expenses of those detained as a result of these incidents. There is, in addition, evidence of popular support being mobilized for protesters from the ranks of nonworkers. This is especially true in the industrial mill towns, such as Mahalla al-Kubra and Kafr al-Dawwar, which are surrounded by more rural areas. Here, almost all residents of the town are relatives of mill workers. During the 1975 uprising at Mahalla, townspeople accompanied workers in demonstrating and raiding the homes of factory managers, and at Kafr al-Dawwar in 1984, townspeople joined workers in battling police. During the Esco protest, committees in several nearby neighborhoods organized to send food to the workers occupying the plant.

In the labor market literature, strikes are commonly envisioned as being called by union leaders. One function of unionists is to gather the relevant information on economic indicators and management's position to permit calculations of expected gains and losses. Unionists are also charged with administering the selective incentives seen as necessary to elicit rank-and-file participation in collective activity, since selfish egoism dictates free riding.[64]

For neoclassical theory, therefore, the informality of these protests raises a central question: how was the free-rider problem overcome? In part, the answer could lie in alternative leadership. The train drivers who struck in 1986 turned to their workers' league, a formal rival to the unions, for organization. Elsewhere, a leadership role played by leftists is evident. Leftist cores have operated historically at the Delta Ironworks in Shubra al-Khayma, the Harir and Helwan Iron and Steel plants, and the MS&W factory at Kafr al-Dawwar, all the scenes of some of the largest incidents cited above. Abd al-Rahman Khayr, a longstanding Tagammu' member, claims credit for organizing the January 1, 1975, demonstration at Bab al-Luq. The 1986 events at Mahalla were led by an independent group, the Workers' Defense Committee, that was formed in the early 1980s under the auspices of the Tagammu' but later split from the party.[65]

But while they are capable of providing organizational skills, leftist groups do not possess the resources to offer any selective incentives of a material type to workers. And many of the protests cited here, while resembling those in which leftists were active, were evidently spontaneous; the Esco occupation during 1985–86 is a notable example. Certainly the widespread walkouts associated with the January 1977 riots exceeded the organizing capabilities of the left.

The absence of formal organization behind these protests does not pose a problem for Marxist or moral economy theorists. Whereas rational choice theory embraces methodological individualism, both of these other perspectives presume the existence of group identities. Marxists believe that subaltern groups will develop bonds of solidarity based on their common oppression, and can find in these incidents evidence of incipient working-class consciousness. Moreover, to the degree that the frequency and scope of labor protest was manifestly higher in the 1980s than in any earlier decade, traditional Marxists can claim that such consciousness is indeed growing as Marx predicted.[66] The problem for Marxism, as before, is the restorative nature of most of these incidents, even those manifestly led by the left. There is also scant evidence of class-based solidarity extending from one industrial region to another, such as a protest in Alexandria to support workers in Cairo or Helwan.

Moral economy explanations see collective sentiments resulting from shared vulnerabilities, expectations, and norms of justice among subaltern strata. Class and community-based categories can be combined in this approach, and the actions shown here may signify participants' blending a class-based identity as workers with a more community-oriented identity as *sha'b*, or common people.[67] In Egypt, Nasirism itself, with its emphasis on the contributions made by workers and peasants to the cause of national development, probably helped to cement such group-based feelings; leftist organizing in the late 1960s and beyond appears to have reinforced existing sentiments of solidarity and norms of cooperation among workers and their families and neighbors. Moreover, through their agitation and distribution of information, leftists may have heightened the anger of workers toward policy-making elites when accustomed patterns were broken.

Increasingly, rational choice theorists are coming to recognize the salience of norms and the prevalence of social bonds which facilitate cooperation among people. A body of game theoretic literature, in particular, seeks to show how cooperation can emerge as a norm after repetitive interactions among the same group of players, in which trust is rewarded and defection can be sanctioned.[68] I remain skeptical of the assumption of innate selfishness which underlies these models. Indeed, the prevalence of trust and cooperative behavior across so many cultures should lead neoclassicists to question this assumption rather than puzzle over the advent of altruistic norms.[69] Still, it is worth noting that Egypt's public sector plants and mill towns, with their stable workforces and populations, respectively, may approximate the conditions under which these models operate.

At the same time, it is not my claim here that norms of solidarity will be uniformly present among all workers. In particular, while anger is seen here as a necessary condition for protest to occur, it may not be sufficient; two plants which follow the same economic policies toward workers might differ in their "protest proneness" according to the workers' willingness to cooperate with each other, and/or their organizing capabilities.[70] Beyond the presence or absence, and the strategies, of leftist organizers it seems likely, minimally, that the size of a plant and the way in which production is structured—both categories of importance to traditional Marxist analysis—would influence whether, and how quickly, feelings of solidarity can develop.[71] Thus, while this study upholds moral economy as the best overall perspective for explaining collective labor protest by workers in Egypt, understanding the spontaneous and informal nature of these incidents is one area where Marxism, moral economy, and new directions in rational choice theory can usefully intersect.

Workers With and Against the State

The moral economy I propose is one in which workers are in a reciprocal, but patron/client, relationship with the state. This belief among workers is manifested in the nature of the protests themselves. It is also evident on the government's part in the pronouncements labor protests invoke from regime elites, as well as in the policy response.

Symbolic Protest

That public sector protests are aimed at agencies of the state is intrinsically obvious. Yet even before the creation of the large parastatals, workers were turning to the government for redress of their grievances; the elevated levels of legal complaints filed in the 1950s, as discussed earlier, is in part a reflection of this. More recently, both government leaders and the ruling National Democratic Party (NDP) have been targeted by protesting public sector workers. For example, after the insurance deductions were increased in 1984, workers at one textile factory in Alexandria demonstrated in their plant, with chants of "down with the NDP." Elsewhere antigovernment slogans were chanted in marches to the local headquarters of the NDP, and several government offices.[72] During the 1988 protests at Mahalla, marching workers carried a large statue of a dog on which they had painted the name Husni Mubarak.

Private sector workers have also targeted the government in their protests, suggesting that they too view the state as the guarantor of their livelihoods. For example, the 1972 walkout by private sector textile workers in Shubra al-Khayma charged the government with reneging on promises to upgrade their benefits; as part of the protest workers blocked a motorcade carrying the Prime Minister. McDermott workers' protests included sit-ins at the NDP offices in Suez and visits to the Minister of Labor at his home. When they were denied the annual production bonus in October, 500 workers occupied the company's headquarters in Cairo, and later staged a sit-in in front of parliament, and when the company refused to implement the spring 1987 court ruling, the workers began a new round of sit-ins at NDP headquarters, and also caravaned from the Suez to the president's palace in Cairo.[73]

Because of the interference of the government in union elections, the fact that most senior union officials were affiliated with the NDP, the historic conjoining of the posts of ETUF president and Minister of Labor, and the frequent unresponsiveness of these union officials to wildcat protests, some workers came to see the confederation itself as an instrument of the state. Hence the sit-in at confederation headquarters by the private sector Atico workers. In numerous other incidents cited here, workers occupied the office of their local seeking official union support for their demands.

But if workers expected the government to ensure that they were compensated justly for their labor, they also exhibited the belief that they had an obligation to the state. In particular, the nature of labor protest in Egypt suggests that workers saw their responsibility lying in production, to contribute to the postcolonial modernization and development of their country. The evidence can be found in the relative scarcity of actual strikes, in favor of protests which exhibited dissatisfaction while not interfering with work. This reflects the fact that workers themselves eschewed actual work stoppages, using them only as a last resort.

The closest alternative to strikes was the in-plant sit-in, during which management was ejected or ignored but workers continued running the factory on their own. This tactic was initiated in the 1963 incident at Kafr al-Dawwar, where the men continued round the clock production for several days to prove their loyalty to the country, but electrified the fence around the plant to prevent security forces or government officials from getting in. The 30,000 Helwan steelworkers who protested in 1971 also maintained production while sitting-in. Harir workers demonstrated *inside* their plant in 1971, and their 1974 protest against forced savings was also a factory occupation. The 1975 sit-in at Mahalla al-Kubra lasted three days, during which time production continued. Likewise, the workers who occupied the Shubra al-Khayma cableworks in 1975 kept the lines going while refusing to allow government or union officials to enter. In the first action of 1984, at the Nasr Pipe Company, the workers actually doubled output during their occupation. When the Esco workers occupied their plant in 1986, they continued working for two days. Only on the third day, when no responsible officials had come, was work suspended.

A second symbolic technique was a boycott on cashing paychecks. It was particularly used in the public sector, where workers say it was effective because it interfered with government accounting procedures. The boycott was first used by the workers of Delta Ironworks in their 1968 protest against salary deductions. By refusing their paychecks, but continuing to work, the workers demonstrated that they remained loyal to the "battle."[74] The next documented use of this tactic was when 2,600 workers at the Tura cement factory refused their paychecks for two days in February 1982. We have already seen that some 10,000 workers in Alexandria boycotted their checks in 1984, and boycotts also occurred at several large factories in 1985 and 1986.

Another form of protest involved sending telegrams to government officials seeking redress of grievances. While milder than the tactics above, such messages can also provide insights into workers expectations of themselves and the state. Thus the initial reaction of workers at Delta Ironworks to Sadat's 1977 removal of subsidies shows both their fear of, and disgust with, the government:

> I suggested that we send a telegram to the officials denouncing them, but the other workers . . . were afraid of denouncing the government. So in the end we decided to send a very satirical telegram. It read:
>
> "To the President of the Republic:
>
> We thank you for increasing prices, and raise the slogan, 'more price hikes for more hunger and deprivation.' May you always be a servant of the toiling workers."

The protest organizers collected signatures and a small contribution from 600 workers in the plant in order to send this message.

The disgruntled railway workers first wired their concerns to the president, the Prime Minister, and other high government officials in 1982, and continued to press their case in this manner over the next four years. The urgent telegrams sent by the league to Mubarak in an effort to ward off the strike is particularly revealing of their moral economy beliefs. It concluded, "We are all waiting here at the league headquarters . . . Some of the trains have already stopped running *in abandonment of our responsibilities*. The situation is getting more serious. It is almost 2 P.M. and at 6 P.M. this evening all of the trains will stop running" (emphasis mine).[75]

Significantly, even in those rare instances where workers have raised aggressive demands in the context of favorable macroeconomic conditions—the type of collective action which conforms to the simplest rational choice model—they used symbolic protest to press their case. Specifically these incidents were a 1982 sit-in by 6,000 workers at the Kalkha fertilizer plant, demanding a lunch compensation and revision of salary and promotion schedules, and a slowdown that same year by railway workers seeking a cost-of-living adjustment, insurance against accidents, an increase in the compensation for uniforms, and elimination of the requirement that drivers be personally responsible for paying compensation to train accident victims. The largest incident of 1983, an aggressive protest demanding a nature-of-work compensation at the Helwan Light Transport factory, was another factory occupation.[76]

While moral economy explains workers' eschewal of strikes by their belief in their responsibilities, neo-Marxism can cast the same phenomenon in terms of the submerging of class to national consciousness. However, this should imply minimally that workers in *infitah* companies would have a greater propensity to strike, with an intermediate level in firms held by private but domestic capital; the data does not seem to confirm such predictions.

Neoclassical models could account for symbolic protest by specifying that workers obtain a positive utility from the feeling of contributing to the national project, although this does stretch the definition of selfishness. Alternatively, or in addition, symbolic protest could be explained merely as a strategy to avoid punishment, provided it could be shown that such actions actually accomplished that goal. The following section will demonstrate, however, that even symbolic protest was usually met with repression.

The Carrot and the Stick

The view that workers who actually or even seemed poised to cease production were harming the interests of the state was actively encouraged by successive regimes in Egypt. It was adopted by the new military rulers in 1952 well before the creation of the parastatals. When workers occupied the Shurbagi Spinning Company in Imbaba in 1953, the military came to disperse them. Some 400 to 500 were arrested and released only after signing an apology "to the state." Later, as we saw above, Nasir explicitly promoted the notion of reciprocal rights and obligations between workers and the state.

Sadat and Mubarak, even while trying to retract the state's role in the economy, perpetuated this philosophy. Sadat condemned the 1971 sit-in at the Helwan ironworks as "an undemocratic act" and blamed both the 1976 bus drivers' strike and the riots shortly thereafter on communist agitation, implying that loyal workers would never otherwise leave their jobs. A decade later, Mubarak vilified the Esco and Mahalla workers' actions as "a lack of nationalist responsibility."[77]

Moreover, despite their differences in overall economic strategy, the Nasir, Sadat, and Mubarak governments pursued very similar policies with regard to labor protest. The actions were quickly put down by a combination of repression and concession. Only the largest incidents were ever mentioned in the official press, and these were customarily blamed on outside agitators. Preventing any escalation of the protest, and maintaining an image of national harmony and worker satisfaction, thus seemed to be far more important to Egypt's rulers than minimizing financial concessions. At the same the consequences of these actions was to reinforce the moral economy and thereby pave the way for future protests.

As an indication of the arbitrary use of power in Egypt, the nature of punishment itself appears to have derived more from regime perceptions of the severity of the incident than from any precise legal framework. When only a small single plant was involved, the immediate handling of the problem was usually left to the discretion of local police, with national security agencies and government figures called in only if the protests could not be quickly broken up. However, protests which involved issues of national policy, large numbers of workers, and/or more than one plant, brought rapid intervention from the highest levels of government.

Thus when the Tanta workers held a sit-in in the early 1960s, the company summoned local police. The men were beaten with rifle butts, and their union leaders were blindfolded and taken to the police station where they were kept concealed. Ultimately, federation officials summoned from Cairo negotiated their release, along with a contract in which workers won restoration of the bonuses as well as other new benefits. But workers who refused their paychecks at Delta Ironworks a few years later were taken immediately to state security police. Their intimidation there had an effect, as Barakat relates, but here too concession was also involved in ending the protest:

> They arrested 22 workers . . . They came back two days later completely silent and went to pick up their checks. We tried to get them to talk but they refused to say a word. After that the stand of the workers began to weaken. But 'Abd al-Nasir in those days was smart—on the same day he sent orders that these deductions shouldn't harm the salaries of the workers and they returned the deductions to us—and that ended the situation.

Around the same time, the Minister of Defense repealed a decision to deduct the transportation compensation and any salary increases due to promotions from workers at another military plant; all back deductions were returned.[78]

Local police sometimes treated workers with particular brutality. Three of the leaders of the 1971 Harir protest were tied to their cell doors and periodically beaten with clubs and whips. Only the intervention of Mohammed Heikel, then editor-in-chief of *al-Ahram*, brought about an end to the beatings and the provision of medical care for the victims, who still bear scars from this incident.[79] After police arrested workers during the 1985 Sigad strike, one woman was beaten in front of her husband, and the arrest of a male worker while his wife was in labor caused her to name the baby "Ifraj" (release). The Sigad detainees were placed in cells flooded with sewer water, forcing them to stand until arraignment.[80] Not surprisingly, though, the volume of arrests appeared to be greater in the more prominent incidents. When security forces stormed the Helwan ironworks plant in 1971, some 3,000 workers were arrested. Seventy-six of the protesters who blocked the Prime Minister's motorcade in 1972 were detained. Significantly, public sector workers have been able to return to their previous employment after release from prison, although those singled out as organizers have sometimes been subject to punitive transfers.

Attempts by security forces to suppress peaceful protests have sometimes contributed to their escalation into violence. The 1975 New Year's demonstration turned into a mini-riot when police attempted to prevent the protesters from marching to a nearby government building; dozens of arrests and a witch-hunt for the action's organizers followed. When security forces raided the MS&W plant in Mahalla al-Kubra in 1975, 2,000 were arrested; 50 workers may have been killed that day.[81] Barakat argues that police initiated the violence in January 1977:

> On the second day we struck at the factory. . . . We went out and demonstrated. They tried a tactic on us—they said they would search us as we were leaving. They took workers in their cars to examine them. There are three bridges near the plant. . . . They dropped workers off at each one, in order to divide us. The result was that there was a demonstration at each bridge, and while each group was marching it was attacked by the police.[82]

Yet each of these events was also met with significant concessions. After the 1971 Helwan sit-in, Sadat quickly promised to investigate and ameliorate workers' grievances. In June 1975, following the series of protests documented above, the employment reform measures were approved for public sector workers.[83] Sadat also approached leaders of the bus drivers' federation to discuss their members' grievances immediately after their 1976 strike. A similar meeting was held between the Minister of Transportation and leaders of the Railway Workers' Federation, and resulted in an increase in their annual production reward.[84]

As we have seen, Sadat also quickly repealed his decision to lift subsidies in the face of the January 1977 riots. Furthermore, soon thereafter he held his first formal meeting with ETUF leaders since before the 1973 war, and agreed to demands not only for wage increases, but also for greater union input into management and governmental decision making. In addition, the Minister of Labor issued a number of new directives aimed at speeding the settlement of individual and collective workers' complaints, and generally improving industrial relations. Periodic wage increases in both sectors continued over the next few years.[85]

Despite Mubarak's moves toward political and economic liberalization, this combination of repression and concession continued during his presidency. At the Helwan Light Transport factory, security forces armed with tear gas, clubs, and electric prods surrounded the plant and broke the occupation, but management did agree to the workers' demand.[86] Three workers died in clashes with security in the 1984 Kafr al-Dawwar riots, and there were over 120 arrests. As a result of the riots, though, Mubarak repealed the price hikes on pasta and cooking oil.[87] During the 1985 incidents in Mahalla, security forces raided the wool factory where workers were holding a sit-in, and arrested 160. The next day, the company closed the plant, putting the workers on forced vacations; another 257 workers were arrested outside the factory.[88] The second Esco sit-in was similarly smashed by the police, who stormed the plant on the third day of the sit-in, arresting over 500 protesters.[89] Nevertheless workers in these cases also won at least part of their demands.

When the train drivers struck, the government called out the army to run an emergency bus service, and the central security police to attack the striking workers. They were clubbed, kicked, and beaten, and over 100 were arrested. Security police continued to hunt down the leaders of the action and those who were arrested were dealt with harshly. Some were denied food and bedding for two days. Unlike any other case cited here, the workers were charged with violations of the emergency laws and arraigned before supreme state security courts, which are empowered to impose indefinite sentences. These courts are supposed to deal only with cases of armed terrorism. The speed and severity of this response would seem to reflect the centrality of the services suspended by the trainmen, and the high visibility of a railway strike relative to the protests at individual industrial establishments. Nevertheless, after the strike was broken the minister finally found time to meet with groups of workers. Most of their demands were met, and some were given financial rewards for agreeing to return to work.[90]

An equally heavy hand was applied to the 1989 occupation of the iron and steel factory in Helwan. An initial sit-in during mid-July ended peaceably with a concession by the company president on the incentive issues, and a promise to look into their demand that the elected workers' representatives be reinstated. But when a larger occupation began two weeks later, after it became clear that the sacking would not be reversed, the plant was stormed by security forces. One worker died and a hundred were severely injured. Over the months that followed, several hundred workers were arrested and two, identified as leaders, were brutally tortured. Moreover, members of a committee of intellectuals formed in support of the workers were also jailed and severely beaten. As in these other cases, the government also made significant concessions to the workers' demands, the list of which had grown during the occupation.

It is instructive to contrast these cases with the government's response to the protests at the *infitah* companies, where workers might arguably have felt less strongly that their labor was contributing to the national project. None of the incidents in foreign-owned firms were met with the same official repression visited upon workers in wholly Egyptian-owned facilities. This suggests that the regime had its own ambivalence about these enterprises, or at least did not want to risk inviting more debate on their role by appearing to side with the companies.

It does appear that punishment was more swift and sure when an actual work stoppage occurred, or seemed imminent. We have also seen that workers sometimes began a struggle with a relatively mild tactic, such as sending telegrams, and then escalated to more pronounced and thereby riskier actions like paycheck boycotts and plant occupations. Rationality can thus account for workers' reluctance to strike as a defensive strategy. But this would leave us without an explanation for the fact that workers did ultimately risk repression, short of assuming again that they assigned precise disutilities, in money terms, to each possible form of punishment.

Such a complex and ultimately authoritarian exercise is not necessary to explain workers' behavior. While positing a commitment among workers to maintain production, the moral economy approach also suggests that their anger toward the state mounts as their demands go unmet, making them prepared to take more radical action and face greater risks until the government fulfills *its* obligations. At the same time, to the extent that repression was more severe for actual strikes, this would only serve to confirm workers' beliefs that their contributions to production were important and worthy of remuneration. In this regard, both aspects of the government's policy toward labor protest served to reinforce the moral economy in workers' eyes. Hence the Western governments, multilateral lending agencies and academicians urging Egypt to adopt market-oriented reforms should have been the first to insist that the long-standing prohibition on strikes be lifted.

In their own right, the struggles of Egyptian workers had a significant effect on the success or failure of various economic reforms in Egypt. We have seen here that protests emanating from the point of production impeded the government's ability to lower subsidies and remove or modify price controls. By resisting cutbacks in various types of supplementary pay, workers also posed a strong obstacle to efforts to cut the fiscal deficit through public sector rationalization.

Workplace struggles have also been used here as a means to make broader inferences about the attitudes of workers toward economic issues. Among the possible and competing explanations of workers' behavior, the evidence most strongly supports a moral economy interpretation, in which workers believe there are reciprocal rights and responsibilities between themselves and the state. Workers contribute their effort in production, and expect the government in turn to ensure their jobs and their livelihoods.

While there is evidence here of a rising level of labor protest, Marxist and neo-Marxist theories are not vindicated. For Marxism, the problem is why more collective struggle, and of a different type, did not occur. If class consciousness entails recognizing state managers as the enemy, workers should have been advancing a program for change, not restoration. At the same time, the private sector, domestic and foreign, should have come under increasing attack as workers pushed for state ownership to be extended there, even as they struggled to capture the state. If the failure of workers to struggle aggressively for a new society does not condemn Marxist theory, we must conclude that the Egyptian left did not successfully articulate for workers a vision for a better alternative future.

Explanations in the rational choice framework which start by assuming selfish egoism on the part of individual workers prove unable to explain why they risked severe repression to engage in collective action for minimal financial gains. Rationality may well be the best explanation for the *opposite* of the phenomena studied here, i.e., for the fact that on any given day, the vast majority of Egyptian workers were *not* engaged in visible collective activity. Indeed it is precisely because the risks associated with collective action were so high in Egypt that it required an emotionally driven *suspension* of rationality in order for protest to occur. Thus the argument here is neither that rationality cannot explain collective action in systems that are both economically *and* politically liberal, nor that Egyptian workers were uniformly and routinely irrational. Rather the claim here is that temporary "moments of madness" were required to surmount the barrier of fear surrounding protest in authoritarian countries like Egypt.

Numerous countries in the Middle East and sub-Saharan Africa have evidenced trajectories of political and economic development resembling Egypt's. It is therefore reasonable to suggest that reciprocal relationships between workers and the state similar to that described here have obtained elsewhere in these regions. The broader claim here, however, is not for the ubiquity of moral economies *per se* but rather for the association between anger, as a *passion*, and labor protest in repressive societies.

Finally, this chapter has examined the role of the formal union movement in relation to rank-and-file protests, and the reaction of the government to them. The state was found to employ a combination of concession and repression aimed at rapidly ending the incidents, with the apparent goal of preserving the impression of regime legitimacy. As we will see in chapters 4 and 5, this same concern led successive Egyptian governments not to pursue certain reforms, or to push them only halfheartedly, out of fear of arousing labor protest.

Except for those who are leftists, union leaders were not involved in the protests enumerated here, and often condemned them. It does not follow logically, however, that opportunistic unionists will in all circumstances side with the state against the rank and file. The next chapter argues that unionists, too, have legitimacy concerns, which cause them to respond differently to issues of national scope than to individual plant protests.

4
The Union Movement With and Against Reform

As we have seen, Egypt since the 1952 coup has experienced two phases of economic restructuring. The first, which started in the mid-1950s and extended through the following decade, was the gradual expansion of the state's role in the economy, accompanied by efforts to politically incorporate workers and peasants. The second, tentatively begun in the mid-1960s and still underway thirty years later, constituted efforts by the Nasir, Sadat, and Mubarak regimes to undo these etatist policies, rescinding by increasing amounts the government's commitments to workers.

Picking up the trail set forth in previous chapters, this chapter explores the response of the trade union movement to the successive government's economic policies. Its central argument is that the union's response to reform initiatives varies over time, as well as across issues. Union leaders' reactions to different policies are shaped by the interplay of numerous factors. These are: the calculations of opportunistic leaders as to what will best insure their incumbency and/or enhance their power and prestige; the perspective of unionists motivated by ideology, and their capacity to prevail over opportunists when their policy responses differ; and, finally, the pressures that unionists face from below, which is a function both of variations in workers' response to the different issues, and in the number of workers affected by any given policy.

The central focus here is on the unions' responses to the Nasir, Sadat, and Mubarak regimes' attempts at economic liberalization, through the late 1980s. Advocates of rapid liberalization malign the government's efforts at reform as piecemeal and halfhearted. I argue here that a key factor behind the successive presidents' tentativeness has been the fear of losing their perceived legitimacy among the working class.

I look first at the 1960s, when the main content of reform efforts revolved around cutbacks in workers' salaries, lengthening the working day, and reducing the number of grievances filed by workers and their locals. During this period the union movement was strictly supportive of the government, in fact going beyond it in its calls for worker sacrifice. Possibly reflecting in part the absence of Marxists in the ranks of union leadership, this phenomenon is also indicative of an embrace of Nasirist philosophy by many federation and confederation leaders. This identity informed the opposition of some unionists to the subsequent efforts to undo etatism.

The material on the post-Nasir years is organized by issue, and aims to show how the response of the confederation and its member federations varied by issue type. I focus on the three issues which proved to be most contentious between the union movement and the state: privatization and public sector reform, the removal of subsidies and price controls, and the opening to foreign trade and investment (*infitah*). The order reflects the descending degree of union opposition resulting from the interplay of the factors cited above.

Nasir's Economics and the ETUF in the 1960s

Beginning in the late 1950s and especially following the socialist decrees, the Nasir regime hiked wages and extended numerous benefits to workers. Nasir's view that unions, as they had been conceived in the past, were no longer necessary after these measures were enacted has been endorsed by some analysts of Egyptian labor. Harbison argues that even in the 1950s, the Egyptian government had appropriated many of the tasks that unions in advanced industrial countries commonly perform. And Goldberg suggests that with the socialist laws and the creation of the public sector, the union movement became altogether "superfluous."[1]

The statistics of declining union membership at the time (cited in chapter 2) can support this claim. But such an argument could not then account for the success of the rivals to the ETUF's basic units; if workers felt no need for representation and collectivity, *all* labor organizations should have floundered. In fact, the union movement did have opportunities to pursue economic demands if it chose to, and its role must be evaluated against these possibilities.

First, although the new laws did deprive unions of the ability to negotiate basic wages in the public sector, a variety of supplements to the base could be enhanced by agreement between union locals and plant management. As we just saw, these increasingly became the focus of struggles in individual plants. In addition, especially during the early 1960s unions could play a role in pushing for implementation of the laws that were resisted by parastatal managers.

In the private sector the scope for union activity was greater. Although virtually excluded from the most strategic industries, the private sector continued to dominate small-scale manufacturing; it was responsible for 80% or more of value-added in such industries as leather, furniture, wood, wearing apparel, and printing.[2] There, locals could negotiate collective agreements covering wages, raises, compensations, and bonuses, while the confederation as a whole could seek extension of the new minimum wage to cover all private sector workers. However, organizing was more difficult in the private sector largely due to a provision in the successive union laws that separate locals can be formed only in shops employing more than 50 workers. In smaller plants, workers desiring to unionize must find other small plants performing similar work, and then band together in a regionally based local. Aside from the logistical difficulties this entails, workers reported that many plant owners deliberately maintained their full-time workforce below 50 to impede unionization and the enforcement of other labor laws.

Measured against this potential, the unions, especially at the upper levels, fell short. Some local union officials were involved in the occasional labor struggles which dotted the 1960s (see chapter 3) as well as in the filing of collective grievances, which remained at relatively high levels.[3] By and large, however, federation and confederation leaders appeared to take no initiative to mobilize workers around economic demands, even though they were now organizationally better equipped to do so. Only one federation, Food Processing, placed special emphasis on organizing private sector workers.[4]

The only economic issue around which the confederation took a stance was the violations of the new labor laws in private sector manufacturing. The ETUF press, when it began publishing, featured periodic exposes on these, and the confederation passed a resolution calling on the government to penalize companies that violated the labor laws at its 1965 labor law conference in Alexandria. There is no evidence, however, to suggest that the unionists ever organized or threatened any kind of militant action to support this demand.

Finally, the union movement could seek to play a role in the policy debates which ensued when the regime had difficulty financing its ambitious development project and generous incentives to workers. In fact, the contradictions with the ASU and the confederation's own internal divisions notwithstanding, the labor movement after 1964 was now better placed to address economic policy issues at the national level. Although there was no provision in the 1959 or 1964 laws for ETUF representatives to participate officially in the discussion and implementation of economic plans, union leaders ran as individuals, with the backing of the confederation, for the workers' seats in the parliament, where they served on various committees.[5] Federation and confederation leaders could also lobby on workers' behalf within the ASU, as well as seek special meetings with regime elites. The union newspaper, which began appearing monthly in October 1965, and then weekly in 1968, provided the confederation leaders with a public voice to advocate for workers. At this stage, however, the unions chose to endorse, rather than oppose, the regime's call for retrenchment.

The Unions and Retrenchment Under Nasir

Nasir's ambitious development project quickly engendered economic imbalances for Egypt. As is typical of countries following import substitution strategies, Egypt developed a widening trade gap. In 1962, within a year after the socialist decrees were promulgated, a balance of payments crisis occurred, prompting negotiations with the IMF. As a result of these talks Egypt devalued the pound and received a standby credit. However, Nasir would not carry through on pledges to cut government spending and curb private consumption, which would have had a more direct impact of the workforce, and risked being perceived by laborers as a retreat from what the government had promised them.[6] In fact, he continued on the same state socialist path after the crisis had abated. The public sector was allowed to expand, and there were further sequestrations of property in 1963 and 1964.[7] There were also additional measures to increase workers' health and living standard standards in the latter year. Under the 1964 Labor Code, previous insurance laws were consolidated and adjusted to increase pensions and injury compensation, and extend health care provisions. Then, early in 1965, wages for public sector workers were switched from a daily to a monthly basis, effectively increasing their minimum wage nominally to 7 pounds/month. The 1962 episode with the Fund marks the beginning of what proponents of Western-style liberalization programs see as the regime's lack of will to implement reform.[8]

However, as we have seen, when the crisis deepened in 1965 Nasir replaced 'Ali Sabri as Prime Minister with the more conservative Zakariya Muhyi al-Din, who reopened negotiations with the IMF. The government cut investment expenditures, implemented some price and tax increases and closed several factories, and after the middle of 1965, no new social legislation was introduced. Nevertheless, Nasir was still worried about the effects of these measures on his image, locally and abroad. He drew the line a year later when Muhyi al-Din concluded an agreement with the Fund that would have required a 40% devaluation of the pound, which could have tarnished the image of steadfastness the regime was trying to project internationally.

Instead, he canceled the IMF deal and dismissed Muhyi al-Din, replacing him with Sidqi Sulayman.

The retrenchment policies had a significant effect on the country's economic expansion. In the three years from FY 1964/5/5 through 1966/7 the average annual rate of GDP growth fell to 3.8%; the average annual expansion of industrial output slipped even more sharply to 2%, with an actual decline experienced in the latter year.[9] Manufacturing employment in the formal sector did not suffer as severely, but also registered a slower growth rate of 7.6 percent.[10] A corresponding erosion in workers' earnings was shown in tables 3.1 and 3.2.

Nevertheless, the regime did appeal to labor for further sacrifice. Shortly after his appointment in 1966, Prime Minister Sulayman summoned union leaders to discuss ways of mobilizing workers to increase output. During the course of the meeting, he complained about the number of workers' grievances being brought to the labor offices, saying "we must put the right of the work before the right of the worker."[11]

Egypt's defeat in the 1967 war with Israel exacerbated its economic difficulties. Nasir once again eschewed an IMF-type stabilization program, especially with regard to devaluation. However, the government did initiate another round of price and tax increases, and there were renewed calls for workers to sacrifice, accompanied by measures which gave them little choice in the matter. Zakariya Muhyi al-Din, then head of the Committee for Planning and Production, issued orders for the work week to be increased from 42 to 48 hours without compensation to workers. Forced savings were increased from 1/2 to 3/4 day per month, and additional retrenchment measures were attempted in some individual plants.[12]

The drive to revitalize the economy put pressure on public sector managers to increase production, and gave new fuel to the hostility harbored by some of them, and by some private sector factory owners, towards workers and their unions. Throughout the latter part of the 1960s some parastatal managers refused to deal with local union leaders and tried to have them fired; local leaders were also subjected to transfer and demotion. In the private sector the harassment of activist local leaders was worse. Furthermore, the problem of arbitrary dismissals, which the socialist decrees had attenuated but not eliminated completely, worsened during this period after a new law issued at the end of 1967 made it easier for managers to evade court recommendations for the reinstatement of workers.[13]

The unions did not counter these measures. Instead, federation and confederation leaders, especially, eschewed aggressive behavior. Moreover, as the dimensions of the country's economic difficulties at mid-decade became clear, the senior unionists took the lead in calling on workers to forego militancy, and to sacrifice some of the privileges they had earlier been granted. In May 1965, the confederation held a conference to discuss the problem and issued the "Charter of Labor for the Take-Off Stage" (*Mithaq al-'Amal fi Marhalat al-Intilaq*). Signed by 288 federation and confederation leaders,[14] its resolutions included the following:

1. Workers should volunteer at least one additional hour per day of free labor.
2. Each worker should save one-half of the cash portion of his/her profit shares, and locals should return five percent of the profit shares earmarked for social services.
3. Calling on the government to create a national savings fund in which workers could participate with one day per month's labor.
4. A commitment not to raise any new economic demands for two years.
5. Forswearing work stoppages for any reason, with any local that violated this commitment to be considered "in deviation of the national plan."[15]

The charter was presented by Ahmad Fahim to the parliament in June 1965. Here Prime Minister 'Ali Sabri, while praising the confederation's initiative, objected that its provisions were too stringent, creating the impression that the state was bankrupt. Parliament enacted a more moderate version in which only one-half day's pay was put into forced savings, and this only for public sector workers. The Labor Code was modified to re-allow paid overtime, but without making it mandatory. It was shortly after this that Sabri was cashiered in favor of Muhyi al-Din.

While the role of local leaders in response to the post-war economic situation is unclear, confederation leaders continued to support the idea of sacrifice by workers for the sake of national reconstruction. However, Ahmad Fahim did adopt a more critical posture than in the past on economic issues. In his 1968 Mayday speech in Kafr al-Dawwar, Fahim charged for the first time that certain public sector managers still had "old" (i.e., presocialist) approaches to industrial relations. At the same he deplored the disparities between the public and private sector and the violations of wage and insurance laws in the latter, and the confederation did push in this period for passage of a draft law prepared by the Ministry of Labor which would give equal legal status to workers in both sectors.[16]

Union Leaders Evaluated

This quiescence on the part of union leaders does not distinguish the 1960s from the previous decade. On the surface, though, it does suggest that the

ETUF leadership embraced the new philosophy of unionism being promoted by Nasir. This impression is furthered by the fact that, after the initial episode with Anwar Salama (see chapter 1), the language of the 1962 National Charter became the basis around which the trade union movement was evaluated, by itself and other concerned parties, throughout the 1960s.[17] In this context, the ETUF's calls for workers to sacrifice beyond what the regime itself was prepared to sanction at the time was probably in part a reflection of the ongoing competition between the confederation and the ASU (see chapter 2).

However, in the authoritarian atmosphere which prevailed in the country, it is reasonable to assume that not all unionists who mouthed the official policy were expressing their true sentiments. Some were likely seeking only personal advancement, and others acting out of fear of repression. For example, while Salama, in his new post as Labor Minister, also became a public champion of the regime's new philosophy, some assert that he retained his private reservations.[18] While Salama's hesitations would have come from his conservative leanings, others more to the left may have silently wanted the ETUF to push more for an extension, and/or better implementation, of the socialist measures.

From below, senior unionists were probably getting mixed signals. At the lower levels of leadership, there is evidence of both acceptance and rejection of the new philosophy, as well as confusion. Kamel interviewed workers in three unionized plants before the June 1967 war. Of 81 workers surveyed, only four felt that the local was doing its job. More than half (46) said that the union did nothing, while the remainder complained that the local carried out only social activities. The workers overwhelmingly felt that their union had little influence over management, and should have considerably more. Asked in what areas they would like to see the local perform better, the strongest responses were regular wages, raises, grievances, and better communication with management.[19]

In evaluations of the trade union movement published in *al-'Amal* and *al-Tali'ah* during the 1960s (see chapter 2), some local and intermediate level union leaders accused counterparts and superiors of failure to adopt the new understanding of unionism. Others, however, bemoaned the loss of the unions' former identity, and in some cases the same individual made both complaints. Ironically, in this discourse those who clung to the notion that unions should be primarily concerned with representing the wage- and job-related concerns of workers were sometimes labeled as reactionary, i.e., in that context, pro-capitalist, elements.

The same series of interviews revealed disagreements and confusion at the base of the union movement. Labor historians as well as unionists and leftists who were active at the time concur that Nasir was immensely popular with the overwhelming majority of workers, and there is some evidence of workers willingly cooperating with retrenchment programs. Although some of it may well be fabrication or hyperbole, *Al-'Ummal* contains numerous interviews with workers promoting the virtues of sacrifice. Also the efforts to reduce the number of grievances did apparently meet with some success, at least in the first year of the crisis, although this could have been the result of pressures on local leaders applied directly by the regime or by senior unionists.[20] Finally, table 3.1 also indicated that workers did commit to longer hours on the job, although the reasons for this cannot be clear. In some cases overtime, although remunerated, may have been imposed by managers. The discrepancies between the hourly and weekly RWI's shown in the table also imply that many workers made up for by declining real hourly earnings by extending the working day.[21]

Nevertheless, the response from below to the calls for extra effort suggests that the regime overestimated its own ability, and that of labor leaders, to win the support of the base for retrenchment. Some workers clearly viewed the calls for sacrifice as an infringement of the rights which they had been promised, and responded in this manner. Anwar al-Salama, then Minister of Labor, told me that the voluntary donations called for in the ETUF's charter could only be implemented with approval from the locals, and that in many cases local leaders, fearing rejection, never brought the proposals before the membership. 'Abd al-Hamid Balal, then an official in the Labor Ministry, cites the ETUF Charter's recommendations as evidence that leaders were out of touch with the rank and file, and charges that none of the proposals were ever put into effect.[22] Finally, as we saw earlier, some workers took wildcat action against the cuts.

The sentiment against sacrifice appears to have been stronger after 1967. Thus, Fahim may have been responding to greater pressures from below when he adopted a more militant economic posture after the war. Alternatively, or in addition, he may have been taking advantage of the more open political atmosphere which followed the March 30 Declaration to express views he had always held but previously suppressed.[23] Regardless, when Fahim died in 1969, his successor officially adopted a policy of "calmness" around economic issues. 'Abd al-Latif Bultiya, who was an affiliate of the WS group, declared repeatedly, "we will have no economic demands until the enemy is vanquished," and put no pressure on the Labor Ministry regarding the problems of workers.[24] The contrast between Bultiya and Fahim illustrates how senior unionists can respond differently to the same countervailing pressures from above and below, demonstrating that union leaders' behavior must always be somewhat indeterminate and cannot, in particular, be predicted solely on the basis of state policies.

The vagueness of this data makes it impossible to generalize about the motivations of the unionists who supported retrenchment. We cannot be precise about the proportion of unionists who genuinely embraced Nasir's policies, versus those who endorsed them out of opportunism or fear. Moreover, the Nasirist camp included individuals like Bultiya and Fahim, who had divergent views on the proper role for unions under "socialism." What is most significant for the decades to come, however, is that a sizable percentage of unionists, including some in the upper ranks of the ETUF, came to identify themselves as followers of Nasirism. As a consequence, when Nasir's philosophy and etatist policies came under explicit attack after his death, these unionists were propelled into unity, in opposition to the government.

Labor vs. Economic Liberalization

Sadat initiated a shift in Egypt's international alliances with the expulsion of the Soviet advisers in 1972. He further pleased the United States by engaging in peace negotiations with Israel after the October 1973 war. Moreover, after several precursors, a corresponding change in the country's economic alliances was confirmed in June 1974, with the passage of Law 43, officially known as the *infitah* (economic opening) law. It opened the country to many Western imports that had previously been prohibited or restricted, and provided generous incentives to lure Western foreign investment back into the country.

The *infitah* law's ideological accompaniment was the exultation of private enterprise in general, and criticism of Nasir's etatist philosophy and practices. What had come to be known as workers' "socialist gains" were widely perceived as under attack. Ironically, these threats served to revitalize the labor movement, restoring a sense of purpose that many unionists felt had been lost during the Nasir years.

Labor vs. Privatization: The Early Battles

The shift in economic and political philosophy promoted by Sadat led logically, and quite quickly, to calls for the public sector to be dismantled. The Egyptian economist Fu'ad Sultan proposed publicly even before the 1973 war that the government shed failing public sector enterprises. Then, in December of that year, the parliament's Plan and Budget Committee recommended partial privatization of successful parastatals. Charged with reviewing the budget presented by Finance Minister 'Abd al-'Aziz Higazi, the committee expressed fears that the deficit financing implied in the budget would be inflationary. But Egypt need not choose austerity as an alternative, their report said, if it could woo investment from neighboring Arab countries awash in surplus capital. Bemoaning the fact that Western nations placed fewer restrictions on Arab investment than did Egypt, the committee called for allowing up to 49% ownership of public industry to devolve to Arabs, slipping in the idea of private domestic participation in this context. The report went on to argue that joint ownership would benefit labor by making parastatals more profitable, thereby directly increasing workers' profit shares, and would enable workers to gain even more if they became private shareholders.[25]

The proposal sparked widespread debate in the ensuing months. In February 1974, the ETUF's executive committee strongly denounced the idea in a formal statement. Charging that the public sector was the embodiment of the socialist principle of collective ownership of the means of production, and that sharing ownership with private individuals negated this concept, the statement warned that "the millions of workers will not permit any attempt to threaten the socialist gains brought about by the July Revolution." Under the din of this and related objections from other quarters, the committee's proposal was never approved.[26]

Gharib confirmed this position in his remarks to an ETUF economics conference in April 1974. Affirming that "socialism is the only solution for progress," he added, "we all agree that the public sector must remain the leader of progress in all realms. It carries the main responsibility for the development plan, and this requires that it be supported and provided with all it needs to enable it to undertake its pioneering role, assuring through this the control of the people over all the means of production."[27]

The sharp ETUF defense of the public sector reflects a combination of the ideological identities of the unionists, rational self-interest, and responses to pressures from below. Both Marxists and Nasirists in the union movement were opposed to privatization out of a philosophical commitment to the public sector. But even unionists who lacked this orientation had reason to oppose the sale of the parastatals, since public sector workers made up the bulk of union membership. Thus the financial resources of the ETUF and most industrial federations, as well as the power and prestige that came with union work, were linked to preservation of the parastatals.

There is little doubt that the rank and file were also attached to the public sector. For one thing, at the time of the initial privatization proposals, parastatal employees had higher average earnings than their private sector counterparts. Tables 4.1 and 4.2 give an indication of the wage differential between the two sectors at the start of the 1970s. It shows that workers in the public sector earned about 1.5 times the wages of those in the private sector, with the difference being more pronounced when only manufacturing workers are considered. A February 1972 ETUF labor law conference had decried the wage disparities between the public and private sectors, and passed resolutions calling for equality between them, and better enforcement of minimum wage laws.[28] The passage of Law 24 of 1972 entitled workers in the larger private sector establishments to the same minimum wages as those employed by the state,[29] and is apparently responsible for the drop in the sectoral differential that year. However, the gap began to widen again thereafter, rather dramatically in manufacturing, apparently because of mobilization for the war.[30]

TABLE 4.1

Public/Private Sector Wage Ratio: Blue-Collar Workers, All Sectors[a]

Year	Public (eg. pias./wk)	Private (eg. pias./wk)	Ratio Pub/Pri[b]	% Chg[b]
1970	419	281	1.49	
1971	423	303	1.40	-6.0
1972	460	409	1.12	-20.0
1973	491	421	1.17	4.5
1974	556	432	1.29	10.2
1975	576	473	1.22	-5.4
1976	663	596	1.12	-8.2
1977	813	744	1.09	-2.7
1978	897	842	1.06	-2.7
1979	1054	1022	1.03	-3.7
1980[c]	1,289	1,204	1.07	3.9
1981[c]	1,576	1,418	1.11	3.7
1982	1,928	1,671	1.15	3.6
1983	2,271	2,120	1.07	-7.0
1984	2,613	2,569	1.02	-4.7
1985[d]	2,800	2,800		
1986[d]	3,100	3,100		
1987[d]	3,400	3,400		

Source: *Annual Survey of Employment, Wages, and Hours of Work* (SEWHW), CAPMAS. I am grateful to Ragui Assaad for providing me with this data.

[a]"Blue collar" is defined by CAPMAS to include all hourly (as opposed to salaried) workers. Only establishments employing more than 10 workers are covered in the SEWHW.

[b]My calculation.

[c]Considered estimates by CAPMAS.

[d]Figures rounded to the nearest pound by CAPMAS; ratio not computed.

There were also marked disparities in benefits between the two sectors which continued to exist after Law 24. Profit-sharing laws were not applied to private concerns other than joint-stock companies, for example, and only joint-stock companies or those with more than 50 permanent employees were required to implement a form of workers' representation in management.[31] As we saw earlier, workers and unionists alleged in the ETUF press and elsewhere that private sector employers violated laws relating to insurance and dismissals, maintaining full-time permanent workers on temporary work contracts to avoid these obligations as well as to keep down their employment numbers. Finally, the access to affordable housing that many public sector workers enjoyed became increasingly important as a housing shortage developed.

TABLE 4.2

Public/Private Sector Wage Ratio: Blue-Collar Manufacturing Workers, All Sectors[a]

Year	Public (eg. pias./wk)	Private (eg. pias./wk)	Ratio Pub/Pri[b]	% Chg[b]
1970	433	268	1.62	
1971	423	290	1.46	- 9.9
1972	466	426	1.09	-25.3
1973	496	331	1.50	37.6
1974	550	424	1.30	-13.3
1975	560	461	1.21	- 6.9
1976	665	588	1.13	- 6.6
1977	777	764	1.02	- 9.7
1978	896	824	1.09	6.8
1979	1,074	917	1.17	7.3
1980[b]	1,327	1,120	1.18	.8
1981[b]	1,641	1,367	1.20	1.7
1982	2,028	1,669	1.21	.8
1983	2,337	2,092	1.12	- 7.4
1984	2,646	2,515	1.05	- 6.2
1985[c]	2,800	2,800		
1986[c]	3,100	3,100		
1987[c]	3,400	3,400		

Source: *Annual Survey of Employment, Wages, and Hours of Work* (SEWHW), CAPMAS. I am grateful to Ragui Assaad for providing me with this data.

[a]"Blue-collar" is defined by CAPMAS to include all hourly (as opposed to salaried) workers. Only establishments employing more than 10 workers are covered in the SEWHW.

[b]My calculation.

[c]Considered estimates by CAPMAS.

[d]Figures rounded to the nearest pound by CAPMAS; ratio not computed.

The benefits available to public sector workers were important to their private sector counterparts as well. These benefits formed a benchmark for private sector workers to struggle for parity. Many private sector workers also aspired to eventually obtain public sector positions, so they too had reason to fear the eradication of the parastatals.[32]

Thus, unionists were getting entreaties from below calling for opposition to privatization. However, they faced countervailing pressures from the regime to support the drive away from etatism. Salah Gharib, who owed his position to Sadat, was especially vulnerable. There were also numerous federations whose membership was less threatened by privatization, particularly those representing mainly civil service workers. It was largely leaders of these federations who were allied with Gharib, and together this group endeavored to dampen the ETUF's opposition to privatization. For example, Gharib tried to obstruct more widespread distribution of the February 1974 ETUF statement. It had appeared originally just in the confederation's paper; only through independent action by left-wing unionists was it reprinted, several months later, in *al-Tali'ah*.[33]

Gharib and his allies also failed to support Fathi Mahmud and the Commerce Workers Federation (hereafter CWF) when they were embroiled in two significant fights against privatization in the commerce sector. The first of these battles began at the end of 1974, when the manager of the publicly owned United Wholesale Textile Trading Company (hereafter UWTTC) signed a contract transferring two small factories owned by the company to the private sector Cooperative for Workers of the Delta Spinning and Weaving Company. UWTTC was a large company formed by the successive nationalization of private stores in the 1960s. The two small factories, 'Azizu and Nublis, had belonged, respectively, to the Hanaux and Salon Verte stores, supplying them with ready-made clothing and other textile products. When these stores were nationalized the factories became part of the UWTTC.[34]

The UWTTC manager, Dr. 'Ali Sabri Yasin, negotiated and signed the sales contract without consulting either his management committee (which, like all public sector companies, had elected representatives from the workers) or the union local. The workers claimed they were told the factories were merely being transferred to the jurisdiction of a public manufacturing company, since their activity was industrial rather than commercial. Upon learning the truth, they refused to work for their new owner, and sent an urgent memo to the CWF. The Federation hired lawyers and raised a court case against Yasin, seeking to invalidate the sales contract; it was the first court case challenging the sale of public sector assets. In their deposition, the lawyers argued that the sale violated those provisions of the constitution which call for the protection and preservation of the public sector and state that socialism is the basis of the Egyptian economy. They further argued that the transfer would deprive them of numerous benefits associated with public sector employment, in particular periodic raises and promotions, annual vacations, and sick leave.[35]

Several months later, Minister of Commerce Zakariya Tawfiq 'Abd al-Fatah decided to fully or partially privatize the publicly owned stores. Shares would be sold in the largest enterprises, and the smaller units were to be sold outright to private entrepreneurs. Once again it was workers in the CWF who were to be affected. When the plans were announced in the spring of 1975, the executive committee of the federation took a position against them, and sent an urgent memo to the ETUF board seeking assistance. The federation decided to sponsor weekly meetings with locals of the stores involved, to invite all the labor delegates in parliament to a general meeting to discuss the issue, and to plead their case with other government officials. Mahmud contacted *Rose al-Youssef*, which publicized the plans via an interview with 'Abd al-Fatah, followed by a lengthy rebuttal from the unionist.[36]

The CWF kept up the pressure on the minister through the summer and fall, and in late October 'Abd Al-Fatah contacted Mahmud and offered to rescind his plans in exchange for an end to the campaign against him; he agreed to a request by Mahmud to announce his decision at a general meeting of the CWF. Thus, one of the first concrete proposals to transfer public assets to the private sector was defeated by concerted union action. But Mahmud had not enjoyed the full backing of the confederation during this victorious campaign. The antipathy of Gharib (and other confederation leaders close to him, such as 'Abd al-Rahman Khidr) to the activities of the CWF was reflected in the complete absence of coverage of these events in the confederation's own newspaper. Mahmud found more sympathy at *Rose al-Youssef*, while Amina Shafiq, labor reporter for al-Ahram, helped him to publicize the issues elsewhere.[37]

Contradictions between the opponents of privatization and the Gharib camp continued as the regime made the first attempts to legislate privatization. In the early summer of 1975, Sadat issued Presidential Decree 262 which authorized the use of share subscriptions to increase the capital of public sector firms. This was followed in July by the passage in parliament of Law 111, which abolished the General Organizations that had been responsible for guiding the performance of public sector firms. The law effectively made the parastatals more autonomous, as well as providing for the presence of "experts" from the private sector on their individual management boards.

Law 111 also set terms for share subscriptions, specifying that they were to be offered first to workers in the affected companies. However, during the parliamentary debate, defenders of the public sector succeeded in adding an article (No. 10) limiting share sales to those public sector firms which already had some private participation, known as "joint companies"(*sharikat mushtarikah*), except in the case of firms incurring losses which had been scheduled for elimination. At the time, there were only 32 such joint firms, as compared to 341 that were strictly state-owned.[38] Moreover, Article 10 specified that the state's overall share in such firms could not be reduced through the subscriptions, although the role of private shareholders in their management could be expanded.[39]

The addition of Article 10 thus constituted a clear setback for the advocates for privatization. It is not clear to what extent union leaders serving in parliament, as opposed to other forces, were responsible for the modification. The prolonged nature of the controversy over privatization does suggest that opposition to the scheme existed outside the labor movement as well, and extended into the government in spite of Sadat's personal support for it.[40]

The ETUF promptly formed a committee to study the new laws and their implications for workers. Its work, according to Sa'id Gum'a, was to be guided by the slogan, "No new capitalism at the expense of the public sector." The committee's report, released by the confederation in August, noted with satisfaction that subscription was to be limited to those parastatals which already had private sector participation, because "public sector companies are not for sale nor for the participation of the private sector." Moreover, the confederation called on the government to ensure that the share of the state in joint companies would not be reduced by stock sales, and that the system of workers' participation in management would not be tampered with in these companies. The statement also upheld the idea of giving workers the first opportunity to purchase shares in them.[41]

However, the report dampened the views of the hard-line opponents of privatization in the labor movement, such as Gum'a and Fathi Mahmud; the former had earlier told *al-'Amal* that workers would rather have their rights in the public sector than the opportunity to buy shares in companies where they worked at the risk of someday being re-classified as private sector employees. The government, he said, should encourage those with savings to invest in new enterprises, not the existing public sector. Because the confederation's statement did not outright condemn subscription sales, it was denigrated by Fu'ad Mursi, a prominent leftist economist and outspoken critic of liberalization, as "the greatest surrender possible." Nevertheless, I concur with Qandil that the ETUF statement stood fundamentally in opposition to the drift of Sadat's government. An indication that pro-regime union leaders were not happy with the document is that it was summarized in *al-Ahram al-Iqtisadi*, but never appeared in *al-'Ummal*.[42]

It was around this time that Gharib's overtures to the ICFTU (see chapter 2) further pitted Mahmud and Gum'a against Gharib, who stood embarrassed and exposed when the two took the issue to the leftist and popular press. Gharib had his revenge in November 1975 by engineering a coup in the CWF, as members of the federation's executive committee voted to withdraw confidence from Mahmud and the other members of the office committee. Gharib recognized the action within 24 hours, issuing a new list of officers and orders to change the signatures on the federation's bank account. He then used his powers as president to launch an investigation into the finances of the federation. As we saw earlier, the move caused a stir of protest within the ETUF, and an investigation by the new Minister of Labor, 'Abd al-Latif Bultiya. But the change was not rescinded. Mahmud believes that higher authorities acquiesced to Gharib's actions against him, because of his outspoken opposition to privatization. Lending credence to that view, one of the first acts of the new CWF leaders was to withdraw the legal case against the UWTTC, thus making the sale of the two small factories final, and forestalling a constitutional challenge to privatization.[43]

These initial privatization skirmishes hold some lessons about the influence of union structure on labor capacity, and its intersection with matters of will. The defeat of the proposed wholesale sell-off in the commerce sector was facilitated by the organization of those workers into a single federation; had they been isolated in disparate locals, or dispersed in several competing federations, mounting a successful campaign against the plan would have required greater amounts of commitment, time, and resources from larger numbers of union leaders. However, given the alliance of the confederation's leader with the regime, the subordination of the commerce federation to the ETUF proved more of a hindrance than a help.

Labor vs. Privatization After the Riots

Salah Gharib's replacement by Sa'd Muhammad Ahmad removed the most staunchly pro-regime unionist from the ETUF's leadership. Although most of Gharib's allies remained in place, Ahmad followed a less partisan style of leadership, seeking to ameliorate rather than exacerbate conflict in the upper ranks of the confederation. At the same time, the Marxist and Nasirist presence at all levels of the ETUF was sharply reduced by the regime's repressive measures after the riots. The combined result was a less fractious union movement, but also a more moderate one.

The pressures from below against economic liberalization may also have moderated somewhat during this period due to the new opportunities for workers to obtain lucrative employment in the burgeoning Gulf economies, which bid up the price of labor in the private sector. As illustrated in tables 4.1 and 4.2, the wage differential between the two sectors began to fall in 1976.[44] In some sectors, such as land transport, workers reported that higher wages could be earned in the private sector as of the mid-80s. Moreover, during the mid-80s, because of its spiraling fiscal deficits, the government made public sector employment more difficult to obtain; although the regime did not formally renounce its pledge to provide jobs to all high school graduates, the wait to obtain such positions became steadily longer. Workers also reported an increase in the low-paying probationary period at public sector jobs; some saw this as a deliberate attempt by the government to discourage young people from working for the parastatals.[45]

Nevertheless, most of the workers I interviewed in 1987–88 continued to prefer public sector employment over private. Even those working in the private sector by choice said they viewed it as a short-term arrangement in order to establishment themselves, and would shift to the public sector as soon as they had married and found housing. The main reason workers cited for preferring the public sector was that it offered greater job security, pensions, and protection from illness and accident. The private sector was frequently accused of violating the labor laws with regard to these protections. A common charge was that employers were refusing to give workers full-time contracts or forcing them to sign undated resignation letters so that they could be fired later. Employers were also accused of lying to authorities about the number of workers they employed, since claiming less than 50 full-time permanent employees facilitated evasion of workers' representation in management and made unionization more difficult as well. These comments suggest that the illegal practices in the private sector reported by the labor and leftist press in the 1960s and early 1970s were never eliminated.[46]

Against this background, the subsequent union/state battles over privatization and public sector reform, or lack thereof, proceeded on two tracks. The first of these, the focus of the remainder of this section, involved regime efforts to pass legislation at the national level that would facilitate selling the parastatals or making them operate more according to market logic. Thereafter I explore the second track, which involved entering individual firms in joint ventures with private capitalists.

There were some regime initiatives to privatize individual parastatals in spite of the limitations set in Law 111. After the law took effect, the Ministry of the Economy hired the Ford Foundation to study ways to manage share offerings, and some wholly owned state firms did receive authorization to sell shares. The process of selling shares developed slowly, however, in part because the stock exchange itself first needed revitalization. It was not until 1978 that capitalization through subscription was budgeted, and these budgets were never realized. In addition, two public sector paper companies were given private sector status as part of an agreement for World Bank assistance in their development, and one joint company was granted private sector privileges at the request of its shareholders.[47] The lack of evidence of an ETUF response to these developments is indicative of the removal, intimidation, or co-optation of the previously most outspoken opponents of privatization in the aftermath of the riots.

The first significant liberalization measure at that time was Law 48 in 1978, which aimed at public sector rationalization. This law upped the minimum wage in the public sector and increased salaries, especially for the lowest-paid workers. However, it allowed parastatal managers to set their own hiring levels, whereby formerly they were required to accept all workers assigned by the Ministry of Labor. Over time, this came to mean that new high school graduates must wait for increasingly longer periods in order to obtain jobs in the parastatals. Thus the measure contributed to a significant reduction in overstaffing in the public sector, whereas the civil service, which was not freed from mandatory hiring, remained bloated 16 years later. In addition, the law gave managers greater flexibility in setting initial wage rates for new hires, although only above the legal minimum. It also provided for greater freedom in promotions, though implementation of this was impeded by the fact that the law failed to specify new standards.[48]

The role played by the labor movement in relation to Law 48 is cloudy. The law itself appears to have generated little public controversy, perhaps because it was associated with wage increases, maybe also because of the intimidation of the left and/or the fact that national attention was focused on the peace process at the time. ETUF literature indicates that the confederation recommended numerous modifications to the initial draft prepared by the Central Planning Agency, only some of which were heeded; the senior unionists then took their concerns to the Cabinet and the Committee on Labor of the parliament, where they elicited further changes. The confederation felt that the final law was still deficient "in application," but maintained that its effect was to provide for more or better compensations, raises and holidays for public sector workers, and better protection against disciplinary actions. However, neither the modifications initially sought by the organization, nor the criticisms of the final bill, were specified.[49]

The confederation did respond rapidly and strongly, however, to the ongoing push for more sweeping privatization. At the national level, the first serious threat came when Sadat appointed a new Minister of Industry in 1980. Taki Zaki promptly introduced a new scheme for selling shares in the public sector, which would have allowed the state to reduce its share in public sector companies to 51%. Ahmad himself had endorsed a version of the proposal that spring, supporting the sale of 49% of failing parastatals through subscription. But Zaki's proposal was angrily denounced by most federation leaders in a meeting held October 13. Amid calls for convening an emergency general assembly of the ETUF, Ahmad proposed that the issue be decided at an upcoming confederation conference on "Promoting Production in the Peacetime Economy."[50]

The other union leaders agreed, and though the conference had been in planning for well over a year, this issue became its main focus. After what was generally described as hot discussion, the overwhelming majority of participants rejected Zaki's proposal "in form and content, in the name of all Egyptian workers." The conference opposed "any trend toward selling the public sector or having private sector participation in it." Ahmad publicly announced the ETUF's position on television at the close of the conference.[51]

Qandil argues that the strong opposition raised by the ETUF was an important factor in the defeat of Zaki's proposal. Subsequent events support this interpretation—three times over the next two years, government officials promoting sell-off plans met with the ETUF leadership seeking their endorsement. In addition, officials of the U.S. Agency for International Development (AID) arranged special meetings with some federation leaders in 1982 hoping to disabuse the latter of their opposition to privatization.[52]

The first new proposal was presented to the confederation in the spring of 1981 by Dr. 'Abd Al-Rizaq 'Abd al-Majid, Vice Prime Minister for Economic Affairs. It called for separating the ownership and management of parastatals by establishing "holding companies" (*sharakat al-qabidah*) in which the ownership of the firms would be vested; the holding companies, in turn, would be open to private participation. In addition, the plan called for public sector management to be given greater leeway in determining promotions and pay scales, and for reduced worker representation in management. 'Abd al-Magid's scheme was rejected in a restricted and unannounced meeting of union leaders, based on detailed analysis of the plan by the confederation's legal department, and was not pursued further.[53]

Sadat's assassination does not appear to have intimidated the proponents of privatization. Another proposal was brought to the labor movement in November 1981, by officials of the Financial Markets Authority, who met with the heads of those federations representing the bulk of blue-collar public employees. Their plan apparently involved establishing a system of branches (*shu'ab*), similar to the old general organizations, to coordinate parastatal activity. Like its predecessors, this plan would have given state managers more leeway in labor affairs, and restricted workers' representation in management (WRM). In addition, it proposed to give management freedom to lower the workers' share of profits (hereafter WSP). The federation leaders involved objected to it for these reasons, also citing the lack of provision for union representation in the branches.[54]

The last attempt in this series came in October 1982, when several ministers met with confederation leaders to discuss a new proposal for subscription in the public sector. According to *al-Ahali*, the leaders concerned agreed to support the plan, which called for private participation to be limited to 49% of the shares in any company, with some modifications: the share price would be 100 pounds, not 1000 as originally proposed, and the federations would have the right to appoint the workers' representatives to the management committees of the jointly owned companies, rather than having them elected by workers. This latter criterion, though obviously a step backwards for workers' democracy, would have resolved the historic objection of some union leaders to the fact that the unions were not formally connected to the WRM system. However, some federations, such as the EEMWF, immediately objected to the new proposal. The left did as well, and was able to publicize the plan and its opposition through *al-Ahali*. Leftists in the trade union movement pressed for meetings of locals and federations to discuss the scheme, and denunciations were reported in the newspaper. Rejections came from a confederation regional meeting in Aswan, two large textile locals in Helwan, and five regional meetings of the Chemical Workers' Federation. Other than this, *Al-Ahali* reports nothing further on the plan; it was, apparently, quietly set aside.[55]

Mubarak himself appears to have concluded, at least initially, that Sadat's association with the privatization drive had contributed to his unpopularity, and sought to distance himself from it. On Mayday 1983, Mubarak publicly committed himself to "no diminution and no sale of the public sector" (*la masas wala bi' al-qita' al-'amm*). He repeated the pledge on numerous occasions over the next few years, and the issue faded for a while from the union and leftist press.

By late 1985, however, the regime was under external pressures to push ahead on public sector reform. Egypt's balance of payments had begun to deteriorate as a result of the softening of oil prices, which adversely affected the country's export earnings, Suez Canal receipts, and expatriate worker remittances. The government approached the IMF to reopen talks that had been stalled after the 1977 riots, and Mubarak appointed Fu'ad Sultan, the initial advocate of privatization, to head up the Ministry of Tourism. The regime also began increasing its connections to recently formed associations of private sector businessmen, and granting them access to top-level economic policy makers.[56]

Proclaiming the need for "awakening and preparation to confront the vicious campaign that some businessmen are launching," the ETUF's executive committee charged that these entrepreneurs were pushing for the sale of inefficient parastatals. This was the impetus behind another ETUF conference on the public sector in February 1986, which affirmed the confederation's support for the parastatals and called for the strengthening of the WRM system.[57] But senior union personnel, and especially Ahmad, now came under pressure to dampen their defense of the parastatals. In 1986 Ahmad, in his capacity as Minister of Labor, prepared a new draft law which would have lowered workers' bonuses and profit shares, and legalized employee transfers. *Al-Ahali* exposed the proposal, and Tagammu' members then began a campaign of petitions and meetings in unions where they had influence to urge ETUF rejection of the law. This movement resulted in withdrawal of the proposal. But around the same time, Ahmad was cashiered from the ministry (see chapter 2).[58]

The government then revived the proposals for selling the parastatals at an NDP-sponsored conference on developing the public sector in early 1987. The opposition parties boycotted the meeting, making it in effect an NDP affair, and selling some public sector assets was one of the conference resolutions. Two months later, *al-Ahali* reported that a new plan for subscription sales, this time involving the banks, was under preparation. Then, over the summer, Tourist Minister Sultan advocated subscription sales to Egyptians as well as foreigners in a magazine interview. A prominent journalist subsequently wrote that most Ministers, especially then-Prime Minister 'Ali Lutfi, supported Sultan's idea but were afraid to say so publicly.[59]

Al-Ahali published interviews with many union leaders, including Mustafa Mungi, the head of the NDP's labor bureau, denouncing the scheme. This time, however, the ETUF itself did not issue a condemnation. It was in the midst of this controversy that Sa'd Muhammad Ahmad resigned as ETUF president. But Mubarak himself was still fearful of arousing too much opposition from workers and other forces. The immediate clamor over Sultan's proposal was quieted when Mubarak once again stated, in two separate speeches, that he opposed the sale of the public sector.[60]

The Controversy Over Joint Ventures

The above clashes over parastatal sell-offs revolved around what I call "one fell swoop" privatization schemes—legislation which would facilitate changing the ownership structure of the entire public sector at once, by mandating widespread subscription sales. From the standpoint of labor/state relations the significance of such schemes, as well as those for public sector reform, is that they threaten the perceived entitlements of the entire public sector workforce at one time, as well as potentially antagonizing private sector workers who aspire to public sector jobs. As such they maximize the pressures from below on union leaders. At the same time, the national scope of the legislation means that it is federation and confederation leaders who are called upon to represent labor in negotiations over these policies, and whose legitimacy is at stake over them.

However, privatization can also be accomplished more gradually, and on a plant by plant basis. In this case, a much smaller number of workers is affected at any one time, and the issue is the immediate responsibility of local leaders. Federation officials may become involved secondarily, but even so, only a small proportion of their constituency is affected. This is even more true for the leaders of the confederation as a whole.

In Egypt's case, staggered privatization was facilitated by the *infitah* laws, which permitted the involvement of parastatals in joint ventures with foreign concerns. Prior to 1974, any company which had partial state ownership was considered part of the public sector and subject to its laws. Law 43 of that year specified that any joint ventures created with firms entering under its auspices would be considered part of the private sector. Public sector firms could thus privatize all or part of their operations by entering into a joint venture with a Law 43 company, subject to the approval of the Investment Authority created under the law.[61] In some cases, public sector firms contributed only assets, such as land, to the new projects. Where a parastatal's entry into a joint venture involved part of its workforce, however, the affected workers would undergo a change of status. Since joint ventures created under the auspices of Law 43 were subject to the same attenuation of WRM and WSP schemes as fully private foreign firms (see below), reclassified workers would suffer diminution of these benefits. For the workforce as a whole, the rapid creation of joint ventures threatened to weaken the philosophical principle that WRM and WSP are crucial to industrial development. The spread of joint ventures would thus mean a direct loss of rights for some workers, and a more abstract takeaway from Egyptian labor as an aggregate.

Through the end of the 1970s, the Investment Authority had approved 956 "investment firms," mostly in the areas of agriculture, livestock, and building materials. In industry, joint ventures involving the public sector represented about 16% of all projects, with public capital contribution averaging 34%. Not all such projects involved the participation of a state-owned industrial firm. In many cases, the public participant was a bank or insurance company; there were also projects in which the state's contribution was real estate. However, as of the mid-1980s, there were at least five industrial joint ventures which competed with the very parastatals that participated in them.[62]

Over time, the joint ventures became a lightning rod for criticisms of the *infitah* itself. Besides the diminution of labor's rights, leftists and nationalists (and some unaffiliated economists) charged that the schemes were undermining the national economy, citing cases where foreign firms had made their participation in projects contingent on cessation of local production of competing products. In other cases, the joint ventures' products shrunk the market for similar goods produced by parastatals or wholly Egyptian-owned private plants, causing them to run into the red. Critics also argued that many of the new ventures were ill-conceived and nonprofitable, and that state participation in them was an unnecessary drain on the Treasury. Some state managers joined in these accusations. For others, however, the projects were a real financial boon; by joining the directorate of the joint venture involving their firm, they were able to earn salaries far beyond what the public sector alone would allow.[63]

Some of these concerns were taken up by union leaders. The confederation formed a committee to evaluate the experience of the projects late in 1979. One year later, at its 1980 conference on the peacetime economy, the ETUF passed a resolution calling for the participation of parastatals in projects with Arab and foreign investors to be limited to new capital creation, i.e., the involvement of existing public sector capital in such projects was rejected.[64] In practice, however, the ETUF did not undertake any campaigns against those projects which violated these stipulations; opposition to specific projects came from locals or federations rather than the ETUF leadership.

The campaigns versus individual projects met with mixed success. The first project which elicited a reaction from labor was a large textile scheme known as al-'Amiriyya, approved by the Investment Authority in 1977. With a projected workforce of 37,000, the $1.3 billion project was to produce yarn, polyester, fabric, and knitwear from a location near Alexandria. The Egyptian participant was to be the state-owned Bank Misr; the foreign partners were to be the Chemtex Co., an American firm, and the Misr-Iran Textile Company, itself a Law 43 concern. The Textile Workers' Federation prepared a detailed report opposing the project on the grounds that the current production of these goods was already sufficient to meet domestic and export demand. The new venture would therefore either drive existing textile firms out of business or find no market for its own wares, thus squandering the state's investment in it. In addition, the report argued, there was an insufficient supply of skilled textile workers to meet the labor demands of the project.[65]

The project was opposed by some powerful other forces as well, including the Ministries of Industry and the Economy, the Federation of Egyptian Industries, and the World Bank. The objections raised by these groups led the Supreme Investment Committee to suspend the project's authorization in August 1978. A special parliamentary committee was formed to investigate and ultimately ruled in favor of the project, albeit with some revisions, but by the end of 1980 all but the polyester production aspect of the proposal had been scrapped.[66]

After 'Amiriyya, all of the controversial projects appear to have been in the engineering sector, thus involving the EEMWF. The first project challenged by the federation was a proposed joint venture between the French-owned Thompson Company and Egypt's Ideal, a parastatal producing refrigerators and washing machines. Approved by the Investment Authority in 1978, the proposal called for Ideal to contribute one of its plants to the new project, in which it would hold 49% ownership. The contract prohibited Ideal's other plants from manufacturing or marketing products which would compete with the output of the joint venture. Critics charged that this latter clause would lead to the idling of these plants and the ultimate bankruptcy of Ideal. They also maintained that the proposal essentially meant selling an Ideal plant to Thompson, thus violating Law 111.

Between 1978 and 1980, the general assembly of the Ideal Company rejected the proposal six times, but it was kept alive by the Investment Authority and the Ministry of Industry.[67] During this period, the EEMWF prepared a report detailing its objections to the proposal. Aside from the dangers to the remaining Ideal plants and workforce, the report charged that workers at the plant to be transferred would lose up to two-thirds of their pay, since the contract called for maintaining their basic wage, but not their incentives, bonuses, and profit shares; these amounted to 200% of the basic salary. The federation also questioned Thompson's commitment to exporting the refrigerators. The proposal was finally brought up for what proved to be a heated discussion in parliament in the spring of 1980, and the project eventually died there.[68]

The dangers to Ideal predicted by opponents of the joint venture were not mere conjecture, but were in fact based on experience elsewhere. Later in 1980, workers in the Alexandria Metal Products Co. (AMPC), an EEMWF local, sent an urgent memo to officials calling for intervention against a joint venture involving that firm with the Wilkinson Blade Company. That project was already underway, and had entailed transfer of two of AMPC's plants to the joint venture. The workers charged that rather than modernize the plants, Wilkinson had deliberately allowed the existing machinery to deteriorate, and then closed the factories, while flooding the market with imported blades manufactured by its plants in Great Britain. The idled workers were transferred to AMPC's other plants, but the company began to incur losses due to the competition from Wilkinson.[69] It is unclear what action was taken by federation officials against the project.

The most militant and enduring struggle was over the Chloride Egypt project, a merger of the British-owned Chloride Company with Egypt's General Battery Company (GBC) and a small amount of private capital. The project was initially proposed directly to Sadat by Chloride's president when the former visited Great Britain in 1975, and Chloride Egypt was formally established by the Minister of the Economy in 1980. GBC consisted of two dry battery factories and one liquid battery plant located in 'Amaraniyya. Under the agreement, the latter plant, which employed 600 workers, would be closed, and the two partners would open a new liquid battery factory. The accord called for 480 of 'Amaraniyya's workers to be transferred to the new plant; the remainder would receive one year's severance pay. Chloride stipulated that the import of liquid batteries into Egypt be banned.[70]

The GBC local, upon learning of the project in the late 1970s, sent inquiries to the unions of Chloride workers in other countries seeking information about the company. The responses they received indicated that Chloride was not a successful firm. When the agreement was signed the local, in conjunction with the EEMWF federation, raised a court case against it, relying in part on this information. The suit further charged that there was no technical necessity for the project. However, the court ruled that it was not empowered to decide the issue, effectively allowing the scheme to proceed.

The issue remained dormant until the fall of 1982, when the new factory was ready to open. Chloride Egypt announced that it would only accept 360 of 'Amaraniyya's workers, 120 less than specified in the agreement, but nevertheless expected the old plant to be shut. The new factory would take some of

'Amaraniyya's equipment, and the rest was to be sold as scrap. The local met and rejected the integration and closure orders. Some workers scheduled to be transferred resisted the orders; the management threatened them by refusing to release their incentive pay. Workers also tried to stop the closure of the 'Amaraniyya plant. Several were arrested, and security forces were called in, under emergency laws, to close the factory at the beginning of 1983.

The local members continued to fight after the closure. They challenged the government to let them take over the plant, and vowed to produce more than Chloride had proposed if provided with one million pounds for modernization. Finally, in the spring of 1983, the Minister of Industry agreed to reopen the 'Amaraniyya plant, transferring ownership, and the remaining workforce, to the National Plastics Company. The plant resumed production of liquid batteries, albeit on a much lower scale. Chloride opposed this arrangement; Former Prime Minister 'Abd al-'Aziz Higazi, who had assumed the management of the joint venture in 1982, charged that the government had yielded to labor pressure.[71]

The GM Story

The other project to generate sustained controversy was a proposed joint venture between Egypt's Nasr Car Company (Nasco) and General Motors. At the time of the deal, Nasco was producing a small passenger car based on the Fiat; the company also made trucks, lorries, and tractors. Egyptian officials had sought the participation of a multinational car firm to upgrade Nasco's facilities for passenger car production at its Helwan plant. GM was one of several bidders, proposing to produce a car based on designs by Opel, its West German subsidiary. The project had a cost of $700 million, of which $200 million would be for plant modernization; the remainder was for the establishment of 12 feeder plants to make parts for the auto. Financing was to come from the Misr-Iran Development Bank, Chase Manhattan Bank, USAID, and Egypt's new Export Development Bank.[72]

The Tagammu' was the first to attack the project, accusing the United States of pressuring Egypt into accepting GM's bid. *Al-Ahali* charged that the company had twice declared that the proposal had been accepted before an actual agreement was reached, the first time coinciding with a visit by Mubarak to the United States, and later during a visit to Egypt by the Assistant Secretary of State for Middle East Affairs. The choice was finally announced on June 16, 1986, when then-Defense Minister Abu Ghazzala was in the United States negotiating a military aid pact. Ghazzala was head of the committee to select among the bids for the project, a break from previous procedure in which the Ministry of Industry and Investment Authority decided. The newspaper charged that the United States had made certain aid provisions contingent on Ghazzala's acceptance of GM's bid.[73]

The Nasco workers had become increasingly concerned as an agreement grew near. Before GM's selection was announced, officials of the EEMWF met with the Nasco local and its elected workers' representatives and formed a standing committee to study the project. They requested to receive all the documents pertaining to it from the Minister of Industry. According to *al-Ahali*, the Minister, angry at his own exclusion from the selection process, was initially hostile to GM and agreed to supply the papers. However, after the choice was made public, he reversed himself and reneged on his commitment. The local then sent telegrams to parliament, the cabinet, and the Minister protesting the decision. When GM invited Nasco's management committee to the contract signing, the elected workers' delegates refused to attend.[74]

The EEMWF council scheduled an emergency meeting for July. In the meanwhile, Nasco's managers, who had also not been consulted earlier about the choice, met to evaluate GM's proposal. On July 14, the managers gave conditional approval to the project, voicing 19 reservations to which they expected GM to respond. The reservations concerned the speed with which the new venture would move to actual automobile manufacturing, rather than just assembly, decision making on imported inputs, penalties to GM in case of contract violations, responsibility for payment of foreign salaries, and the name to be placed on the autos. The federation endorsed these conditions. Then, at the general assembly of the EEMWF in October, the union members rejected the GM project "in form and content."[75]

The precise effect the project would have had on Nasco's labor force is unclear. *Al-Ahali* charged that GM planned to use robots on the assembly lines, and would render more than half the workers on passenger car production lines superfluous.[76] Mr. Rif'at A. Rahman, then Director of Government Relations for GM Egypt, hinted at something similar to me. He stated that there was so much excess employment at the factory that many workers actually moonlighted during working hours. Of 12,000 employees at the plant, he said, only about 2,500 were really necessary. He attributed the workers' hostility to the project to this; even though those who would work for the joint venture would receive higher wages, he said, all the workers were concerned for their jobs. However, Mr. Rahman did not actually state that GM would lay off workers, and the union did not explicitly raise this as a central concern.[77]

Between July and December, there were negotiations between GM and Egyptian officials over Nasco's terms, and no work on the project was begun.

Nasco threatened to renew production of Fiats if the delays continued, and *al-Ahali* reported that Nasco had resumed negotiations with the unsuccessful bidders on the project. The paper also claimed that the Minister of Industry admitted in parliament that he had been pressured by GM. When no work had begun by the end of December, the agreement with the Investment Authority became inoperative. Misr-Iran bank applied for and received a renewal, but in January GM announced that it wanted to review the agreement. The company cited the fall in the value of the Egyptian pound, changing conditions of production in Egypt, and restrictive laws on operation and expansion. This request was rejected by the Cabinet, which agreed to reconsider the contract only after work on the project had begun.[78]

Al-Ahali reported on March 4 that Nasco officials, with the support of the Ministry of Industry, refused to conduct any further negotiations with GM, and had set to resume Fiat production. Two weeks later, the paper said that GM had submitted a new proposal to the Investment Authority, prompting Nasco to formally requested abrogation of the contract between Egypt and the multinational. The final cancellation of the project by the Ministry of Industry was not celebrated by *al-Ahali* until October 21, 1987. However, MEED reported on June 20 that the scheme was "frozen," and on August 29 that it had "collapsed." The left attributed the defeat of the proposal to the opposition of labor and Nasco's managers; GM's reservations, listed above, were portrayed as a smokescreen to deflect attention from the unpopularity of the project.

The full story of GM will probably remain ambiguous for some time. It does seem likely that in this case *al-Ahali* exaggerated the significance of labor's opposition to the project. At a minimum, the part played by Nasco's managers seems more important; it also appears that there was growing resistance to the project from the Ministry of Industry. Indeed, while it may have been GM, rather than Egypt, that ultimately withdrew from the agreement, there is some evidence that the tide among Egyptian officials had turned against joint ventures involving public sector industries. In October 1986, the ministry warned public sector managers against such schemes, and set restrictions limiting the companies to technical participation only, and only on the condition that the project's products not compete with the output of existing public sector firms, except in food production.[79] I found no evidence of any new controversies over joint ventures thereafter.

On the whole, then, the record of labor opposition to joint ventures was a mixed one. At least some projects which violated the ETUF's stipulations did not encounter resistance from labor, and the confederation itself did not play a key role in the battles which did occur. Given this, it is difficult to give labor full credit for any restriction of joint ventures. On the other hand, the story of 'Amariyya and the controversial projects in the engineering sector do show that labor resistance, where it did occur, was able to at least modify the plans for certain projects, and protect some of the rights of threatened workers.

Leftists within the union movement see the ETUF's inactivity around joint ventures as a sign of the leadership's conciliation toward the regime. But this cannot explain why the senior unionists did speak out against the plans for subscription sales. While the co-optation of the senior unionists cannot be denied, I believe this difference also reflects the fact that an individual project's real or potential effect was limited to a circumscribed group of workers, whereas subscription sales threatened to "privatize" a large segment of the working class at one time. This means that more of the confederation's membership base was put at risk by the latter, which carried implications for the legitimacy of the ETUF leadership if they did not speak out.

The fact that not all joint ventures were opposed by the locals or federations involved is also indicative of different orientations among unionists at these levels. It is likely that the bulk of joint ventures which did not arouse specific reactions from labor did not involve worker reclassification. The concentration of opposition to joint ventures in the EEMWF can also be attributed to structural factors, since firms in this sector were entering joint ventures in proportionally greater numbers than other industries: at the end of 1981, a study prepared for the Consultative Assembly reported that while 25 percent of all parastatals were involved in or planning to join a joint venture, a full 33% of the state-owned companies in engineering were already in joint venture schemes. This, in turn, may stem from the fact that the ratio of public to private capital in engineering at the outset of *infitah* was relatively higher than in most other industries.[80]

I would argue, however, that the EEMWF's activism versus joint ventures also reflected the presence of a nationalist/leftist tendency in the leadership of the federation and its associated locals. As will be recalled, Sa'id Gum'a, who had served as president of the EEMWF since 1971, was one of the militants in the earlier battles against privatization. Although he did join the NDP, he remained close to Nasirist unionists, especially Fathi Mahmud. Moreover, Niyazi 'Abd al-'Aziz, who served as the union's cultural secretary from 1976 until 1987, was a leader of the Tagammu' in its early years, and even after leaving the party he remained one of the few federation leaders who refused to affiliate with the NDP.[81] The Nasirist identities of its leaders were also plainly evident from the adornments in the federation's offices. This would render the EEMWF more prone to fighting joint ventures than federations where Nasirist proclivities were not so strong.

Privatization in the Tourist Sector

Nationalist philosophy also played a role in the privatization battles fought in the tourist sector. Here, a third type of privatization was being pursued—management leasing. In this case the issue was handled on an industry-wide basis between the government and the federation representing hotel and tourism workers.

When the economic reform program was launched, the overwhelming majority of Egypt's large hotels are actually owned by a small number of public sector companies, but the most successful of these had historically been managed by private hotel companies under long-term leasing arrangements.[82] The Nile Hilton, for example, was owned by Misr Hotels, while the Cairo Sheraton and the various Oberois belonged to the Egyptian General Company for Hotels and Tourism. However, another public sector tourist company, the Egyptian Hotels Company (EHC), assigned the running of its seventeen hotels to a management company which it also owned. The EHC hotels, which include some famous facilities that had been nationalized in the 1960s, such as the Shephards in Cairo, the Winter Palace in Luxor, and the Cataract in Aswan, did not perform well and deteriorated.[83]

When Fu'ad Sultan became Minister of Tourism in September 1985, he stressed the need to privatize the management of these hotels, and made the like-minded Bahi al-Din Nasr chairman of EHC. Together they aggressively sought to lease the EHC hotels to foreign companies, under contracts that required the leasee to renovate the facilities during an initial grace period. When Nasr took over the EHC, the company had 7,200 employees, with a ratio of employees to room of approximately 3.5:1. The new leasing arrangements committed the foreign firms to maintain a ratio of 1.4:1, about the world standard.[84]

The arrangements thus threatened to cause lay-offs. Before the first deals were concluded, Nasr and other Tourism Ministry officials met with leaders of the Hotel and Tourist Workers' Federation (HTWF), and reached an agreement whereby workers laid off under new management agreements would remain on the payroll of the EHC. They would continue to receive their basic salary, but not get incentive pay and bonuses; which amounted to about 33 percent of their previous earnings. Nasr also offered early retirement to about 1,800 workers. The EHC reported in 1988 that its workforce had declined to 3,500, including 500 workers on basic salary, and that it had realized considerable savings from the employee reduction.[85]

The federation, while feeling that its agreement with the EHC protected its members, remained opposed to the new leasing deals. Mustafa Ibrahim

Mustafa, its president, charged that the new agreements were not beneficial to Egypt, precisely because the EHC was now required to keep unproductive workers on its payroll. Union leaders also maintained that EHC undervalued the property it leased, and that the land's worth was increasing dramatically over time. Since the government's participation in these deals, and hence its share of profits, was based on the land's valuation, the HTWF charged that the EHC was loaning Egypt's assets to foreign firms for a song. Thus nationalist concerns were paramount, as Mustafa confirmed when he told me, "Our opposition to privatization is not based on workers' rights (because we have protected them) . . . It is based on economics and on nationalism."[86]

Meanwhile, Sultan began to push aggressively for selling some EHC properties to foreign companies outright. However, in the winter of 1988, he caused a stir by announcing the intended sale of the San Stefano hotel in Alexandria. A coalition of leftists, nationalists, and other unionists joined the HTWF in fighting this scheme. The issue was hotly debated in parliament, where opponents again brought up Mubarak's pledge not to sell public assets. The president ultimately upheld his promise, and the sale was canceled.

On the whole, privatization appears to have proceeded more rapidly in the tourist sector than in manufacturing. In part this would seem to reflect a greater commitment to it from the Mubarak regime; the president's selection of Sultan, known as a fierce advocate of privatization, to head up the ministry is indicative of this. But even here there were limits to how far the regime was willing to go in confronting opposition from labor, nationalists and leftists. The sale of the San Stefano was blocked, and the government's deal with the HTWF attenuated the benefits of the leasing deals for the state treasury.

Defending Subsidies and Price Controls

Prices supports first became an issue for the unions in the mid-1970s, when Sadat initiated discussions with the IMF, seeking assistance with Egypt's balance of payments difficulties. For workers, the removal of subsidies and/or price controls means immediate inflation. Thus, like one-fell-swoop privatization schemes, price reform would adversely affect the entire membership base of the ETUF at the same time.

For unions, the issue presents several policy options. First, obviously, union leaders could support such measures as necessary to reduce the government's fiscal deficit and restore market mechanisms, with accompanying calls for workers to understand the need for belt tightening. Alternatively, unions can either oppose the removal of price supports, or try to protect their members by insisting that workers' wages keep pace with inflation. In Egypt, the unions' initial position was outright opposition, but over time this evolved into a combination of the other responses.

The issue of price reform confronted the confederation in the fall of 1976, just after Gharib's ouster. In the face of rumors that the government was negotiating with the IMF, one of the first acts of the newly elected ETUF slate was to convene another national conference in December 1976, focused on the issue of wages and prices. The discussion was hot, with many speakers pointing to the failure of the government to implement decisions of prior conventions. The resolutions called for a system of cost of living adjustment in wages and pensions, lower taxes on the poor (to be compensated for by a higher tax on luxury items), and the maintenance of all subsidies and price controls. Sayyid Mar'ai, then Prime Minister, who opened the conference, promised to carry its recommendations to Sadat.[87]

It was a month later that Sadat decreed the lowering of subsidies on numerous items, sparking the January 1977 riots. Because the official press would not cover the angry response of the confederation, it is sometimes claimed that the ETUF did not support the protests.[88] But union leaders were obviously under considerable pressure from below to condemn the price increases; moreover, they took the decision to raise prices so soon after their December conference as a political insult. The confederation called an emergency meeting on the night of January 18 and issued a strongly worded denunciation of the government's actions, demanding to meet with Sadat. Their statement read in part:

> The ETUF holds that this decision is an affront to the feelings of the masses, a serious disregard of the trade union movement and a suppression of the right which was guaranteed to it, in Trade Union Law 35 of 1976, Item 17, to express its opinion, as the leadership of the Egyptian working class, on draft laws and by-laws and decisions related to the organization of affairs of work and workers, to participate in discussing draft economic and social development plans, and to defend the rights of Egyptian workers and protect their collective interests.
>
> We consider the decision to raise prices a torpedo aimed at the demands of the Egyptian trade unions which were defined at the ETUF's conference on wages and prices held at the end of December 1976 . . .
>
> For this reason the ETUF rejects in form and content the increase in prices and demands its repeal. And we have decided to meet with the President of the Republic to place the entire matter in his hands so that he can order the repeal of these decisions which the toiling masses consider a new burden. . . .
>
> Long live the struggle of Egypt's workers. Long live Egypt.[89]

Further, and unlike the aftermath of the January 1975 demonstration (see chapter 3), the confederation now organized to assist the arrested workers. The ETUF's Committee on Union Freedoms visited the workers in detention, as well as their families, to provide reassurance and financial assistance. In addition, the confederation assigned lawyers to defend the workers and ensure that they were not mistreated in prison. Finally, the individual federations were instructed to contact the plants under their auspices to ensure that detained workers continued to receive their salaries.[90] It is significant as well that the confederation publicized these policies, unlike the clandestine support that had been given to victims of repression in the early 1970s. This approach is indicative of the greater level of unity the confederation achieved under Ahmad's leadership, and its initially more defiant stance toward the regime.

When Mubarak reinitiated efforts to liberalize the price structure and reduce the burgeoning fiscal deficit in the early 1980s, his approach was more gradualist than Sadat's attempt had been; regime officials referred to this policy as reform in stages (*daragat*), and explicitly defended it to multilateral lenders as necessary to forestall further outbreaks of mass protest.[91] The government also adopted a strategy of reform by stealth. Rather than lowering the prices of some subsidized goods, their quantity or quality was lowered—this was especially the case with the popular loaf of bread (*'ish baladi*), which Egyptians say shrunk several times during Mubarak's rule—or the availability of goods was reduced, causing long waiting lines and periodic absences of basic staples from the market.[92]

In this context, the ETUF's response proved to be more conciliatory. Confederation leaders at first cooperated with the regime's efforts simply by not speaking out about the reforms. Then, in 1984, they quietly endorsed a plan to increase workers' paycheck deductions for insurance just as another round of partial subsidy removals was hitting. As we have seen, this sparked the riots in Kafr al-Dawwar which forced some backing off of these measures by the regime (see chapter 3).

The ETUF's position in this case was not discussed through the ranks of the union movement, and generated some controversy within it. With its limited presence in the unions at that time, the left relied on its press to call attention to the ETUF leadership's conciliation and tried to push them to play a stronger role in defending workers. It was *al-Ahali* which exposed the confederation's support for the insurance deductions. After this embarrassment, and in the face of the upsurge in plant-based protests which followed it, union leaders did become somewhat more outspoken. While expressing a willingness to consider subsidy modifications targeted at the affluent,[93] the ETUF now insisted on continuation of the subsidies for workers and the poor.

In the main, however, the confederation now directed its efforts to seeking wage increase to compensate workers for the rising prices. The call for formation of a high council on wages and prices was revived, and repeated by Sa'd Muhammad Ahmad at all his official appearances as ETUF president. At the 1985 Mayday celebration, Ahmad called on Mubarak to turn the Mayday bonus into an actual raise for workers, and the confederation called for a new increase in the minimum wage at its general assembly held a few months later. The ETUF leaders stressed this issue in particular during a meeting with Prime Minister 'Ali Lutfi after the July 1986 train drivers' strike.[94]

There was never any threat of the confederation itself attempting to mobilize workers behind these demands, however, and unionists soon came under countervailing pressure from the government to tame them. At the 1986 ETUF annual meeting, held in late fall, the meeting hall was full of uniformed and plainclothes police sent by the authorities. Discussion in the committees was hot, but the final resolutions of the assembly did not criticize any government policies, and failed to include many recommendations passed by the committees such as the call for release of the train drivers. The economics committee was chaired by Ibrahim Shalabi, then president of the Textile Workers' Federation, who demanded that wages be increased by no less than 50% if subsidies were lifted. Shalabi himself later excluded this from the committee's final recommendations.[95]

Thus, in contrast to its steadfastness on privatization, the ETUF's oppositional stance on price reform weakened under Ahmad's tenure. Several factors, I believe, account for this attenuation. First, privatization invoked fears of increasing foreign control of Egyptian resources, and thereby tapped into the nationalist sentiments of unionists in a way that price reform did not. Relatedly, although subsidies and price controls were an important component part of Nasir's commitments to workers, they also benefited other strata of the population, including the affluent, who were not part of the ETUF leaders' constituency. Thus, while co-optation into, and pressures from, the regime no doubt played a role, even sincere union leaders could feel that they were still representing their members' interests by pushing for wage indexing rather than resisting price reforms per se. Third, the removal of price supports did not constitute an institutional threat to the unions, as privatization did.

Finally, there was some diffusion of the pressures from below on unionists around this issue. Although, as we have seen, there were numerous incidents of spontaneous protests related to workers' eroding earning power, these were all contained in specific areas. There was no repeat of the widespread rioting of 1977. In this regard, Mubarak's strategy of gradual and obfuscated reform was somewhat effective.

Nevertheless, the fear of mass insurgency erupting spontaneously, as it had in the past, underlay the continued caution with which the government approached reform, and the specter of renewed riots was invoked by the regime in its negotiations with the IMF, resumed in the mid-1980s. When the two parties finally concluded an agreement in May 1987, it did not call for the bread subsidy to be lifted; the Fund's demands revolved around increasing energy prices, phased devaluation of the pound, and other cutbacks in government expenditures. The terms were generally considered lenient relative to other IMF deals, and interpreted as an indication of Western creditors' fears that imposing too much economic hardship on the Egyptian people would cause political problems for the Mubarak regime.[96]

After the agreement was signed, there was about a 20% increase in the price of sugar, cigarettes, and several other items. But almost simultaneously, Mubarak announced a 20% exceptional raise for all those on the public payroll, with tripartite negotiations to follow for a similar increase in the private sector. While the pay raise did preempt any recurrence of the 1977 riots, it also counteracted the effects of the price increases on the fiscal budget, and contributed to Egypt's failure to meet the deficit-reduction terms of the IMF accord. Though it was never formally revoked, the May 1987 agreement basically became a dead letter the following year because of the Fund's dissatisfaction with the pace of Egypt's reform. In spite of the conciliation of the ETUF's leaders, latent opposition from the labor movement is indicated as a cause here.

The Labor Movement and the *Infitah*

Sadat's opening to foreign trade and investment was not an issue that directly affected or threatened workers real wages, or job security and benefits, in the way that privatization or price reform did. However, because it was linked philosophically to these other reforms, and was the opening salvo in his campaign against Nasirist economics, *infitah* initially became a battleground for the Marxists and Nasirists in the ETUF. This resulted in a confederation policy statement critical of the opening. However, the confederation itself never actively tried to undo the *infitah*. During the 1980s some individual federations raised concerns about the opening's effects on industries whose workers they represented, but here too there was no organized oppositional efforts. And unlike the issues discussed above, the labor movement's concerns about *infitah* do not appear to have produced any noticeable effect on regime policy.

Law 65 for foreign investment, passed in 1971, was a precursor to the 1974 *infitah* law. It drew in some 250 project proposals, although only 50 of these were accepted and none had been initiated by the time Law 43 was introduced.[97] Caught up in the 1971 elections and the ensuing leadership changes, the ETUF leadership did not react to its passage. However, the union movement did respond to a little-known working paper issued by a joint committee of the ASU and the parliament in August 1973. Entitled "World Changes and Their Effect on the Path of National Action" (*al-Mutaghayyarat al-'Alamiyah wa Tathiruha fi Masar al-'Amal al-Watani*), it was the first written recommendation for an economic opening. The ETUF distributed the paper to the rank and rile for discussion, and in late August a series of meetings were held in which each federation summarized the feedback of its members. A report on these responses was prepared by a subcommittee of the confederation leadership and submitted to the authorities.[98]

The report contained the germs of what was to become the dualistic position of the trade union movement on the *infitah*. It recommended that constraints be placed on the opening process so that it would not affect the socialist path or place the country "under the control of world monopolies." There was, however, no outright rejection of foreign capital. Under a section entitled "strengthening Egypt's economic capabilities for the sake of the battle, new construction, and raising the standard of living," the report called for issuing laws to protect foreign and Arab capital seeking to invest in Egypt, widening the scope for international economic agreements and expanding licensing for oil exploration and joint ventures in petrochemical manufacturing.[99]

Clarification of this conditional acceptance of *infitah* came several months later, in the ETUF's 1974 statement about privatization. The gist of the document was discussed above; relevant here, though, is the confederation's affirmation that its opposition to selling the public sector "does not prevent it from welcoming any initiative from private local, Arab, or foreign capital to participate in the projects of development plans on the condition that such projects are coordinated in the framework of the general national plan and under the umbrella of existing laws."[100]

However, there were stronger reservations about *infitah* within the ranks of the ETUF than this statement suggests. This became apparent in April 1974, when Sadat issued another working paper proposing a course for the country in the aftermath of the war; it came to be known as the October Paper. While hailing the public sector as the backbone of the economy, the document criticized past neglect of domestic private enterprise, and called for barriers to its expansion to be removed. An open-door economic policy was promoted on the grounds that Arab and other foreign capital would provide the finance and technology for development.[101]

In a discussion of the paper sponsored by *Al-'Amal,* both Fathi Mahmud and Hassanayn Muhammad Hassanayn, president of the Media Workers' Federation, expressed strong reservations about rehabilitating the private sector. Hassanayn further charged that the ranks of the trade union movement were disunited and disorganized, and that Gharib was doing nothing about it. Also, several middle-level unionists who served in parliament raised sharp critiques of the private sector during a discussion of the disparities between the two sectors in April. However, Salah Gharib told *Rose al-Youssef* that the unions would treat the October Paper "like a constitution that we adhere to at all levels. We will adjust the programs at the Workers' Educational Center and its institutes to the October Paper; it will be the basic material for all the lectures to prepare the cadre of workers."[102]

As a result of these disagreements, no formal confederation response to the October Paper was issued. There was also no official ETUF response to the passage of Law 43 two months later, even though the law exempted foreign firms from affording certain benefits to workers incumbent on the Egyptian private sector. Some form of profit-sharing and worker participation in management was required, but the details were left largely to the discretion of the management.[103]

It is difficult to assess to what degree this debate within the ETUF leadership reflected the sentiments of the rank and file. The confederation did not organize mass discussion of the October Paper as it reportedly did for "World Changes." The October Paper was submitted to a popular referendum in May, and approved overwhelmingly, but this may not reliably indicate true support for the document; in authoritarian societies such polls can easily be orchestrated or manipulated. On the other hand, it is doubtful that widespread initial opposition to *infitah* could have been thoroughly camouflaged. Some Egyptian leftists suggested to me that Sadat's popularity was quite high in the aftermath of the October war, and Egyptians were eager to believe his promises that the West would bring new prosperity to the country.

It is also likely that workers would not feel threatened by the *infitah,* in its narrow sense, in the same way that they did by privatization. Neither the opening to foreign investment nor the influx of foreign products in and of themselves violated the formula of reciprocal rights and responsibilities Nasir had initiated. While new enterprises created with foreign capital might fail to provide labor with the gains that advocates of *infitah* promised, diminution or elimination of the public sector threatened to take away benefits that workers had already realized.

In lengthy interviews conducted by *Al-Tali'ah* during 1976 workers generally endorsed the goals of the opening.[104] In this regard, then, the leftist unionists who were completely opposed to Western investment appeared to be acting according to their own ideologies rather than representing their constituents. The workers interviewed also stressed, however, that the public sector should not be undermined, and displayed a considerable distrust of the private sector. This suggests that Gharib, in his unwavering support for Sadat's encouragement of the latter, was also acting out of personal motivations rather than responding to the concerns of the base.

Some workers did express disappointment that the new policies failed to fulfill the regime's promises, however, and there is other evidence that as it was implemented, opposition to *infitah* spread somewhat among ordinary workers. In part because new foreign goods did find a market among workers, the first consequence of the opening was a widening of Egypt's trade gap. The influx of imports also contributed to inflation, which was exacerbated by the demobilization of soldiers during the disengagement process. Real wages fell in 1975, and remained below the prewar level the following year as well (see table 3.3). At the same time segments of Egypt's upper class, previously inhibited by Nasirist rhetoric, took to lavish displays of wealth encouraged by Sadat's own ostentatiousness. This reappearance of obvious class differences violated the spirit of egalitarianism promoted under Nasir. As we know, 1975 and 1976 saw a small but militant strike wave, as well as a riot on January 1, 1975 that erupted out of a labor demonstration. During these protests workers expressed their outrage over the increasing manifestations of class divisions in the society.

The leftists in the union movement who organized the January 1, 1975, demonstration intensified their push for a clear confederation stance against the *infitah* afterward. The result was an ETUF conference on the Economics of Labor, held April 6–9 of that year. Gharib, who had resisted convening the conference, opened it by presenting a statement in support of the government, insisting that foreign expertise was necessary to speed the development process in Egypt. The conference then heard papers on the economic opening from a number of academics and other intellectuals. Among them was 'Abd al-Mughni Sa'id, the long-time Labor Ministry functionary who had published one of the early critiques of *infitah* in the leftist journal *al-Katib*.

Another critic argued that infitah was in contradiction with socialism. This position was challenged by others who cited moves toward economic opening in a number of Eastern bloc countries. All the speakers concurred, however, that the influx of foreign capital into Egypt should be subject to control and supervision, in the context of national development plans and priorities.[105]

The committee charged with formulating resolutions on the *infitah* adopted this unifying approach. While recognizing "the necessity of *infitah* for economic growth," the committee insisted that *infitah* should serve the country's development plans, and not vice versa. Further, they said, the opening should be planned, all entering foreign capital should be productive, and Arab capital should be given preference over foreign capital. Other resolutions called for the maintenance of the national alliance, preservation of workers rights, and maintaining the public sector as the key to the economy.[106]

These points were ultimately formalized in the Charter of Ethics for Union Work (Mithaq al-Sharf al-Ikhlaqi lil-'Amal al-Niqabi), approved by the general assembly of the confederation in October 1976. The statement expressed support for the *infitah* and the encouragement of Arab and foreign capital, but only so long as it "does not disturb the socialist principles on which our economic and social system is based, nor contradict the national interest. The influx of foreign capital "should not take control over the basic structure of our national wealth, and must be in accordance with our development needs." The remainder of the document committed the union movement to struggle for the protection of the "workers' gains of the July Revolution." While affirming that the public sector represented the backbone of the economy, the document did pledge the unions to support the private sector, which should be kept free from exploitation and monopoly.[107]

Leftists and Nasirists in the ETUF considered the statement a victory. Although the Charter of Ethics represented a compromise among the various forces in the trade union movement, the reservations about *infitah* it expressed were much stronger than the confederation's original statement; I join Egyptian scholars in interpreting the Charter as essentially a rejection of government policies.[108] This would seem to reflect both that opposition to *infitah* in the unions had grown during the three years which elapsed between the two policy statements, and the related diminution of Salah Gharib's authority. However, the long delay in issuing the statement was evidently the work of Gharib. Its publication thus marked the first formal statement of the ETUF under Ahmad's leadership and indicated the greater unity that was achieved among union leaders after Gharib's ouster.

Union Leaders and *Infitah Under Ahmad*

The Code of Ethics remained the basic ETUF policy statement on *infitah* through the 1980s. In the late 1970s the confederation also adopted the critique raised by some nationalist economists that the opening had proven to be mainly "consumptive," in other words, flooding the country with imports rather than bringing in new investment in production. This critique was later adopted by Mubarak himself, whose regime committed itself to seeking more direct foreign investment. However, the latter threatened the public sector with more local competition. In response, Ahmad issued a statement calling for no licensing of foreign firms, either alone or in conjunction with a local partner, to make goods for which local demand was already being satisfied by public sector products. This demand was also raised by the nationalist opposition forces, but there was no policy response from the government.[109]

Around the same time, out of growing concern that the *infitah* not undermine established workers' rights, the confederation insisted that Law 43 firms should be required to abide by the same stipulations for WRM and WSP that were incumbent on the Egyptian private sector, although it is not clear how many wholly Egyptian-owned private firms were subject to this law at the time.[110] And, in a reflection of nationalism, the ETUF demanded that local firms, both public and private, be entitled to the same privileges being granted foreign firms; this latter demand was also being raised by Egyptian entrepreneurs. Law 159 of 1981, a new private companies law, did extend Law 43 privileges to new Egyptian-owned joint ventures, but this included extending the exemptions from WRM and WSP laws to them, not making them more stringent for foreign firms. In this respect, the law constituted a defeat for the ETUF, although the confederation does not seem to have conducted a vigorous campaign against it prior to passage.[111]

At the next level down, some federations attempted to influence particular policies or projects which exclusively affected their membership or associated industries. The earliest case of this, to my knowledge, was the EEMWF's demand that even strictly private sector joint ventures in car production be halted, fearing that these could lead to the closing of the public sector Nasr Car Company (Nasco). This concern arose after the Investment Authority approved a joint venture for car and truck production involving GM, Mercedes, and Ford. Later, the Textile Workers' Federation called for a ban on imports of ready-made clothing, or the imposition of tariffs on them; the textile industry began declining when such imports were permitted.[112] Similarly, the Commerce Workers' Federation requested a ban on the entry of bagged tea, and restrictions on the import of loose tea, to protect the local tea producer, Shamtu.[113]

But while federation leaders may have lobbied government officials around these issues, and/or attempted to pressure the regime through press exposes, they did not initiate any mass campaigns behind their demands. Also, unlike privatization and price reform, these issues did not generate protest at the base. In sum, as with Law 159, there does not appear to have been a great deal of agitation around these issues, and none of the efforts that were expended appear to have elicited a policy response from the government.

This chapter has had the twofold purpose of, first, demonstrating and accounting for variations in the labor movement's response to different components of economic reform and, second, showing that these differences have in fact had an impact on government policy. The emphasis has been on the formal policies of the ETUF and its member federations to the government's economic liberalization initiatives, but I have shown how these policies, in turn, are influenced in part by the reactions of workers, unionized or not, at the base.

The union movement's response to reform measures can range on a continuum from strong support to strong opposition, where the intensity reflects not only the vehemence of the officially stated position but also the presence or absence of efforts to actually mobilize workers behind them. Supportive positions can reflect a genuine belief by unionists in the correctness of the policy, or opportunist calculations of potential personal gains from compliance or losses due to government retribution against opponents. In contrast, where unionists challenge the government, personal philosophies or concerns about legitimacy with the base are indicated. Given the impossibility of establishing precisely what lies behind individual behavior, the analyst can only rely on personal judgment in accounting for the different union responses illustrated here. A third dimension of labor's response is the degree of unity within the trade union movement around a given policy stance. Disagreements within the ranks of the ETUF weaken the unions' ability to affect change through either voice or veto.

The material presented here shows only one instance of *active* union *support* for government policies which constituted a retraction of workers' "socialist gains." This occurred in the 1960s when an ETUF congress called almost unanimously for labor sacrifice in the face of the country's economic difficulties and some unionists, at least, did seek to mobilize workers behind this program. While the trade union movement was virtually unanimous in its position, however, there is evidence of dissent among the rank and file. Fear of retribution and/or ambitions for advancement were doubtless factors behind the embrace of retrenchment by some union personnel, but the evidence also suggests that a significant core of unionists were acting out of genuine devotion to Nasirist philosophy.

In the 1970s and '80s, in contrast, union leaders either *resisted* economic liberalization policies or gave them *passive* support, with even the latter generating disunity in the confederation. The strongest and most unified opposition from the ETUF came in reaction to one fell swoop privatization schemes, and the sweeping subsidy cuts of 1977. In the latter case, the senior unionists were responding to manifest and intense anger from the rank and file. In the former, this potential anger was presumed, but also preempted by the defeat of the policy proposals. Both were issues of national scope, and are illustrative of the ways in which the existence of a singular trade union confederation, even under authoritarian controls, can serve to defend the interests of workers.

A key question here is why the ETUF remained vigilant against privatization through the 1980s, whereas its stance on price reform softened, when both issues were clearly of major concern to the base and the opposition forces within the labor movement. I have suggested that the difference can be explained in part by the institutional threats to the confederation's power inherent in privatization. In addition, the unions could claim to be defending their members' interests versus the inflation generated by price reform through seeking wage indexing, whereas there appeared no ready compromise position on privatization.

At the same time, it is instructive to compare the ETUF's reaction to both these issues to the privatization proposals for individual industries, i.e., the 1975 plan in the commerce sector, and the 1980s' scheme in tourism, or individual plant privatization via joint ventures. In these latter cases, although opposition among the workers affected was seemingly intense, the overall ETUF posture was weaker. I hold that this is because fewer workers were threatened, making it less problematic for opportunist *confederation* leaders to ignore their concerns. In this vacuum it fell to the federations to organize resistance, but their campaigns lacked the advantages of centralized resources and mobilization that confederation support would have offered.

In the case of the *infitah*, the ETUF's policy was forged in the heyday of Nasirist and Marxist influence in the union movement; to the degree that it voiced hostility to foreign investment and products, this appears to reflect more the clashes between the leftists and the pro-regime unionists than the sentiments of the base. The lack of strong opposition to *infitah* from below can explain why the NDP unionists did not feel compelled to do any more around this issue than release occasional pronouncements of their concerns.

It is not the claim here that labor opposition, or the perceived threat of it, is the sole factor shaping government economic policy. Nevertheless, there is a crude inverse correlation between the intensity of labor opposition and its breadth, in terms of numbers of workers involved, and the regime's actual implementation of reform measures during this period. Economic liberalization went furthest precisely in the areas where labor opposition was overall weakest: the *infitah*, and individual plant privatization through joint ventures. In contrast, price reform was at first rescinded and then imposed gradually and by stealth, and even then with some retraction and concessions to labor. One fell swoop privatization was proposed and quickly withdrawn several times in the face of strong labor opposition.

Given this, it is difficult to avoid the conclusion that Egyptian policy makers have been particularly fearful of an organized national challenge from workers. All three presidents proved ready to back off of schemes for economic reform in order to prevent manifest displays of labor opposition. This in turn suggests that privatization in Egypt would remain politically problematic unless its proponents recognized the need to protect the interests of the public sector workforce that felt threatened by these policies. It was exactly this issue that became the stumbling block to further economic restructuring in the 1990s.

5
Toward a New Era in Labor/State Relations

Clearly, the project of economic reform in Egypt, through the 1980s, was protracted and fraught with political controversy. At the core of this conflict lay the determination of Egypt's labor movement to preserve the commitments the government made to workers in the 1960s, manifested especially in the modus operandi of the public sector. While many workers and union leaders were prepared to accept other aspects of economic liberalization, preservation of the public sector and the protections to workers it embodied became the unifying call and the raison d'être of the trade union movement.

Under growing pressure from international creditors, especially the IMF and the World Bank, the Mubarak regime in the first half of the 1990s moved more decisively to rationalize, and ultimately dismantle, the public sector. Sweeping legislation to facilitate privatization was passed in 1991, as Law 203. It spawned the proposal of companion legislation that would grant workers the right to strike in an explicit quid pro quo for greater management prerogatives in hiring, firing, and wage setting. Combined with the ongoing retraction of government from the economy, the proposed new labor law would signal the final withdrawal of the government from the Nasirist moral economy. As of this writing in the fall of 1996, however, the bill had not been passed, and remained the focus of controversy between the regime, the trade unions, and leftist opposition forces. Egypt was poised before a new era of labor relations, but still stuck at an impasse.

The following two sections trace the interaction between Egypt and the multilateral agencies that shaped the new reform program. In them I demonstrate that the Mubarak regime, in resisting but gradually responding to the pressures to privatize, repeatedly undermined its own credibility. At every stage

Mubarak made pledges to Egyptians on which he later reneged. This, and the widespread popular perception that Egypt's structural adjustment program was being imposed by the West, eroded the regime's legitimacy, and meant that the reforms were being pursued by an increasingly authoritarian government.

Labor's response to these developments is described and analyzed in the succeeding sections. Succumbing to heightened pressure from the regime, the confederation's leadership endorsed the privatization legislation. At the same time, as public sector managers faced growing pressures to make their firms more profitable and thereby attractive for sale, rank-and-file protest increasingly took an anti-privatization character. Faced with these countervailing pressures, the unionists drew their battle lines around the legal framework under which the newly privatized firms would operate.

The union movement which confronted these challenges was increasingly divided at the top over structural issues. Ultimately, the career unionists won government support for legislation that ensures their prolonged incumbency and tightens their control over the base, measures the regime apparently felt necessary before it could contemplate legalizing strikes. Nevertheless, the conflicts exposed the careerism of the senior ETUF personnel, and gave new impetus to the movement for union reform.

The final section examines the proposed new labor legislation which emerged from this interplay of forces and anticipates its consequences for the labor movement and the future of labor/state relations. Paradoxically, in losing the fight to preserve the public sector, the union movement would gain an increasing importance to the lives of workers. With the government no longer acting to guarantee raises, benefits, and job security, much more of workers' livelihoods will depend on the outcome of collective bargaining between labor and management. This outcome, however, would focus growing attention on the hierarchical functioning of the trade union movement and the government's interference in its operations, and seems destined to increase pressures for more democratic unionism.

A Regime Change of Heart

In 1988, in the wake of the controversy over selling the San Stefano hotel, President Mubarak pledged once again to neither sell nor shrink the public sector. By the end of 1995, his government had restructured the public sector to promote greater efficiency, and prepared a lengthy list of firms to be fully or partially privatized; several had already been sold. To Egypt's creditors, and many prospective local and foreign investors, the pace of reform remained too slow,[1] but for workers the dramatic change over time in the government's official rhetoric constituted a series of broken commitments. Efforts to ameliorate or preempt labor opposition lay behind the regime's various pledges, as well as the hesitancy in reform implementation which so frustrated its proponents.

The government's new openness to privatization was a direct result of pressures from the IMF and the World Bank, emerging after the effective collapse of the standby agreement that Egypt had signed with the IMF in May 1987.[2] That accord was followed by the lowering of subsidies on several popular consumer items, such as cigarettes and tobacco, resulting in a 20% price increase. But to compensate workers, the government had almost immediately declared a 20% raise for all civil service and public sector employees. The multilaterals saw this as a sign that the regime was not committed to the reform package, and indeed the raise did contribute to the government's failure to meet its deficit reduction targets; the schedule for unifying the exchange rate was also not met. In the spring of 1988 the Fund refused to release the second tranche of the loan and, without a seal of approval from the IMF, Egypt was precluded from negotiating a new debt rescheduling agreement with the Paris Club when the existing arrangement expired that June. By the fall of 1988 the country had already fallen into arrears on some payments, and the balance of payments situation continued to deteriorate.

Negotiations for a new agreement with the Fund formally opened in 1989. By that point, however, the IMF was no longer exclusively concerned with its standard stabilization package (the customary terms of which had been considerably softened in the 1987 accord) but had now joined the World Bank in pushing for structural adjustment, including liberalizing the management of public sector enterprises as well as promoting privatization.[3] Moreover, because of the 1987 events, the Fund now insisted that many reforms be implemented *prior* to the signing of a new agreement.

Consequently, while insisting during the negotiations that the Fund should modify its stipulations, the Mubarak regime gradually softened its stance against privatization. In July 1989 Mubarak announced that he did not object to the sale of *loss-making* parastatals, so long as heavy industry remained in the hands of the government. Shortly thereafter, the government submitted a new letter of intent to the IMF, pledging to reduce its stake in joint ventures with the private sector, and to sell off some small public sector concerns.[4]

Following Mubarak's July speech the Egyptian media opened a lengthy debate on the issue. The government-owned press featured frequent exposes on losses and inefficiency in the parastatals and gave prominence to the views of Tourism Minister Fu'ad Sultan and other staunch advocates of privatization. An array of leftists and nationalist intellectuals spoke out for reform, rather than liquidation, of the public sector. The debate culminated in December with a speech to parliament by Prime Minister 'Atif Sidqi, again endorsing the sale of unprofitable concerns while also calling for enhancing public sector performance by allowing managers to link wages to productivity.[5]

The ruling National Democratic Party then published a working paper on privatization which was to serve as a basis for new legislation the following year. The NDP paper went further than Sidqi's speech in several significant ways. It called for government ministries to divest themselves of land and real estate, and for the sale of enterprises owned by local governments. It also endorsed the sale of shares in all public sector enterprises, providing that the government's equity did not fall below 51%, with workers to be specifically encouraged to participate. In addition, it advocated freeing public sector concerns to sell properties as needed to enhance liquidity or support purchase of new capital. Thus it did not limit privatization to only unprofitable ventures.

Emboldened by the shift in policy, Sultan accelerated his efforts to privatize the tourist sector; on June 30, 1990, the first public auction of a state-owned asset took place with the offering by the Egyptian General Company for Tourism and Hotels (Egoth) of an 80% share in its Cairo Meridien Hotel. The terms of this offering were different from what Sultan had proposed for the San Stefano: the buyer had to commit to financing a 700-room expansion of the property; Egoth was to maintain at least 20% of the equity in the complex; and it pledged to invest the proceeds from the sale in other publicly owned tourism projects. Because of these stipulations, the auction generated less opposition that the proposed sale of the San Stefano had. Nevertheless, the contrast between the success of this effort and the earlier failed initiative was a clear sign that the winds had shifted.

In April 1990 Mubarak appointed former Minister of Defense 'Abd al-Halim Abu Ghazzala to spearhead the new privatization drive. In addition to overseeing the sale of the companies owned by the regions, Abu Ghazzala was also to focus on identifying loss-making industrial concerns that could be sold. His appointment was soon followed by the announcement of invitations for a $5 million, five-year USAID contract for consultants to the privatization program, for which the bidding was intense.

Yet the regime was still proceeding cautiously, and efforts to avoid antagonizing labor were manifest. In talks with the World Bank in February 1990, government officials indicated that a phased approach to privatization would be followed, in which the first project would be the sale of companies owned by provincial governments, of which some 2,000 were identified. Of these, 30 companies from the regions of Manufiya, Fayum, and Damietta were identified for participation in a pilot project in which they would be sold via subscription. The concerns in question were mainly small poultry and cattle raising operations,[6] which were unlikely to have unions, and all lay outside the major cities where labor's strength was concentrated.

At the same time, in the ongoing negotiations with the IMF (which had rejected Egypt's September 1989 letter of intent as insufficient) and the World Bank, Mubarak's spokesmen frequently warned that rapid reform would increase the suffering of Egypt's lower classes, and could thereby lead to political unrest. In response to these concerns, the Bank proposed the creation of a "social fund" to which Egypt's creditors would contribute, to ease the burdens of reform on the poor. Bank officials suggested that part of the money be earmarked for retraining public sector workers displaced by privatization.[7]

In addition, the regime formally embraced the idea of employee share in ownership (ESOP) plans as a way to make privatization more popular. Part of Abu Ghazzala's charge was facilitating the purchase of shares in companies by employees. Shortly before his appointment, the state-owned National Investment Bank announced plans to develop a program for worker stock purchases.[8] Earlier, USAID had initiated an effort to sponsor privatization by providing a loan for workers in a public-sector tire factory in Alexandria to purchase equity in that firm.[9]

Mubarak had to contend with other opposition to privatization as well. Any widespread sell-off evoked fears that Egypt's strategic assets would once again fall under the control of foreigners, and engendered resistance from nationalists. Within the regime these concerns were championed by Minister of Industry Dr. Muhammad 'Abd al-Wahab. In the summer of 1990, 'Abd al-Wahab announced only limited plans for privatization in industry. These were, first, allowing the private sector to expand its share in joint ventures established before the *infitah*, and selling the government's holdings in those set up under the auspices of Law 43. Second, loss-making public enterprises, and small units owned by, but peripheral to the operations of, larger parastatals would also be sold. Finally, he would permit the sale of land and buildings owned, but not used, by the parastatals, and the leasing of unused productive capacity. Mubarak himself repeatedly sought to assure the nationalists that the country's most strategic and largest parastatals would not be sold.[10] But as negotiations with the multilaterals continued through 1990, his speeches gradually indicated that he no longer held objections to selling profitable parastatals in nonstrategic sectors, such as food processing. Contrasted with 'Abd al-Wahab's statement, this indicated that there were still disagreements with the Cabinet over the extent to which the government should commit to privatization.[11]

As 1991 opened, the government was moving ahead on a range of reforms sought by the IMF. Prime Minister 'Atif Sidqi reported in February that some 851 small establishments, mostly those previously owned by the governates, had been sold. Although some Western analysts considered the figure inflated, it is clear that at least several hundred such enterprises were transferred to private owners.[12] By May of that year, Egypt's exchange rate had been unified, subsidies on a wide range of goods had been reduced, and interest rates were increased. The government then prepared and submitted to parliament the legislation necessary to proceed with privatization of larger industrial establishments. Known as the "public enterprise law," it was needed to override and to replace Law 111 of 1975 (see chapter 4). The IMF signaled its approval of these measures by signing a new standby agreement. A generous debt rescheduling accord with the Paris Club followed shortly thereafter, on the heels of the substantial debt relief Egypt had already been granted as a result of its role in the Gulf war. As part of these deals, a $600 million Social Fund was created with Western and Arab Gulf financing. Egypt's balance of payments position improved considerably as a result of this debt relief.

The public enterprise legislation was passed, as Law 203, in June 1991. In addition to paving the way for full or partial privatization of parastatals, the law also sought to make public sector firms operate more according to market considerations by removing the management of existing public sector concerns from the government ministries. Instead, the government's interest in parastatals would be overseen by several dozen holding companies whose members would be appointed by the government, and who were expected to manage their portfolios according to market principles. Each of the existing public sector concerns would be assigned to one of these companies.

While the holding companies themselves would be wholly state owned, their minimum share of equity in the subsidiary firms was set at 51%; the remaining 49% of the assets could be purchased by the private sector through subscriptions. The resulting joint ventures would still be considered part of the public enterprise sector. However, the holding companies could also decide to maintain only a minority share in a subsidiary, and in this case the firm would be considered a private sector joint venture subject to the terms of the private companies law (see Chapter 4).

Finally, the holding companies were empowered to divest themselves of a subsidiary completely, or close it down. The government further committed itself to identifying an initial group of parastatals to be completely liquidated or sold, since the process of choosing the holding company board members, and allowing them sufficient time to familiarize themselves with the operations of their subsidiaries, would necessarily delay the selection of firms to be offered for full or partial privatization.

All of the reforms leading up to the IMF package were enacted with little press attention and fanfare, and the parliamentary debate over Law 203 was brief. And while the Mubarak regime was moving forward on economic liberalization, it had turned its back on further political reform. Parliamentary elections held at the end of 1990 were boycotted by most opposition parties because of the restrictive electoral laws. In the spring of 1991, Mubarak extended the longstanding emergency detention laws for another three years, and shortly thereafter he rejected appeals from the opposition for constitutional reform.[13] He explicitly justified these moves on the grounds that economic restructuring had to take precedence over political liberalization.

Nevertheless, law 203 did encounter resistance from the labor movement. Significant pressure was exerted on some union MP's to win their votes, and some nevertheless refused to support the bill. While the details are discussed below, it is essential to note here that some ETUF leaders insisted on several critical amendments to the initial draft before they would endorse the legislation. One was that the prevailing requirement that 10% of company profits be distributed to workers remain intact; in fact a previously existing annual cap on the amount of profit shares workers could receive was removed. Secondly, the final version specified that existing national labor legislation, with it protections against arbitrary firing and mass lay-offs, and for health and accident insurance and pensions, would continue to apply unless and until overwritten by a new labor law.[14]

These modifications enabled both government and union spokesmen to maintain that workers' interests would not be harmed by privatization. After Law 203 was enacted, "no diminution of workers' rights" replaced "no diminution of the public sector" as the official propaganda slogan. But it was another pledge that the regime would soon betray.

Slow Progress on Privatization

Shortly after Law 203 was passed, the lucrative AID contract for advising the privatization program was awarded.[15] Over the course of the next six months, the government busied itself mainly with naming the board of directors of the holding companies and grouping the existing public sector firms into them. Numerous prominent private sector businessmen were named as directors, including 29 members of the Egyptian Businessmen's Association. The regime also set to drafting the necessary legislation for the expansion of the capital market to facilitate the sale of stocks.

However, the list of firms to be privatized did not appear, and although the World Bank signed a structural adjustment loan with Egypt at the end of 1991, and the IMF approved release of the second tranche of the new standby credit, both agencies made it clear at the time that they were dissatisfied with the pace of the structural reforms. In addition to the slowness in developing the Capital Market law, which was not passed until May 1992, and in the identification and valuation of the candidates for privatization, the multilaterals complained that the investment climate in Egypt posed a hindrance to privatization. The job security regulations most valued by Egypt's workers, and preserved by Law 203, were specifically mentioned.

Accordingly, in October 1991, the regime quietly established a committee to begin drafting new labor legislation that would replace Law 137 of 1981 as well as the additional regulations binding only on the public enterprises. The committee consisted of representatives from the ETUF, businessmen's organizations, the Ministry of Labor, and the local legal community. At the suggestion of Ahmad al-'Amawi, then president of the ETUF, the International Labour Organization (ILO) was also invited, and provided funding for the endeavor. Its role was to ensure that the new law did not contradict Egypt's participation in international agreements.[16] Since the long-standing ban on strikes had already been proven to contradict Egypt's signature on the International Human Rights Agreement (see chapter 2), the regime's consent to ILO participation signaled its willingness to consider legalizing strikes in some fashion as a quid pro quo to labor for liberalizing hiring, firing, and promotion regulations. Discussion of this explicit exchange increasingly appeared in the discourse of regime officials and union leaders, especially as the opposition press leaked news of the committee's supposedly secret deliberations.

The government also created a Public Enterprise Office (PEO), charged with supervising the privatization program, under the leadership of Prime Minister 'Atif Sidqi. In January 1992, the PEO announced that the government's share in 23 industrial joint ventures would be completely sold, and invited local and international auditors and consultants to bid for contracts to help value the assets of the firms selected. At the same time, the regime implemented several other significant reforms, including raising the price on petroleum products, cigarettes, alcoholic beverages, telephone calls and railway tickets, and reducing the size of the popular subsidized load of bread. The government also announced that it had decided to begin the process of selling wholly publicly owned firms with profitable concerns. The original list of such firms, which would be revised over the course of the next year, included nine textile firms, several construction companies and food processing plants, and two of the country's largest fertilizer manufacturers. Several prominent tourist facilities were added to the list in the spring. Thus the regime went back on its earlier pledge to sell only loss-making enterprises.

Nevertheless, the multilaterals found new reasons to disparage the reform effort. The valuation of the companies to be sold was proceeding slowly, and the agencies complained that because the transfer of assets to the private sector had barely begun, public sector firms were placing extensive demands on the credit system, crowding out the private sector; they also objected to delays in the liberalization of trade regulations which protected the public sector from foreign competition. Privately, officials of the multilaterals acknowledged that one reason for urging haste in the reform program was to deprive its opponents of sufficient time to mobilize resistance.[17]

To demonstrate its displeasure, the IMF delayed for three months the scheduled second review of Egypt's progress under the May 1991 agreement, thus jeopardizing implementation of the second phase of Egypt's debt reduction accord with the Paris Club, and the World Bank delayed release of the second tranche of its structural adjustment loan. As a result of these pressures, the government lifted import quotas and/or tariffs on 30 items.[18]

At the same time, the expanding scope of the privatization program widened the public questioning of it. The New Wafd Party, traditionally representing Egypt's liberal capitalists, charged that it was corrupt public sector managers, not workers, who were responsible for the losses in that sector. Thus they agreed that workers should not be harmed by economic reform. The Wafd also argued, for nationalist reasons, that only small companies should be sold, joining other opposition forces in raising concerns about the sale of profitable enterprises to foreigners.[19] This prompted a statement from Fu'ad 'Abd al-Wahab, the head of the PEO, that priority in all sales would be given to Egyptian purchases, and that no unit would be wholly owned by foreign interests; 'Abd al-Wahab also reasserted that oil companies, the Suez Canal, and other strategic assets would not be sold.[20]

The year 1993 saw further progress toward privatization, though its slow pace continued to pique the multilaterals. The process of shifting the selection of firms to be privatized from the PEO to the holding companies was set back because of a government decision to restructure the holding companies; almost a third of the subsidiaries were shifted in an effort to prevent the development of monopolies and distribute resources more equitably among the holding companies, whose number was reduced from 27 to 17 as part of this process.[21] Mubarak's renomination for the presidency in the fall of that year served as another reason for delay as anticipated and then actual Cabinet changes rendered officials hesitant to make decisions.

Nevertheless, some holding companies did select firms for sale; added to the privatization list as a result were the two bottling companies that held the franchise for Coca Cola and Pepsi Cola, some additional tourist properties, nine firms in the engineering sector, and two cement factories. The PEO began supervising the review of applications by bidders to prequalify for share purchases, and the government specified a strategy for full parastatal sales. This entailed finding one large anchor to take 40–60% of the equity of each firm, with an additional 10–30% to be sold to the public, and the remainder of the shares to be distributed among the firm's employees. At the end of the year, the two bottling companies were sold under these terms.[22]

The bottling companies' purchasers agreed in writing not to reduce the existing workforce of the firms, and the government continued to proclaim that its privatization program was geared toward protecting workers' jobs and livelihoods.[23] Yet throughout 1993, public sector managers were under pressure to make their firms more profitable, and their first line of attack was workers. Numerous firms had resorted to hiring temporary workers, and now began to lay these off. Other establishments simply declared that large numbers of their full-time, permanent employees were redundant, and therefore not entitled to supplementary wages.[24]

In the meanwhile, there was little progress in the effort to develop a new labor law. Union leaders, under the spotlight of the opposition press, continued to resist the retraction of job security or other traditional benefits enjoyed by public sector workers. The government for its part sought to strictly contain the right the strike; cracking down during this period on the Islamic associations and their growing influence in the professional associations,[25] the regime obviously feared that legalizing *any* form of collective protest could have a snowball effect. The government was also coming under new pressures from employers' representatives, some of whom also opposed lifting the strike ban. Indeed, because smaller domestic entrepreneurs had historically evaded the labor laws (see chapter 4), they stood to gain little from the right to strike/right to fire exchange. It would increase their management prerogatives legally, but not actually, while they would lose some claim to government assistance in suppressing labor protest.[26] Some private sector businessmen also complained that the privatization process showed bias toward foreign investors; they accused Coca Cola and Pepsi Cola of unfairly withholding information to gain an advantage in the bidding for their franchises. One EBA member raised the old anticolonial slogan, "Egypt for the Egyptians."

The government itself expressed dissatisfaction with the low value that foreign auditors were placing on the assets to be sold, and complained that proceeding too rapidly with the sell-off would further depreciate the value of the properties. In the summer of 1993, then Public Affairs Minister 'Atif 'Ubayd publicly repudiated World Bank pressures to speed up the process, stating "the adviser will not be driving the car, but might be willing to visit us in the hospital if we crash it."[27] Under Cabinet changes in October, 'Ubayd became the new Minister for Public Enterprises, the office charged with supervising the privatization program.

For their part, the multilaterals kept up the pressure for more extensive reforms, using delays in performance reviews as their main weapon. The IMF postponed final review of the standby credit, originally scheduled for December 1992, citing slippage in the privatization schedule as one key reason. When a favorable review was finally issued in March 1993, the World Bank agreed to release the second tranche of its structural adjustment loan, originally scheduled for June 1992. Nevertheless, both agencies expressed disappointment with the pace of privatization. Thus, before the IMF granted Egypt a new, extended finance facility in September 1993, the government had to agree with the World Bank on a revised package of structural reforms, including privatization of state-owned banks and insurance companies, which had previously been considered strategic assets and hence excluded from the process.

This "ritual dance" between Egypt and the multilaterals continued in 1994 and 1995.[28] Toward the end of 1994, the regime declared that numerous firms had begun selling shares to raise capital while 12 additional firms had been fully privatized, and announced another group of enterprises to be sold.[29] The government also stepped up pressure on loss-making ventures by removing the managers of four large parastatals, including the Misr-Helwan Spinning and Weaving Company. These measures resulted in accelerated efforts at some plants to cut workers' wages and benefits and/or reduce the workforce through attrition and dismissal of temporary employees.

At the same time, the regime continued its efforts to ameliorate labor opposition. There were once again no lay-off clauses specified in the sales agreements, and the government was vigorously promoting the establishment of shareholders' collectives among workers in the public enterprise sector. PEO Minister 'Ubayd, at a large conference on private sector development in October 1994, publicly acknowledged the government's efforts to win the support of union leaders and their rank and file before stepping up the pace of privatization.

The government may have exaggerated even the slow progress of the program; in the spring of 1996, *MEED* reported that only one sizable concern, the Pressure Vessels Company, had been sold after the two bottling firms, and that only 16 parastatals were offering shares. In any case, IMF and World Bank officials remain dissatisfied with the government's commitment to privatization and still held a club over the regime's head: the third tranche of debt relief under the terms of the 1991 Paris Club accord, scheduled for the summer of 1994, required confirmation of satisfactory reform progress from the multilaterals. The IMF withheld its seal of approval; moreover, the extended finance facility signed in 1993 effectively lapsed in 1995 following the repeated refusal of the Fund to approve its first review. Although the primary bone of contention between the Egyptian government and the Fund had been the latter's demands for further currency devaluation, more rapid privatization was a second major issue; it rose to the forefront in the spring of 1996 when the Fund dropped its demand for devaluation. Accordingly, the government announced new plans for more extensive and rapid privatization.

Meanwhile the World Bank, while supporting Egypt before the Paris Club, did repeatedly criticize the government's gradual approach to reform, mentioning in particular the regime's failure to revise the labor market regulations. But the project to reform the labor laws stalled as the Mubarak regime continued its retreat from political reform. Beginning in 1994, members of the Muslim Brotherhood, previously tolerated as moderate Islamists, were being rounded up. The detentions accelerated before the fall 1995 parliamentary elections. Under new regulations, Islamists were now brought before military courts where they had fewer legal rights and faced stiffer punishments. The regime also became more repressive toward informal labor protest, which increasingly took an anti-privatization character (see below). In this climate, the idea of permitting even a very restricted right to strike was evidently too hot for Mubarak to handle.

Unions and the State in the 1990s

The government's gradual reversals on privatization came at a time when the ETUF was more divided at the top than it had been a decade earlier. Although Sadat had earlier resorted to separating the posts of ETUF president and Minister of Labor, it proved for him a temporary move; Mubarak maintained the split. The competing careerist ambitions and power quests between 'Asim 'Abd al-Haqq, in the Labor Ministry, and Ahmad al-'Amawi, at the ETUF's helm, diverted the attention of the senior unionists from economic issues and weakened the ETUF's oppositional stance.

The 1980s ended with al-'Amawi and other senior unionists hostile to 'Asim pushing to deny the ministry the right to supervise union finances, and agitating against his mandatory retirement decree (see chapter 2).[30] As the 1991 union elections approached, al-'Amawi moved to restrict the ministry's prerogatives by proposing new union legislation that would embody these changes. In addition, and as evidence that ensuring their prolonged tenure was a primary motivation of the unionists involved, the draft called for extending the term of union office from four years to five, allowing unionists promoted to senior management positions not to forfeit their leadership posts, and enabling long-term incumbents to run for federation office without the nomination of their local. The legislation was drawn up by an ETUF committee and disclosed to the press, which reported that the upcoming elections would take place under its auspices, before it was circulated to the federation leadership or the locals.[31] This further exacerbated the internal disagreements within the ETUF, however, and the draft remained dormant.

As a consequence, 'Abd al-Haqq's decree remained in effect during the 1991 elections, and impending retirement caused a relatively large number of federation officials not to seek reelection.[32] Some who did run were rejected by workers at the local level. All told, there was about a 40% turnover in federation leadership, with new blood entering mainly at the secondary level while more experienced officers moved up. This enabled some opposition elements to enter into, or advance in, the federations, and reflected in part growing turmoil at the base as economic liberalization intensified (see below). Nevertheless, 'Abd al-Haqq's desire to flush some of the entrenched senior union personnel is also indicated in this outcome.

Retirements also created some vacancies on the ETUF board, but these were all filled with tested NDP unionists. As before, the entire slate was chosen by default, with each federation nominating only the one representative to which it was entitled. Al-'Amawi became president again.

But although the ETUF board was solidly regime loyalists, clear divisions soon emerged within it. Shortly after the elections al-'Amawi resumed his efforts to have new union legislation passed which would permit him to remain in office after turning 60. The previous year's draft law was revived and submitted to the NDP's parliamentary committee, again without prior consultation with federation or local leaders. Some of the bill's provisions remained controversial within the ETUF, and opposition to it was encouraged by 'Abd al-Haqq. In an effort to preserve unity within the confederation, al-'Amawi quietly withdraw the legislation from consideration again.[33]

Nevertheless, this was not the end of the game. In a move that reportedly caught all the players off guard, al-'Amawi ran for and won a leadership position in the International Confederation of Arab Trade Unions during its convention in Libya in the spring of 1992; this required that he vacate his ETUF post.[34] Al-'Amawi informed Mubarak of his resignation when the two intersected at a Libyan airport. A journalist who witnessed the exchange reported that Mubarak was both surprised and disappointed, and offered to override the retirement rule on al-'Amawi's behalf if he would stay on, but the latter declined.[35]

At the helm of the NDP, Yusuf Wali quickly drew up a short list of candidates to replace al-'Amawi. 'Abd al-Haqq's favorite was his long-time ally Khayri Hashim of the Communications Workers' Federation, a leading proponent of ETUF relations with the ICFTU and thus, indirectly, the Histadrut. Hashim immediately began lobbying the other service federation leaders for support. However, the post was ultimately conferred on Sayyid Rashid from the Textile Workers' Federation. The reasons for the choice are unclear: some attribute it to the desire of the Interior Ministry to have an industrial unionist at the ETUF's helm; others intimate that Mubarak had been influenced by al-'Amawi, who favored someone on good terms with the other Arab confederations, during their airport meeting.[36]

But Rashid was also nearing retirement age. Anticipating that Hashim and 'Abd al-Haqq would move against him, he again revived the draft new union legislation and had an ETUF committee prepare it for submission to parliament.[37] At the same time, he arranged employment for himself in the private sector (where the mandatory retirement age was 65), to circumvent the union laws. In the spring of 1993, as Rashid's sixtieth birthday approached, the attempt to oust him came anyway. Relying on 'Abd al-Haqq's 1987 decree, which focused on age (60) rather than employment status, Hashim mobilized a majority on the ETUF board to declare Rashid's presidency illegal, and claimed the post; 'Abd al-Haqq quickly recognized him.

Rashid fought back, appealing to his own allies in the confederation, NDP and parliament; within the ETUF, his main backers came from the industrial federations. Rashid's supporters formed a "Union Salvation Committee," which kept the official press informed of developments and attacked 'Abd al-Haqq publicly in the ETUF's newspaper for the first time, and Rashid was implicitly backed as well in the opposition press. Prime Minister 'Atif Sidqi sent the issue to the parliament's legal committee, and reinstated Rashid pending its ruling, but Hashim would not honor that decision.

The controversy raged for several months, and tore the ETUF apart to the degree that virtually no official work was done at a time when the privatization program, with all its incumbent threats to workers, was heating up; even the government-run newspapers could not refrain from commenting on the unprofessional preoccupation of unionists with these squabbles.[38]

Several months later, the legal experts ruled in favor of Rashid; around the same time, after Mubarak was confirmed for a third term as president, he removed 'Abd al-Haqq in a cabinet reshuffle, and replaced him with Ahmad al-'Amawi. Subsequently, at Rashid's initiative, 'Abd al-Haqq's membership in the Textile Workers' Federation was revoked. However, Hashim and his allies retained their ETUF and federation posts in an agreement, brokered by al-'Amawi, to restore unity to the confederation. As part of this deal, not only did the mutual recriminations cease, but the formation of the Union Salvation Committee and its disclosures to the press were denied.

With both Rashid and Hashim loyal to the regime and the ruling party, the reasons for Mubarak's ultimate backing of Rashid can only be speculative. Plausibly, he felt that this would buy greater cooperation of the industrial unionists with the economic reform program, especially privatization; there is some intimation that Rashid explicitly promised to support a lower annual raise for workers in exchange for Sidqi's backing.[39] But as they had earlier with Ahmad, some of the leftists in the Tagammu' and Nasirist parties implicitly supported Rashid, partly in the belief that he would be more inclined to resist privatization.[40] If ETUF relations with the Histadrut was the prize that Hashim and 'Abd al-Haqq were offering, the outcome suggests that further normalization of relations with Israel was not then a priority for Mubarak.

For our purposes, the main significance of this affair is the spotlight it threw on the functioning of the ETUF and its interactions with the government. The blatant opportunism of the senior union personnel it revealed generated new interest in the institutional structures of union/state relations. Egypt's labor left used this opportunity to intensify its campaign for trade union reform.

In the height of the conflict between Rashid and 'Abd al-Haqq, *al-'Arabi* published an article in which labor leaders from left parties considered splitting the confederation.[41] This appears in retrospect to have been more of a means of pressuring the government to support Rashid than a reflection of changed thinking; Fathi Mahmud later denied to me that he had ever contemplated the idea of dual confederations.[42] However, other opposition forces promoted a new spate of publications advocating reforms.[43] Leftists affiliated with

Sawt al-'Amil (Voice of the Worker, VOW) opened a Center for Union Services (Dar al-Khidamat al-Niqabiyah) in Helwan and, several years later, Shubra al-Khayma. Besides producing pamphlets and books, these centers sponsor regular forums on union-related issues and help workers in need of legal assistance for job-related matters. VOW members also see their activities as working to create alternatives to the ETUF at the base level.[44] And a mimeo by the National Committee to Fight Privatization, distributed by VOW activists, advocated a return to the bottom up clauses of the 1976 union law, before its "loopholes" were adjusted by the senior ETUF personnel (see chapter 2), as the means to ensure union democracy.[45] Amin 'Izz al-Din, the Nasirist labor historian, began to agitate for eliminating dues check-off completely, despite having been one of the earliest proponents of this system. He maintained that eliminating the check-off would compel union leaders to pay much more attention to the rank and file, or risk dramatic reductions in membership.[46] Islamic forces associated with the Socialist Labor Party (SLP),[47] who had begun to contest union elections in 1987, now also took up the banner of union democracy, advancing proposals similar to those advocated by the left.[48]

The result of the wider public debate on the unions' functioning was to increase the vulnerability felt by senior ETUF personnel, who intensified their push for revisions to the union laws which would further insulate them from rank-and-file pressures. This demand was realized in 1995, as part of the ongoing machinations over the new labor law (see below). But far from quieting the calls for union reform, it appears more likely to aggravate them when its effects are combined with more extensive privatization.

The ETUF's Changing Stance on Privatization

The ETUF's endorsement of Law 203 represented a significant turnaround from its earlier opposition to privatization. Union acquiescence was achieved in part through heightened levels of coercion, combined with a continuation of the regime's customary co-optation techniques. However, the confederation's ongoing cooperation was also purchased through concessions to labor's concerns which enabled some unionists to claim that privatization would not harm the interests of their membership.

The first harbinger of a weakening in the confederation's stance against privatization appeared in May 1989, when the confederation issued a joint statement with the Egyptian Businessmen's Association (EBA). The document called for a number of reforms in the operation of the public sector, including freeing it from subordination to government administration, separating its management from the state, enhancing its access to credit, lifting price controls on its products, allowing individual public sector firms to adopt their own regulations, and linking wages and incentive pay to productivity.[49] Although the statement did not call for a sell-off, and its recommendations were in fact supported by many of those who were arguing for preservation of the public sector, it did signify a marked departure from the previous ETUF stance toward the businessmen's group, which was seen as the strongest nonstate domestic force urging privatization. In 1986, the confederation had issued a statement condemning the developing close relationship between the EBA and the government's economic planners, and charged the businessmen with seeking to undermine workers' gains.[50]

Al-'Amawi had acted independently in issuing the joint statement, and his actions brought an angry response from a number of other senior unionists, particularly those representing federations with membership primarily in the public sector. The heads of the Textile, Engineering and Metalworking, Military Production, and Woodworking and Construction federations, along with several others, demanded an extraordinary meeting of ETUF leaders to discuss the statement. Some charged that it had been signed in spite of their rejection of it, while others protested that they had been denied an opportunity to comment on a draft. Under the leadership of leftist activists, a number of industrial locals also condemned the statement, blaming al-'Amawi for failing to insist that opposition to privatization and defense of workers' rights be included in it.[51] The Tagammu' and Socialist Labor parties promoted the campaign against the statement in their weekly newspapers.

Although no action appears to have been taken in response to these protests, Al-'Amawi's tone did change somewhat during the course of the privatization debate which ensued later that year; he argued, in particular, that any increased freedoms given to public sector managers needed to be balanced by more freedoms for workers, and that the British model of privatization was not applicable to third world countries like Egypt. At the same time, several of the same federations that had condemned the EBA joint statement likewise rejected the privatization proposals emanating from the NDP and the government, and pushed for an official ETUF policy. The result was the formation of an ETUF committee on public sector development, which agreed to back some liberalization of public sector management provided that workers' rights were not harmed, but rejected the idea of holding companies and "any attempt to sell the public sector by any name."[52]

Thus, formally, the confederation remained in opposition to privatization. However, as the government's plans took shape over 1990 and 1991, there was no organized and unified ETUF campaign to disrupt them. Rather, opposition was maintained only by a minority of federations, mainly in industry; it was spearheaded by the Electrical, Engineering, and Metal Workers' Federation (EEMWF), historically the most militant of the confederation's member groups, and home to several of the few leading unionists not then affiliated with the NDP. But these unions apparently lacked the backing not only of their counterparts in the service sector, but also of Al-'Amawi, making them more vulnerable to attack by the regime.

There is clear evidence that the government did attempt to intimidate union opponents of privatization. In April 1990, as the Mubarak administration was formulating its sell-off policy, the EEMWF invited other federations to a meeting to discuss the government's proposals. At least several unionists were contacted by government or security officials and urged not to attend, and while the meeting was in session, security police surrounded the ETUF headquarters. This had never occurred before in the confederation's history. The SLP claims that the police declared the meeting illegal and threatened those who crossed the barricades with arrest, in addition to monitoring the proceedings. The meeting resulted in formation of a "Committee for the Defense of the Public Sector," which included representatives of the same federations that had opposed al-'Amawi's collaboration with the EBA. This committee issued a statement calling privatization proposals "a bomb threatening social peace and stability, which will lead to serious social explosions," and demanded that al-'Amawi lead the ETUF in more forthright opposition to the new schemes. But the meeting's attendance fell short of expectations because of the repression.[53]

The more militant federations continued to pass resolutions over the following year, but the ETUF took no further steps against privatization. The confederation did not, in particular, condemn the sale of the regional cooperatives, nor attempt to block them. Moreover, during the period between the proposal and actual enactment of Law 203 other senior unionists left the opposition. The EEMWF was the only federation to mobilize internally against the law, holding a series of open meetings and ultimately rejecting it unanimously in a federation congress. The federation demanded public hearings on the draft and that it include the right to strike for workers, and called for marches to parliament and Mubarak's palace to protest the law.[54]

There is no record that any such demonstrations actually took place. However, the regime did attempt to mollify, as well as pressure, the recalcitrant unionists. A series of meetings were held between them and government officials before the vote in parliament, and the draft was revised to protect workers from layoffs or reductions in the workers' sharing in profit schemes. As a result, some unionists who had participated in the "Committee for the Defense of the Public Sector," now declared their support for the bill, claiming that it would actually benefit workers; several union MP's also denied that the law would really lead to ownership transfers.[55]

The ETUF's formal endorsement of Law 203 after its passage prompted charges of treachery from the opposition press. There were, nevertheless, 25 union MP's who voted against Law 203, indicating that privatization remained a controversial subject within the confederation in spite of the regime's combination of techniques.[56] And the endorsement was not without its reservations, expressed at a special conference on the law held at the Workers' University in August. The unionists present insisted on the ETUF's continued right to be consulted about all laws affecting workers, and to participate on any committees formed to evaluate the functioning of the new public enterprise sector. They called for clear principles to be established for the selection of key decision makers for the sector, and for sound technical, financial, and administrative practices in the assessment of public sector resources potentially for sale. Finally, they demanded that a fund be established to compensate any worker harmed by closure or consolidation of public sector firms.[57] Neither the official nor opposition press took much notice of the event, however.

However, the contradiction between yielding to pressures from the regime, and maintaining some semblance of credibility with the rank and file, did not go away, and after the 1991 trade union elections the ETUF's leaders adopted a more oppositional posture again. The corporatist system protected senior union personnel from losing their posts in the elections—changes at the top occurred only due to retirements—but they appear to have been chastened by some significant NDP losses at the base (see below).[58] The most promising change for opponents of privatization was the election of Niyazi 'Abd al-'Aziz to the EEMWF helm. As we have seen, 'Abd al-'Aziz had formerly been affiliated with the Tagammu' and then joined a Nasirist opposition party. He was credited by the SLP with having spearheaded the EEMWF's campaign against Law 203, and his candidacy was also implicitly supported by some leftists. 'Abd al-'Aziz was later co-opted into the NDP, and some leftists charged that he, too, became an opportunist, but he did use his new post to push for a stronger ETUF stance against privatization.[59]

Although the ETUF's endorsement of Law 203 was not reversed in the elections' aftermath, the unionists did begin to draw more public and private battle lines around the maintenance of workers' customary rights and benefits. Thus, while Al-'Amawi urged the newly constituted federation boards to support law 203 and cooperate in drafting the new by-laws of the subsidiary firms, he also publicly pressed for the government to guarantee alternative employment to any workers laid off as a result of privatization. The confederation also decried the attenuation of the workers' representation in management (WRM) system that would occur in firms that were fully or partially privatized under Law 203. Under the terms of the Private Companies Act, such firms were required to have some form of WRM, but not the high proportion of elected workers' delegates imposed on the public sector. In raising this issue, the ETUF also sought, unsuccessfully, to realize its longstanding desire for greater input into the selection of workers' representatives to the management councils, who were elected by workers at the same time as, but independently from, union leaders.

These concerns were heightened in January 1992 when some prominent advocates of liberalization, including former Prime Minister 'Ali Lutfi, called for new WRM elections to be held in the entire public enterprise sector as if all its firms had already become joint ventures. Although the ETUF's leaders vigorously opposed and successfully defeated this initiative,[60] it is clear that the WRM system had already been weakened. There were no provisions for WRM on the holding company boards, and with the progress of privatization the proportion of elected workers' representatives on individual company boards would be reduced.

Sayyid Rashid, upon assuming the ETUF presidency in the spring of 1992, proved to be more outspoken than al-'Amawi. Despite having voted in parliament for Law 203, Rashid soon issued a statement condemning the closing of any public sector company and/or the layoff of any employee, charging that such measures would cause social instability. He also accused the government of mishandling the Social Development Fund, challenging the regime to demonstrate any positive effects from Social Fund expenditures, in particular with regards to providing job retraining for workers.[61]

The confederation also reacted strongly, and with at least a public facade of feeling betrayed, when the holding companies began to target subsidiaries for closure that summer. The list included six companies in the construction and contracting sector, and two in commerce: the Eastern Cotton Company and the Nile Company for Agricultural Exports. With the president of the Commerce Workers' Federation (CWF) threatening that there would be strikes and sit-ins over the decision, which would affect 7,500 workers, the ETUF called an emergency meeting in mid-July and sent urgent memos to Mubarak and 'Atif Sidqi. The confederation charged that the decisions violated Law 203, since the general assemblies of the companies (on which union representatives sit) were not consulted, and in general that proper investigation of the companies' financial situations had not been done. A week later Rashid informed the ETUF that the government had agreed to postpone the closures until proper studies had been conducted, and the CWF delegated a committee to explore means to improve the companies performance. However, a year later a confederation adviser reported that the Nile Company for Agricultural Exports had been liquidated.[62]

By the end of 1992 it was clear that the confederation, despite having endorsed Law 203, was at odds with the regime over its application. Another emergency meeting of the confederation had been convened in September, in response to a charge by EEMWF president 'Abd al-'Aziz that Law 203 had been transformed into "a weapon against the Egyptian economy and its workers." Rashid agreed at that meeting to set up a special ETUF office under his leadership to confront the problems arising from implementation of the law, and urged each federation president to do likewise.[63] At its general assembly held in December the confederation passed a resolution condemning privatization and called for legalizing strikes if any workers did lose their jobs as a result of public sector diminution. The meeting also condemned the price increases resulting from lowered subsidies, and renewed the call for a high council on wages and prices to ensure that the former keep pace with the latter. Significantly, and for the first time in its history, the delegates refused to send the customary telegram of support to the government, and the decision rejecting this gesture was unanimous.[64]

Nevertheless, while ETUF leaders warned publicly that massive layoffs could produce a "social explosion," they made no effort to mobilize the membership behind these implicit threats. Their actions, as in the past, were confined to public statements and behind-the-scenes lobbying of other regime elites. Moreover, as we have seen, much of 1993 saw the ETUF divided by, and preoccupied with, the challenge to Rashid's presidency. It was not until the end of that year that another series of meetings was convened to discuss the application of Law 203. These resulted in a 22-point memo sent by the ETUF to 'Atif 'Ubayd, the Minister in charge of the PEO; Rashid, 'Ubayd, and Ahmad al-'Amawi, who had become Minister of Labor in cabinet changes that October, met to discuss the document. As before, the confederation's chief complaints were that workers were being threatened with layoffs, and that the holding companies and subsidiary management councils were not properly consulting union officials before deciding matters with deleterious effects on the workers.[65]

One apparent outcome of this meeting was the addition of federation officials to the boards of holding companies for firms whose workers the unionists represented. This enabled unionists to claim that they were carrying on the struggle against privatization in other venues.[66] It also appears that, whatever their public demeanor, the senior unionists privately recognized that they had already lost the battle for preserving the public sector and its workforce, and now came to redefine their raison d'être as winning the best terms for workers who would be displaced or reclassified. Thus, while labor leaders' pronouncements continued to decry layoffs, a discussion of tradeoffs increasingly found its way into their public statements as they negotiated these with the regime in closed-door committees.

In October 1993, just as the first industrial sales under Law 203 were being finalized, it was disclosed that ETUF delegates had participated in a committee to devise ways of eliminating redundant public sector workers. Along with representatives from the Ministry of Labor, several insurance companies, and the Social Fund, which would finance the endeavor, this committee agreed upon three options:

1) encouraging early retirement with full pensions. This would mean lowering the legal retirement age to 51 for men, and 45 for women;
2) offering a lucrative severance payment to workers between the ages of 37 and the minimum for retirement. This compensation would be earmarked for the establishment of a small business;
3) providing job retraining for workers between 18 and 36 years of age.

By the end of 1994, the government had begun offering these alternatives to workers in targeted establishments.[67]

Although the work of this committee apparently proceeded amiably, the debates in the committee for devising the new labor law were heated. It was here that the terms under which those who remained in the workforce would be employed were being fought out. When listening sessions on the proposed legislation were held by the Ministry of Labor and the ILO at the end of 1993, the ETUF complained that all the work on the draft to that date had gone on in smaller working groups, and that their team had never been shown the full text of the draft prior to the hearings. Charging that this was done by intent to weaken the confederation's position, the team issued a sharply worded critique of the proposals emanating from the committee. They accused the government of being biased in favor of the employers' team, whose suggestions "seek to restore the labor relations prevailing before the 1923 constitution." The statement warned that implementation of such suggestions would "lead to a dangerous imbalance in the basis of social peace," arguing that the labor law should be viewed as a "social and human law before it is an economic one."[68]

Thus, while the confederation was resisting the repeal of parastatal workers' protections, by 1994 ETUF opposition to privatization per se had receded. A year later, Sayyid Rashid declared that Egyptian workers supported privatization, in light of Mubarak's promises that it would not hurt them. And in 1996 former ETUF president al-'Amawi, then at the helm of the Labor Ministry, criticized the opposition press for questioning the veracity of Mubarak's slogans.[69]

Some of the ETUF's member federations proved more recalcitrant, however. As in the past, the most militant were the CWF and the EEMWF. When the management council of the Egyptian Company for Chemicals and Minerals agreed to sell 46 of its branches, and actually concluded some of these sales without public announcement, these two federations filed three court motions against the management.[70] In 1996, the EEMWF accused the Babcock and Wilcox Company, which had purchased the pressure valve parastatal at the end of 1994, of violating the terms of its sales agreement by laying off workers and forcing others into resigning. The federation sent an urgent letter to al-'Amawi about the company, and shared its contents with opposition parties who publicized the case.[71] The federations representing workers in the printing, and chemical and pharmaceutical, industries were also relatively outspoken.[72]

The Pressures from Below

While those federation and confederation leaders who remained opposed to privatization confined their efforts to issuing public denunciations, lobbying the regime from within, and perhaps taking cases to the courts, leftist forces tried to organize resistance from below both through direct organizing at the plants, and indirectly through their publications. These efforts served to keep the base informed about government policies and the positions of ETUF leaders, thereby putting pressure on both regime and union officials. At the same time, rank-and-file workers acted spontaneously in various locations when assaults on their wages, benefits, or job security appeared imminent; increasingly after Law 203 was passed, such takeaways were linked to the plans for privatization. Workers' struggles won some immediate rollbacks and likely contributed to the regime's hesitancy to push through the new labor law. As countervailing pressures from the multilaterals intensified, however, the regime offered fewer concessions to rank-and-file protest, and its response to them became more repressive and violent.

Sensing that it implied a weakening in the confederation's posture against privatization, the left campaigned against the joint statement signed by the ETUF and the EBA in 1989. Meetings were organized in most of the major industrial areas. The largest was in Helwan, attended by about 80 workers, and resulted in the formation of the Committee for Defense of the Public Sector and National Industries. This committee filed a legal case seeking to nullify the statement, but the courts refused to hear it. Throughout the country, locals where leftists held leadership positions sent telegrams to the ETUF headquarters denouncing the statement. A similar campaign was undertaken by the Socialist Labor Party, although its influence at the base was less extensive. All of the denunciations were publicized in the SLP and Tagammu' newspapers, contributing to the controversies in the upper layers of the ETUF.[73]

The Tagammu' took up a similar campaign before and immediately after the passage of Law 203. Leftist workers organized a meeting in Helwan in March 1991, and drafted a statement criticizing the repeal of benefits and job security protections present in the original draft of the law. They collected signatures on the statement and sent it to the parliament. Significantly, the petition maintained that any attempt to institute capitalist labor relations in the public sector had to be accompanied by legalizing strikes, the demand later embraced by ETUF negotiators on the labor law committee. In June, leftists in Alexandria began organizing a Popular Committee for Combating the Sale of the Public Sector; three workers were arrested distributed the literature of this committee.[74]

In their fall, 1991 union election campaigns, Marxist, Nasirist, and SLP candidates made opposition to the privatization program an issue. As further evidence that some leftist pressures for union reform did have concrete results, the OSP was not used against any candidates in these elections, although there was no formal government denunciation of the prosecutor's past practices. Tagammu' affiliates and other leftists made some significant gains in this poll, doubling their overall presence to about 400 local positions. Although this still represented only a small portion of the total number of local slots available, the leftists were concentrated especially in the public sector industrial unions, and claimed that they now held about 25% of these seats.[75] The results represented some significant victories against entrenched NDP incumbents. At the Helwan Iron and Steel factory, the defeated included Sulayman Idris, former president of the local and vice president of the EEMWF. Opposition elements were also able to maintain secondary leadership positions in the Military Production and Textile federations.

The committees formed to combat Law 203 appear to have disbanded that fall as their activists were subsumed by the election campaign, but several were revived after a hiatus; in April 1993 leftists affiliated with the Tagammu' established Committees to Defend the Public Sector and National Industry in both

Mahalla al-Kubra and Helwan.[76] The efforts of these groups, and of opposition forces in individual plants and locals, now focused on trying to prevent the sale or consolidation of particular plants. Blaming the problems of the public sector not on inefficiency but on indebtedness and management incompetence (and, in some cases, embezzlement), they proposed cancellation of some portion of public sector companies' debt, and rescheduling of the remainder with 10 years' grace. They also attacked the work of the holding companies and the government's choice of members for their boards, charging that the regime had deliberately empowered businessmen interested in purchasing the firms with making the decision to sell them.[77] The opposition press gave wide play to such stories, although no particular victories seem to have resulted.

The most significant of these efforts was the National Committee to Combat Privatization. It was established in November 1993, by more than 50 opposition unionists, mostly local leaders but some holding second-tier federation positions; all of the industrial federations, and all regions of the country except the south, were represented in its membership. The committee published a lengthy newsletter, *al-Tadamun* (Solidarity), detailing the government's plans for privatization to that point, and alerting workers to the dangers inherent in the proposed new labor law. It was also *al-Tadamun* which published the ETUF team's criticism of the government's conduct in the deliberations over the law.[78] Relatedly, in the spring of 1996, some 150 unionists and other labor activists from the opposition parties, representing most industrial federations and governates, held a one-day hunger strike at the headquarters of the Arab Nasirist Party to protest the government's new push on privatization.

The lack of a strong rank-and-file response to these particular initiatives is not inconsistent with the attitudes and behaviors of workers described in previous chapters. As we have seen, workers have traditionally been moved to protest in Egypt only after some negative change to their accustomed standard of living and working conditions. Law 203, while it made such disruptions imminent, did not in and of itself create them. In this sense, the government's 1990s' strategy of separating passage of the legal framework for privatization from the actual sales, and the sales themselves from any setbacks to wages and benefits, appears to have been successful in preempting widespread protest. However, workers did react strongly to takeaways which occurred as a result of the privatization program, indicating that the regime's expressed concerns about the potential for greater unrest with more extensive privatization, if perhaps exaggerated, were not imaginary.

An incident in the summer of 1992 served as a harbinger of the turmoil to come, as well as an indication that opposition to privatization was not limited to public sector workers; many of those employed privately were still hoping to join the government's payroll in the future. It occurred in 10th of Ramadan city, one of several new industrial areas set up by the government in remote locales to both ease the population pressure on Cairo and encourage private enterprise. When the Minister of Labor suddenly decreed that full-time private sector workers should have their names removed from the waiting list for government employment, several hundred workers resigned from their jobs in protest, asking to be rehired as temporary employees so they could remain on the list.

Significantly, the enterprises in 10th of Ramadan offered higher wages than government jobs, and complied with all the insurance laws often evaded by other private sector entrepreneurs; in seeking reclassification as temporary employees the workers sacrificed these benefits. Asked to explain their actions in this context, some workers mentioned the fact that government positions enable workers to take prolonged leaves of absence for employment abroad, a sentiment seized upon by privatization proponents as evidence of widespread redundancies and moonlighting in the public sector. However, most of the workers cited the harsh conditions and remote location of their employment; several of the women interviewed in *al-Ahram* complained that subsidized housing near the factories was available only to male employees.[79]

The government's response to this incident is an indication both of how divorced it was from mass sentiment, and of the contradictions inherent in its attempt to win popular support for the state's retreat from the economy. Both the Minister of Labor and the city's employers expressed shock at the workers' reaction, even though the preference of private sector workers for public sector or civil service jobs had been evident years earlier. The government accused the workers of laziness for seeking less demanding positions, and of selfishness for refusing to yield their spots on the waiting list to others who lacked employment. At the same time, the official press lauded the 10th of Ramadan entrepreneurs for creating jobs and exhorted the workers to trust the country's future to private enterprise, or in other words to the selfishness of capitalists.

Meanwhile, workers in dozens of plants across the country were subjected to sudden reductions in various components of their supplementary pay, and in some cases the threat of layoff, as the holding companies began the work of reviewing their subsidiaries' finances and operations. More spontaneous labor protest was the result. As in the past, the actions were primarily informal. But although senior union leaders did not organize these incidents, and in many cases the local leaders were also not involved, they did serve to bolster the ETUF's warnings that the government's policies were likely to cause social disruptions.

The first case to attract national attention was the Southern Company for Land Reclamation, whose 1500 employees were members of the Agricultural Workers' Federation. The Agricultural Ministry issued orders in 1992 to liquidate the company and sell half its land; the remainder would be distributed to employees upon their formal resignation from the company. However, roughly half the workforce had been working under nonpermanent employment contracts for periods ranging from two to seven years. Although lacking a formal job grade, they had been receiving all the customary benefits and pay of permanent public sector employees. After the liquidation decision, these 750 workers were terminated and denied pensions, severance pay, and the opportunity to purchase land. They began a hunger strike in January 1993, which lasted at least ten days and caused several workers to be hospitalized. But their plight went unheeded; the company maintained that as temporary employees, the workers were not entitled to the benefits.[80]

At the General Battery Company, workers held a sit-in at company headquarters in April 1992, protesting the sudden curtailment of their incentives which resulted in a 50% decline in take-home pay. They also complained of an end to compensation for work clothes and an increase in occupational diseases as a result of cutbacks in protective apparatus. The local began sending telegrams to the Ministry of Industry warning of a strike if the situation persisted. Although they won some concessions, the following year the company was marked for closure, and another campaign of telegrams to the Prime Minister and president was begun.[81]

Workers tried to prevent privatization of the Habi Company, which manufactures eyeglasses, by filing a suit against its holding company. They also tried to discourage potential buyers by circulating rumors that the factories would be burned if any sales went through. However, many of the resources of the plants, including the land of the branch in Shubra al-Khayma, were sold while the court case was under appeal.[82]

At dozens of other establishments, covering every major industrial region of the country, workers boycotted paychecks and/or occupied their plants in protest of declining pay as a result of cutbacks in incentives and other supplementary wages. The workers routinely charged that company mismanagement, rather than lack of productivity, was responsible for the parastatals' poor performance. In most cases, at least partial concessions were won. Toward the end of 1994, however, the government's response to these symbolic protests became more repressive. This resulted in the death of at least six workers or relatives, and the wounding of more than a hundred others, at the MS&W plant in Kafr al-Dawwar; these are the highest casualties from a single factory protest in Egypt's modern history. As we have seen, Kafr al-Dawwar's 80,000 industrial workers[83] have a long tradition of activism and workers' solidarity, and some labor activists see the incident as a sign that, with increasing pressures from the multilaterals for more rapid privatization, the regime wanted to test just how far it could go with the impending new labor legislation.[84]

The story reveals the same pattern as previous violent incidents as described here: a long build-up of workers' grievances, management promises remaining unfilled, and then a symbolic protest that turned bloody at the provocation of the security forces. Its roots date back to May 1993, when a new director, Fathi Muhammad 'Ali, was appointed head of the management council. 'Ali quickly worked to co-opt the trade union leadership, by transferring its members to administrative positions where they were rumored to be receiving wages and bonuses almost six times as high as the average plant worker. He also appointed a board of 17 advisers, also earning exorbitant salaries, to help supervise his new policies.

The situation began to decline when 'Ali, with the approval of the local leaders and new board of advisers, began to implement a series of policies designed to drive wages down. Among these were: 1) surprise administrative and security searches whereby minor infractions such as being momentarily away from one's station or tardiness would be punished by a 3-month loss of incentive pay and bonuses; 2) pay cuts for excess sick leave at the rate of one day for the first day, 3 days for the second day, 6 days for the third day, etc.;[85] 3) laying off approximately 1,500 temporary workers, and transferring some permanent workers to distant branch plants; 4) attempts to evict over 200 families, some of which had been residents for twenty or thirty years, from company housing when they no longer had a head of household working for the factory; and 5) the demolition and nonreplacement of other company housing, despite a severe housing shortage.

Over the following year, the workers sent numerous appeals to the Textile Workers' Federation, the ETUF, the Minister of Labor, and to Mubarak himself. They eventually won a decision by the government's Central Accounting Office declaring the management policies illegal. However, 'Ali was allowed to remain in his position and some of these policies were implemented nevertheless. The rank-and-file's anger festered and then, on September 27, 1994, they learned that the local had agreed to a company proposal which included a 3-month loss of incentives for workers taking 1 day of sick leave or being caught sleeping or absent from their work site. The infuriated workers chose their own representatives who conducted negotiations with management on September 28 and 29. This resulted in circulation of an administrative announcement saying that the 3-month penalty had been withdrawn. However, on September 30 'Ali announced that he had not authorized that statement and would not rescind his decisions.

A sit-in began that evening, with workers from incoming shifts joining until they were prevented from entering the factory the next day; about 7,000 of the plant's 22,000 employees participated. The workers demanded the removal of 'Ali, the reinstatement of the laid-off temporary personnel, the return of all workers who had been transferred, and the nullification of all the policy decisions of the previous months, with back pay to all workers who had suffered losses as a result of them. They also called for the dismissal of the new advisory council and permission to elect a new union local; during the sit-in, three local leaders were beaten by workers, and the rest fled.

The next day, as families began collecting food and sending it to the workers, the local units of the state security police (SSP) were out in full force, and special forces and reinforcements were brought in from neighboring towns. At dawn on October 2, the SSP surrounded the nearby village of Nazir, which had often been the starting point for workers marches; they also cut the rope to a small ferry to deter workers from other villages from joining the protests.

The confrontation at the factory gates began when the SSP tried to turn back families bringing food, then confiscated the food and threw it in a nearby canal. When the families protested and started chanting, the SSP fired in the air, the families responded by throwing bricks, and the SSP then attacked with live ammunition and tear gas. Workers watching from inside the factory then opened fire hoses on the SSP in an effort to defend their families, and were fired on as well. At about the same time it was announced that school was canceled for the day, and families went to collect their children and get them out of the way. One of those killed was a nine-year-old boy who was shot while his father was leading him home from school. Some older school children went to see what was happening, joined in the chants, and were also fired upon.

Later during the day management announced that it would accept the workers' demands and also called for an immediate one-week vacation. As the occupation broke up, 71 workers and family members were arrested either in the street, at home, or at the hospital. Another 22 were arrested from their homes over the next two weeks. Some 75 people were hospitalized, including eight individuals who lost one or both eyes, and one who lost the use of his legs; many others were afraid to seek medical treatment for fear of arrest.

Subsequently, most of those arrested were released after paying moderate fines. A government-appointed committee investigating the incident found that the union did not play its proper role in defending the workers' interests, and that their demands were legal. 'Ali was put on an indefinite leave of absence as a result, but with the workers' other demands only partially addressed, labor activists reported several months later that the situation in the village was still tense.

Shortly thereafter, the regime took preemptive action to prevent a similar outbreak at the MS&W plant in Mahalla al-Kubra.[86] There, a committee of labor activists from the opposition parties had begun agitating around several economic issues. They were seeking an increase in incentive pay, the opening up of promotions, which had been blocked for several years, and compensations for meals and housing, since few workers there had company housing such as that available to their counterparts in other industrial areas. The activists distributed leaflets calling for workers to raise these demands at a meeting at the local's headquarters on October 20, where the local leaders rejected them. The next day, workers demonstrated, marching from the local's headquarters to the management offices, and adding demands for the abrogation of Law 203, the resignation of the company president, and the release of the workers still detained at Kafr al-Dawwar. The SSP was called in and blockaded the factory.

After this, local leaders promised to investigate the demands, but when they did nothing the activists began gathering signatures on petitions to withdraw confidence from the local. On November 1 workers gathered to protest again, marching through the streets of the town with wooden coffins on which they had written the names of the company president and the head of the local. At the same time, activists distributed leaflets calling for a strike on November 5. On that morning, the SSP began arresting the ringleaders, so the planned protest was averted. The detainees were charged with handing out illegal pamphlets and, rather ironically, with opposition to the country's socialist system! Two months later five of the 12 activists detained were still in prison.

As at Kafr al-Dawwar, the regime also made some concessions to workers' demands. However, lawyers representing workers in both incidents say that the government was conceding less to labor protests at individual plants in the 1990s than it had in previous decades. They saw in this, as well as in the violence at Kafr al-Dawwar and the preemptive repression in Mahalla, a sign that the regime's position toward labor protest was hardening as it moved toward adoption of a new labor code. But such a change could only serve to belie the government's pledges that privatization would not harm the interests of workers, further undermining the credibility of the regime.

The Proposed New Labor Law

At the end of 1994 the various parties involved in the negotiations over the new union law finally agreed on their sixteenth draft. It would facilitate layoffs in the public sector and newly privatized firms, and generally reduce the benefits and protections available in the public sector to those which actually prevailed in the private sector. As signaled for several years, it would also legalize strike for the first time in Egypt's post-coup history, but this quid pro quo for the sacrifice of job security was sharply constrained.[87] The bill was then expected to be passed in the spring of 1995,[88] but its submission to parliament was put off until after parliamentary elections in the late fall of that year. As of the fall of 1996, the new legislature had not yet enacted it, and the draft bill was continuing to generate public controversy.[89]

The proposed new legislation requires that firms must obtain government approval for any mass workforce reductions. This provision also existed in the previous law, but in the past, such approval was almost never given. Events since the passage of Law 203, reviewed above, indicate that the government would now be amenable to layoffs. The new law verifies this by stating explicitly (article 198) that it is an employer's right to adjust the workforce according to economic conditions.

The draft law would also considerably increases an employer's freedom to dismiss workers individually. The 1981 legislation required that employers consult a tripartite committee representing management, unions, and the Ministry of Labor before firing a worker. Although the decisions of this committee were only advisory, dismissed employees could appeal its rulings before turning to the courts, which in turn could require an employer to continue paying the worker until the case was settled.[90] The new legislation—at the insistence of union leaders—makes committee decisions to reinstate workers binding. However, it expands the committee to include two judges, which will probably make rulings in favor of workers less frequent.[91] While a worker could still appeal a negative committee decision to the courts, there are no longer specific provisions for ongoing salary payments, thus making it unlikely that workers would be financially able to appeal dismissals.

The draft maintains the prevailing requirement that all workers must be employed under written contract, which can be of indefinite ("permanent") or specified duration. However, it widens the range of serious worker infractions for which employers are entitled to break a contract. Moreover, while Law 137 specified that a temporary contract, if renewed, would acquire indefinite status, the new law would permit multiple renewals. This makes it likely that no temporary worker would ever achieve the security of permanent status, and that no new worker would be hired indefinitely. Finally, the law (Article 203) enables employers to effectively force workers to resign by empowering them, "for economic reasons," to lower the contractual wage and/or require employees to perform different jobs. This article, included at the insistence of the businessmen's organizations, nullifies the very essence of the work contract concept. Nasirist and leftist delegates to parliament vowed to challenge it there,[92] but with the NDP's crushing majority, this opposition could only be token.

As we have seen, many private sector employers had been evading the old protections against firing through techniques such as using only temporary contracts and terminating workers briefly before renewing them, or forcing workers to sign blank contracts or undated resignation letters upon hiring. The resulting disparity in job security between the two sectors was the major reason why qualified workers preferred public sector employment.[93] Even though public sector managers had increasingly resorted to temporary hiring to enhance profitability, such employees could still aspire to permanent positions. The new law, by facilitating individual firings as well as mass layoffs, clearly aims at eliminating this basis for long-standing labor opposition to privatization.

The negotiations over the new labor law were intended to be secret, and were not officially acknowledged until the fall of 1993, two years after they had begun. However, the opposition press, especially the left, learned about and publicized them early on. Under this spotlight, ETUF leaders insisted that they would oppose any diminution of workers' established rights.

Two concessions from the government at the end of 1994 enabled the unionists to claim adherence to this principle. Article 2 maintains Law 137's stipulation that all workers should receive a minimum 7% annual raise, and actually removes a previous cap on the pound value of this increase. In addition, Article 4 specifies that workers' current wages and benefits must not be reduced by the law's application. Thus, although the law eliminates the rigid job grades and pay scales which used to constrain public sector managers upon hiring and promoting workers, only new hires would be affected. Some leftists charged that this "grandfather" clause aims at dividing the interests of the current public sector workforce from those of future entrants, thereby ameliorating opposition to the law.[94]

The last-minute inclusion of these two articles suggests that the law's *immediate* goal is to facilitate layoffs and dismissals; greater wage flexibility would be achieved only gradually, as older workers retire and are replaced. Nevertheless, Article 4 is ambiguous and may not provide the full guarantees it seems to promise the current workforce. It appears to be contradicted by Article 203. Moreover, it does not specify which of the many supplements to workers' basic wages—including "exceptional" raises, "bonuses" for religious holidays and the start of the school year, "compensations" for meals, uniforms, and difficult or dangerous work, "incentives" for meeting monthly production targets, and profit shares—are covered. As shown here, attacks on these had repeatedly prompted wildcat public sector protests over the preceding ten years. Nor are pensions and insurance specifically mentioned, and the Egyptian government had already implicitly interpreted these out by enacting new health insurance legislation which increased workers' contributions for medication; the Prime Minister was empowered to order further increases by decree.[95]

The long-awaited provisions for legalizing strikes require that two-thirds of a *federation's* leadership endorse a walkout at individual plants. Thus the law denies decision-making power to the locals, where militants had been most successful at gaining influence. The regime and employers' associations also insisted on other provisions which would render legal strikes rare. They are still prohibited while contracts are in effect, during mediation and arbitration, and in vital services. Moreover, union leaders are required to give employers and the government 15 days written notice, as well as a justification, before declaring a strike.

These provisions are obviously intended to constrain labor protest and maintain the control of the NDP-affiliated senior unionists over the base. This was confirmed by the regime's decision, early in 1995, to finally approve the changes to the union law that al-'Amawi and Rashid had been advocating. These were passed as Law 12 of 1995, and published in the Official Gazette, March 30, 1995. This legislation would serve to entrench the then-current crop of senior union personnel by allowing them to remain in office past retirement age as well as by extending the term of union office from four to five years. Moreover, the latter provision was allowed to take immediate effect, thereby postponing for a full year the union elections that had been scheduled for the fall of 1995. The law further makes it likely that federation and confederation leaders will be divorced from the base by allowing them, but not lower-level officials, to continue to hold union office even after promotion to senior management positions in their workplace establishments.[96]

In numerous ways, the new union law tightens the control of the federations over the locals and makes it more difficult for new blood to enter the ranks of union leadership. Workers are now ineligible to run for union office until they have completed a full year of local membership, and a full term, now five years, of service on a local executive board is required before a worker can run for federation office. In addition, federation leaders are given enhanced power to expel individuals from union membership, while the limited provisions for their accountability to the membership are further eroded.

Ironically, but perhaps not coincidentally, only a month after Law 12 took effect the Egyptian High Constitutional Court issued a ruling on the restriction, initiated with the 1964 union law and maintained in its successors, that only 20% of the leadership of a union local could come from professional workers (see chapter 2). The court held that this provision was unconstitutional, and hence that all the union bodies elected on its basis—the entire union structure—was null and void. But rather than immediately holding new elections to conform to the court's ruling, as the regime had done on several occasions with the parliamentary poll, the government declared that the provision of Law 12 which postponed the upcoming union elections took precedence, thereby forestalling any electoral challenge to the entrenched NDP unionists who had helped to draft the new labor legislation, and presumably permitting it to be enacted before the rank and file would again go to the polls.[97]

The new union law thus signaled the regime's confidence in the loyalty of the current crop of NDP unionists. At the same time, the prolonged delays in issuing the new labor law, and the government's foot-dragging on privatization itself, would appear to indicate recognition by regime elites that the loyalty of the senior unionists alone is neither a guarantee of, nor a substitute for, the support of workers at the base. The regime tied its own credibility to the continuation of workers' perceived entitlements, so that the rollback of these could only erode its legitimacy. Yet the unfolding of events through the summer of 1996 suggested that the government would soon succumb to multilateral pressures to enact the law, while pushing forward with the privatization program.

The Outlook in 1996

Implementation of the new labor law, seen here as likely in the near future, is in fact apt to foster greater interest in unionism. This is because it increases the importance of unions to workers' livelihoods. Its elimination of fixed pay scales renders the starting wage of future public sector employees subject to negotiations between unions and management, and it explicitly relegates supplementary wages in both sectors to collective bargaining. The irony of this should not be lost: with their surrender in the battle for preservation of the parastatals, which was the ETUF's rallying cry for two decades, the unions stand to regain the raison d'être that was largely lost when the public sector was first created.

Yet it is not clear that the ETUF is organizationally suited for the challenges that lie ahead. The single, centralized, hierarchical labor confederation which helped workers fend off a reversion to neoliberal capitalism is not necessarily the best formation to help them protect their interests while living under it. The collective bargaining provisions in the new labor legislation will focus greater attention of the internal functioning of the confederation. While they seem to imply a greater role for locals, it is actually the federations whose influence these provisions enhance, since the ETUF's hierarchical structure denies locals the right to negotiate independently by requiring that federation officials must approve all contracts. The stipulations for legal strikes further augment the powers of the federation boards.

It follows that as the performance of locals and federations become more important to workers' paychecks, the hierarchical functioning of the ETUF will come under increasing challenge. The question is what route any changes in the ETUF's structure can, or should, take. Though the left has been able to effect some changes in union regulations, its long-standing efforts to reverse the flow of power within the confederation while otherwise maintaining its present structure have never succeeded and are unlikely to do so, because they rely on the acquiescence of the very senior unionists who benefit from the existing hierarchy. Leftists and other sincere unionists advocating structural reform would do well to question the theory that maintaining a single national confederation is always in the best interests of workers. Either separate, independent federations, or some form of competitive unionism, seems more likely to permit rank-and-file supervision of leadership and restrict the ability of the government to intervene in union affairs and vet labor militants.

In the meanwhile, there is little reason to believe that the prolonged incumbency of the present senior unionists, combined with the tightened control over the base the regime recently granted them, will succeed in eradicating wildcat protests. Virtually all of the rank-and-file actions of the previous decades either occurred spontaneously or were organized outside of the formal union structure. Such protests are apt to continue as workers are faced with deteriorating earnings and an unresponsive union leadership. The recent incidents at Kafr al-Dawwar and Mahalla confirm that continuing steps toward privatization will result in further outbreaks of labor discontent.

Moreover, it is more likely that such wildcat actions will now take the form of strikes rather than symbolic protests that do not jeopardize production. This is because the government itself has legitimated the idea of striking as a quid pro quo for its withdrawal from the economy. Leftist labor activists who were themselves devoted to the moral economy in the past are now more willing to call for strikes, and say that workers share these sentiments; the leftists' recent strike call at Mahalla serves as an example of this.

The government's response to these wildcat actions will remain repressive. Like his predecessors, and despite his initial steps toward political liberalization, Mubarak has shown little proclivity to tolerate public displays of discontent, and became increasingly authoritarian in the 1990s. But the retraction of the government's long-standing pledge to be the guarantor of workers' rights, and the transfer of Egyptian assets to multinational corporations, do more than undermine the basis for workers' commitment to the cause of national production. They also erode the previous developmentalist rationale for crushing worker protest. The continued repression of labor will therefore increasingly expose the regime as subservient to foreign capital, and concerned primarily with its own survival.

Conclusions

The primary purpose of *Labor and the State in Egypt* has been to examine a phenomenon—the labor movement in Egypt—and to explore how it both affected and was affected by the policies of successive authoritarian regimes in Egypt over the forty-four years since the Free Officers coup ushered the military into power. Toward this end I have consciously applied a blend of different theoretical approaches to the study of labor, drawing from institutional-, interest-, and identities-based perspectives. At the same time, I have also sought to use the Egyptian case to add to comparative theory, in the middle range.[1] In particular, this study has made broader generalizations about labor/state relations in other authoritarian developing countries. These concluding comments summarize, in turn, the findings of this study in these two areas.

Labor and the State in Egypt has also reflected my normative concerns with the plight of developing countries' workers in an increasingly globalized and capitalistic international economy. The chapter concludes with some final remarks on this issue.

Workers, Unions, and the State in Egypt

From the late 1950s onward three successive presidents—Gamal 'Abd al-Nasir, Anwar al-Sadat, and Husni Mubarak—have sought to manipulate and control the behavior of Egypt's workers while claiming to represent them and enjoy their unqualified support. The nature of the three regimes in regard to labor has been different, however. Nasir built an etatist economy enshrined in populist rhetoric, which encouraged workers to believe in a system of reciprocal rights and responsibilities between themselves and the government. Sadat and

Mubarak launched efforts to retract the state's commitments to labor established under this Nasirist moral economy, while endeavoring to convince workers that orthodox economic reforms are in their best interests.

The vehicle of labor control favored by all three presidents, after some initial hesitation on Nasir's part, was repressive controls on a hierarchically ordered trade union confederation. Yet this structure was not one imposed from above on reluctant unionists. Rather, significant segments of the labor movement, including the communist forces who played a key role in establishing many of the unions, advocated the singular, centralized structure as the best way for unions to advance workers' interests. At the same time, trade unionists opposed to confederation were able to interact with its opponents among regime elites to obstruct and delay corporatization.

Nor were the corporate structures of union/state relations, once established, static or uncontroversial. State corporatism generated conflicts within the union movement, along with continual pressures for an end to government interference in union leadership selection and revision of the hierarchical controls confederation leaders have over the lower bodies. While the basic structures of state corporatism remained intact, these pressures succeeded in winning important modifications at several critical junctures. Again, disagreements within the regime were reflected in the trade union movement, and vice versa, making institutional policies the outcomes of complex political maneuvering among these various forces.

The main victim of the government's intervention in union elections was the Egyptian left. But despite being effectively purged from the ranks of union leadership in the late 1950s and again twenty years later, Egypt's Marxist forces were able to recover both times, and to regain an ability to influence union affairs. Since the advent of *infitah*, an uneasy coalition of Marxists and Nasirists in the labor movement has been the main force behind criticisms of the confederation's operating laws. Nevertheless, a majority of these leftists remained committed to the idea of a singular confederation, and this is part of the explanation for its tenacity.

As a mechanism of labor control, state corporatism failed its regime proponents in two important aspects, but succeeded in one other. First, corporatist controls proved unable to prevent outbreaks of militancy at individual plants and industrial communities. Workers have frequently protested changes to their wages and working conditions outside of formal union channels, at times relying on alternative forms of organization. These struggles almost always resulted in significant management concessions to workers' demands which, when occurring in the public sector, impeded government efforts to narrow the fiscal deficit by squeezing parastatal workers. At the same time, the increasingly brutal repression the ruling regimes applied to quell these protests served to reinforce workers' beliefs in the contributions they make to society, thereby perpetuating their sense of entitlements and undermining the government's efforts to alter their perceptions of the state's obligations to labor.

Nevertheless, the failure of federation, confederation and sometimes even local leaders to lend their support to these localized struggles, and to back them with the resources of the unions, shows where state corporatism served its purpose to the regime. The local protests remained contained, isolated and often largely unpublicized, the stuff of local legend but not of national political discourse. The vision of a working class mobilized nationally to support workers at individual plants—the dream which drove the original leftist project of confederation—never materialized.

Corporatism also failed in guaranteeing the support of senior unionists for the regime's national economic policies. Such support was forthcoming during the 1960s, when Nasir initiated limited retrenchment efforts. But despite the co-optation of virtually all federation and confederation leaders into the ruling party under Sadat and Mubarak, and the perfunctory ETUF endorsements of the government's foreign and domestic political policies, the confederation's reaction to those presidents' economic liberalization policies ranged from reluctant and tepid support to open opposition. Labor resistance obstructed privatization and, through pressures for cost-of-living adjustments to wages, attenuated the benefits to the fiscal budget of lifting pricing reforms.

One could argue that corporatism has actually worked to the regime's advantage in these areas as well, given that the confederation failed to actually mobilize workers around these concerns. Egypt has never seen a general strike or even a union-sponsored national march or demonstration around economic issues. The closest phenomenon to this was the 1977 riots against subsidy reform, and these were spontaneous protests whose demands trade union leaders took up after the fact. Nevertheless, the existence of the ETUF can serve to remind regime elites of the prospect of national mobilization, and its specter has apparently been sufficient to wring concessions. Fear of visible labor opposition—actual, or only threatened—is indicated in the limitations of Nasir's retrenchment and the Sadat government's failure to effect price reform and privatization; the Mubarak regime has publicly attributed its gradual and tentative approach to these reforms to its concerns over labor discontent. In contrast, the liberalization measures which went forward most successfully—the opening to foreign trade and investment, and exchange rate unification—were those which engendered the least opposition from labor.

Thus, *Labor and the State in Egypt* has not attempted to refute per se the assessment of state-centered theorists, who have attributed Egypt's hesitancies over economic restructuring to a lack of elite will. Rather, I have argued that a critical factor underlying this lack of will was the successive leaders' fears of losing their presumed and proclaimed mass legitimacy by antagonizing labor on a national scale. The strength of the push for reforms has varied indirectly with the intensity of labor opposition to them. This makes understanding the dynamics of labor's reactions to economic policy initiatives—and hence, society-oriented scholarship—critical to assessing the reasons for their success of failure.

The dynamics of labor opposition to reform shown here began with ordinary workers, reacting with anger to violations of their perceived entitlements under the Nasirist moral economy. Careerist union leaders concerned with their incumbency and advancement made decisions about supporting or neglecting workers' concerns based on the anticipated effects these decisions would have on the unionists' patronage from regime elites. The logic for support was strongest where the concerns emanated from broad numbers at the base, such that neglect would have exposed the unionists to widespread charges of misrepresentation, eroding their legitimacy and thereby damaging their usefulness to the regime. Other unionists approached policy issues from an ideological perspective, with both Marxist and Nasirist philosophies coming into play.

This combination resulted in privatization, with its associated retraction of the benefits enjoyed by Egypt's public sector workers, becoming the major sticking point in Egypt's economic restructuring program. The ranks of the labor movement were more unified around preserving the public sector than any other issue, for a variety of reasons. First, privatization would adversely affect the interests of broad numbers of workers both inside and outside of the ETUF. Second, it threatens to weaken the unions institutionally, since organizing in the private sector is more difficult, and the unionized public sector workforce would be diminished by layoffs. Finally, defending the parastatals served as a rallying cry for Marxist and nationalist forces in the unions who oppose the prospect of Egypt's industrial resources being turned over to multinational corporations. The salience of labor's position was manifest in the Mubarak regime's five-year long reluctance to revise its labor laws which, in the face of the 1991 privatization legislation and the successful implementation of the various other components of IMF- and World Bank-designed stabilization and structural adjustment programs, remained a significant obstacle to shrinking the public sector, and one of the last vestiges of etatism.

Workers and Unions Under State Corporatism

Egypt has been treated here as part of a particular category of developing countries, those with authoritarian political systems that repress labor. As the current wave of political liberalization gathers steam, the number of countries which reside under this tent is diminishing. Nevertheless, Egypt certainly does not inhabit it alone; it is joined minimally by its Arab neighbors to the east and west, and by numerous countries to its south. Moreover, the fledgling democracies of Asia and Latin America are by no means immunized against reversions to authoritarian rule.

In this context, the broadest comparative lesson this study offers is simply to reaffirm that societal forces matter, even where they are precluded from partisan political participation. Authoritarian institutions designed for mass control do set the framework for popular behavior, but they do not and cannot determine it. Moreover, the design of these institutions themselves cannot be presumed to merely reflect elite will, but may be the product of multiple state/society interactions, as the evolution of Egypt's corporatist labor laws illustrates.

Among these excluded societal forces, workers are a particularly important group. Developmentalist regimes institute labor controls as a means to both promote and supervise the industrialization process. Yet in doing so, they bring labor into the political game and signal to workers that they have cards to play. State-centered scholarship necessarily overlooks the dialectical nature of labor/state relations in the political economy of these countries.

While rejecting the notion that state corporatism enables authoritarian ruling elites, through co-opted senior unionists, to regulate the behavior of ordinary workers, *Labor and the State in Egypt* has upheld the historical institutionalists' emphasis on the importance of established rules and procedures to political processes. I have focused on the ways in which the institutional structures of union/state relations—in particular the conglomerate of regulations which constitutes corporatism—shape the capacity of unions to respond to different policies, and hence frame the choices available to different labor actors. But in contrast to those who propose that corporatist systems uniformly either strengthen or weaken labor, I have stressed the importance of an issue-based approach. This study argues that, in an authoritarian context, a single, hierarchical confederation enhances the capacity of labor to respond to national issues at the expense of struggles at individual plants, and, secondarily, at the industrial level.

What constitutes a national, industrial, or local issue will vary from country to country, of course, depending on how economic policies affecting workers are set. Moreover, while the Egyptian union system closely resembles Schmitter's state corporatist ideal type, different organizational forms may operate elsewhere, with several confederations representing different industries or localities, or only industrially or regionally based federations in the absence of any confederative structures. While the same underlying logic would apply in these cases, the capacities of labor with regard to different types of issues will vary with these organizational differences. Thus specific claims regarding labor responses to the various economic issues studied here are made in the particular context of Egypt, and not presented as generalizable propositions. The broader argument is that in any given country, labor capacities are not uniquely determined by the structure of unions, but rather vary, on an issue by issue basis, with the interaction between union structures and those of economic policy making.

My exploration of local-level labor action demonstrates the propensity of workers to engage in spontaneous and informal protest in spite of the constraints on unions. The nature and frequency of these protests supports a moral economy interpretation of workers' attitudes, i.e., that workers believe themselves to be in a relationship of reciprocal rights and responsibilities with the state.

The moral economy approach is set against both traditional Marxist and some neo-Marxist perspectives. Organizing by leftists was shown to play a role in encouraging collective action, suggesting that Marxist workers succeeded in promoting some kind of class-based identity amongst their colleagues. But the struggles themselves were shown to be overwhelmingly restorative in nature, rather than representing some emerging new consciousness. No alternative to the etatism of the past, enshrouded in socialist rhetoric but nevertheless capitalistic in nature, emerged from either the Egyptian left or the workers' movement.

Stylized versions of rational choice are also contradicted here. Rather than the dispassionate calculation of costs and benefits proposed by rational choice theorists, I showed that an emotional response to policy changes—anger—lay behind these wildcat protests. This anger, shaped by collectively held norms of elite responsibilities toward workers, enabled them to overcome the fear of punishment that made the potential costs associated with these actions so high.

The argument that anger can precipitate protest presumes that workers can simultaneously maintain both individual and group identities. It is a shared feeling of oppression—based on a group identity as workers, or perhaps more broadly as subaltern people—that enables workers to move together from common anger to collective protest. Thus this claim minimizes the significance of the presumed "free rider" problem as a barrier to collective action.

Nevertheless, *Labor and the State in Egypt* does treat union leaders as individual, rational actors. Union work attracts those whose group identity as a worker develops so strongly that they are motivated to seek to lead labor struggles. It also appeals to individuals who are seeking a ladder up from the common oppression workers suffer. But whether it is their intended goal or not, union office pulls both categories of leaders out of the daily lives, experiences, and passions of workers, and imposes a different set of decisions on them. In making these decisions according to consistent preferences, whether motivated by ideology or careerism, union leaders are broadly rational. It is not possible, however, to predict what decisions unionists will make when faced with the dilemma of countervailing pressures from above and below. This is what makes the outcomes of union/state interactions over economic policy as well as legal/structural issues indeterminate, and necessitates that solid, ideographic work be done by those who are interested in further studying these phenomena.

Finally, the Egyptian case has implications for the ongoing debates about the sequencing of economic and political liberalization.[2] The Sadat and early Mubarak years are both illustrative of the dilemmas inherent in simultaneous reform programs. Sadat's early retraction from his limited political liberalization is often attributed to popular disaffection for his peace overtures with Israel. But it could just as arguably be traced to the January, 1977 riots after he agreed to IMF-imposed subsidy modifications. Mubarak in the 1980s promised to liberalize gradually in both spheres, but at the end of the decade neither democracy nor a significantly freer market economy was in evidence.

The 1990s saw Mubarak publicly turn his back on further political reforms, and rule in an increasingly authoritarian manner. Mubarak justified his retreat from political liberalization as necessary to prioratize economic restructuring. The United States implicitly endorsed this approach by pushing further economic reforms while remaining noticeably silent on the growing repression of Islamist forces and labor protests.

Defenders of this "delayed democratization" strategy point to the East Asian "miracle" countries as evidence that authoritarian rule is necessary to push through economic liberalization, while success in the latter realm will lay the basis for democracy down the line. As a single-country case study of Egypt, it is beyond the scope of this book to present a fully developed critique of this rationale. Yet there are numerous reasons to be skeptical that the East Asian model can be readily applied to the Egyptian case. I sketch these out here in the hope that future comparative research will explore them more thoroughly.

First, because the global economic marketplace is more competitive in the 1990s than it was when the "four tigers" began their transformations, greater sacrifices by the citizenry of reforming countries may now be required to achieve the same results.[3] In addition, the call for such sacrifices is more problematic politically in countries like Egypt, whose inward-looking growth period was more oriented toward popular welfare and linked to an ideology of labor entitlements. Finally, the gap in wages and working conditions between Egypt's public and private sectors appears to have been greater than that in the Asian miracle countries, and during the etatist period illegal practices in the Egyptian private sector were notorious. This would render Egyptian workers less prone to believe free-market rhetoric which paints private-sector entrepreneurs as heroes.[4] The combination of these factors makes popular resistance to structural adjustment more likely in the Egyptian case, and hence the level of repression necessary to implement reform higher. Such elevated levels of repression increase the risks of political instability, and reduce the likelihood that incumbent authoritarian elites will democratize.

Thus, I concur with Armijo et al. that the proper sequencing of economic and political reforms is historically contingent; there is no universal model.[5] Egypt's history makes it unlikely that any authoritarian regime there will be able to achieve or maintain legitimacy while implementing economic reforms that are widely seen as a Western imposition. Democratizing first, even if it would mean long delays in the march toward the global marketplace, is the wiser as well as more moral strategy.

Workers, Unions, and Economic Liberalization Under Authoritarian Rule

Western capitalist democracies long ago established social safety nets in the form of unemployment insurance and welfare payments. With increasing globalization of the economy, these protections have come under attack from conservative forces who see them as impediments to competitiveness. Progressive forces can legitimately debate the design of welfare programs and whether they have been the most effective means of achieving their humanitarian goals. But the principle of protecting society's vulnerable members from the market failures inherent in capitalism should never be sacrificed.

Etatist economies like Egypt developed different mechanisms of protection. Their social safety net was not cash grants, but rather took the form of employment in a burgeoning and inefficient public sector, and controls on the prices of basic necessities. Like some of the welfare modifications currently in vogue in the United States, the orthodox economic programs imposed on

Egypt and other developing countries by international creditors have resulted in slashing this safety net with no guarantees of protection for the millions of people they have kept afloat.

As explained in the Introduction, the current rationale for the acknowledged suffering of parastatal and other organized, urban workers under structural adjustment programs is the belief that the inward-looking development strategies and excessive government intervention in the economy these countries have experienced resulted in stagnation and widespread poverty, especially in rural areas and among some segments of the informal sector. Formal sector workers' "transitional pain" is held necessary to alleviate the poverty of the lowest income strata, in the short run, with the expectation that market-oriented reforms will eventually result in economic growth beneficial to all income groups.[6]

While there is no gainsaying the considerable macroeconomic distortions that Egyptian etatism caused, I remain skeptical that economic liberalization will bring Egypt's lower classes the salvation its advocates have promised. There is growing evidence that the United States' and Great Britain's embrace of neoclassical economics in the 1980s served to widen income gaps.[7] Studies of Latin America and Africa reveal that successful structural adjustment programs have in many cases resulted in a similar outcome.[8] Labor's opposition to economic reforms in Egypt has been treated here in a positive light because it foregrounds this issue, and encourages a continued search for viable but more humanitarian alternative economic models.

In the same vein, union organization has been approached here from the standpoint of whether it impedes or promotes the ability of workers to defend their interests. The answer has been contradictory because of the different nature of the issues workers have confronted. The model of a single, hierarchical confederation, under authoritarian government, unquestionably served to impede labor struggles around individual plant issues. At the same time it helped to stave off, over a twenty-three-year period, the elimination of the public sector and the accompanying retraction of the protections that parastatal workers enjoyed and private sector workers coveted.

In the fall of 1996, the prolonged conflict between workers and the government over labor market liberalization was at an impasse. Yet the most likely scenario appeared to be eventual passage of the new labor law, which would promote privatization by removing these protections for workers while maintaining the repressive controls on union behavior which had historically accompanied them. In the increasingly decentralized economy that would emerge, the detrimental effects of a hierarchical confederation on individual local-level struggles would become primary over its capacity for centralized representation. Under these circumstances, Egypt's leftist forces would do well to reconsider their long-standing support for this model of union organization.

The proposed new labor law embodies the Mubarak regime's vision for the future of labor/state relations. From the government's apparent reluctance to grant even the very narrow right to strike the law entails, it is evident that the informal, localized protests which have historically accompanied threats to workers' livelihoods will continue to be suppressed. By calling attention to the Mubarak regime's brutal suppression of labor protest, this book seeks to promote greater international awareness and condemnation of it, in the hope that this will contribute to its eradication.

Notes

Introduction

1. Waterbury (1983), pp. 17–20.

2. The critique of pluralist studies is developed in the introductory and concluding essays to Evans, Rueschemeyer, and Skocpol (1985), the seminal work arguing for state-centered scholarship. Among recent studies of Egypt, I would include in the state-centered category Waterbury (1983), Springborg (1989), and Hinnebusch (1985). Earlier scholarship tended to focus on ruling elites rather than state structures per se. However, many of the younger generation of Middle East political scientists have focused their research on societal groups in defiance of this disciplinary trend. On Egypt, this study joins those by Singerman (informal sector), Brown (peasants), Vitalis (1995, 1996: businessmen), and Wickham (professional associations). See also Bellin and Alexander on North Africa.

3. Based on an interview with Niyazi 'Abd al-'Aziz, president of the Engineering, Electrical, and Metal Workers' Federation, Cairo, January 1, 1995.

4. The literature is vast. The most comprehensive recent studies are those sponsored by the Overseas Development Council (ODC): Nelson et al (1989); Nelson, ed. (1990); and Haggard and Kaufman (1992). These studies conclude that labor has limited ability to impede reform, except when it is joined by broader societal opposition. Related works by the authors involved in the ODC studies include: Haggard and Kaufman (1989a), pp. 209–43; Haggard (1986, 1990); Callaghy and Wilson (1988); and Nelson (1988, 1991).

5. See esp. the introductory chapter in Nelson (1990), pp. 3–32.

6. Nelson herself (1991) attempts to account for cross-national variations in labor militancy, defined by wage demands, among developing countries. I take issue with some of her conclusions later.

7. See Waterbury (1993), esp. p. 235; Haggard and Kaufman (1992), pp. 28–30, and (1994), p. 10; Nelson (1992), esp. 229–33; and the World Bank (1995a), esp. pp.

103–08, and (1995b), esp. pp. 15–27. Richards (1991, 1992) is one of the strongest academic proponents of orthodox reforms for Egypt.

8. See, e.g., Wade (pp. 372–75); an excellent overview of the debate on the sequencing of political vs. economic reforms can be found in Armijo et al. (1994).

9. For a very well-documented and cogent critique of the neo-liberal reform argument, see Chaudhry (1993).

10. This is tantamount to the outcome of "embedded liberalism" favored by Thomas M. Callaghy (1989), p. 129.

11. This analysis is associated most closely with the article by Frobel, Heinrichs, and Kreye. See also the review of NIDL theory in Southall (1988b), pp. 1–17.

12. See esp. the works of Guillermo O'Donnell. Evans' work linking multinational penetration of Brazil to the political repression and exclusion of labor bears close resemblance to O'Donnell's theory. Related arguments can be found in the works by Hobart Spalding, Robert Kaufman, Cardoso and Faletto (1979), and Almeida and Lowy. See Schamis for an important review and theoretical revision to this literature.

13. The seminal work is Schmitter (1974).

14. Mericle (1977), p. 304.

15. Youssef Cohen (1982), p. 46.

16. Galenson (1962), p. 9.

17. Waterbury (1993), pp. 235–42. The labor movement is worth only a few scattered pages in his 1983 study.

18. The term "pessimistic" is taken from the discussion in Southall (1988b), pp. 28–30, who in turn appropriated it from an earlier work by Richard Hyman.

19. For evidence from other countries, see *inter alia*, the articles in Southall (1988b), and the works by Bayat (1987), DiTella (1981), Kraus (19796, 1977, 1979), Roxborough (1981, 1984), and Waterman (1975).

20. Haggard and Kaufman (1989b).

21. Roxborough (1981), esp. pp. 88–89.

22. Youssef Cohen (1982).

23. A useful comparative study of labor resurgences is Valenzuala (1989). For more on labor in southeast Asia, see the works by Deyo.

24. For a recent application to Egypt, see El-Shafei.

25. See esp. the works of Frederick Deyo.

26. Haggard and Kaufman (1989a), pp. 225–26. To the authors, this constitutes an "intermediate" level of political strength for labor. The weaker situations are characterized by labor penetration and repression, such as in BA regimes, and here labor is portrayed, as above, as unable to effectively challenge stabilization programs. Conversely, the stronger situations are equated with societal corporatism in Europe. In these cases, peak associations have cooperated with wage moderation programs in exchange for programs which enhance job security and provide social safety nets. On this point, see below.

27. al-Sayyid (1983), Qandil (1985), Bianchi (1989). See also the article by 'Abd al-Mun'im.

28. This argument is simply an application of the long-standing wisdom of the power analytic literature, which has established that configurations of power—in

this case, the power of labor vis-à-vis the state—differ according to issue types. See works listed here by Baldwin, Lowi, and Frey.

29. See esp. the works by Pizzorno and Panitch, and the literature reviews in Golden (1993), p. 440, and (1992), pp. 312–15.

30. See the discussions in Lange (1984), Przeworski (1985), esp. chs. 4 and 5, and Golden (1993).

31. For a postmodernist approach to defining Egyptian workers, see Lockman.

32. The gap between moral economy and rational choice appears to be narrowing, with the recent acceptance by some rational choice theorists of the salience of norms. See the book edited by Cook and Levi. For a review of, and contribution to, the literature on norms vs. rationality in the international politics field, see Klotz. For attempts to reconcile Marxism and rational choice, see Przeworski (1985), and Elster (1985).

33. The remainder of this section restates, with some modifications, my argument in Posusney (1993).

34. With regard to agrarian societies and peasant rebellions, moral economy arguments propose that peasants are primarily concerned with ensuring a subsistence income; they will avoid risks to, and particularly resent elite actions which threaten, this subsistence. See Scott (1976), and the discussion in Popkin. For an application of peasant moral economy theories to Egypt, see Brown.

35. E. P. Thompson (1971), Charles Sabel (1982), pp. 128–36; Swenson (1989), pp. 11–108; Kraus (1979), pp. 272–73.

36. Assaad and Commander (1990), p. 11.

37. Assaad (1995b), table 3 (n.p.).

38. See, e.g., Waterbury (1993), p. 126.

39. See the discussion in Hansen (1991), p. 116.

40. This discussion draws on, but ultimately departs from, the juxtaposition of the moral economy vs. rational choice debate in Swenson (1989), pp. 13–15.

41. See, *inter alia*, the works by Elster (1983), esp. pp. 1–42; and Hindess (1984), Riker (1990), and Simon (1985).

42. For the debate about selfishness within the rationality school, see, e.g., the works by Sen and Elster (1983), esp. p. 11. A growing number of economists are challenging the selfishness assumption, incorporating the idea that the behavior of both workers and managers can be shaped by beliefs about fairness. See, e.g., North (1981), esp. chs. 1 and 5; Frank (1988); and Kahneman, Knetsch, and Thaler (1986).

43. The seminal work is Olson (1971).

44. For a review of empirical studies which support this argument, see Weintraub; Bruce E. Kaufman; and Shalev. See also Kennan for a more recent overview of this literature.

45. Nelson (1991), pp. 43–48.

46. Goldberg (1992), pp. 152–58.

47. Goldberg also contradicts his own logic here, since he posits that workers with relatively secure jobs did engage in collective action during the pre-coup period. Also, the cost and benefit structures associated with joining unions and participating in collective protest, either union-sponsored or informal, are different.

Goldberg's article mistakenly groups these together in one category of collective action to which the same logic supposedly applies.

48. Their considerable evidence in support of this theory comes from studies conducted in advanced industrial economies. That Egyptian workers' behavior conforms to it suggests that loss aversion is not only a product of western cultures.

49. Kahneman makes this argument in his article with Thaler and Knetsch.

50. Elster (1990).

51. See Geoffrey Brennan's response to Elster in Cook and Levi (pp. 51–59, esp. p. 53), and the introductory essay to that volume. Elster's own argument that norm-driven behavior is also irrational because it is not "outcome oriented" (p. 45) does not make sense to me. Clearly, people who are motivated by norms are placing a priority on earning or maintaining the respect of their peers, which seemingly affects their own self-esteem; their behavior is outcome driven in this respect. Moreover, Elster himself has elsewhere argued, using game theory, that individually rational workers can develop feelings of solidarity—a norm.

52. Possibly unique to Egypt, paycheck boycotts are explained in chapter 3.

53. Scott (1976, p. 4; 1985, pp. 242–47, 320–26) suggests much the same for peasants in his discussion of agrarian rebellions.

54. For an overview of culturally-oriented neo-Marxism as well as the structuralist approaches discussed below, see Marshall.

55. See, e.g., Mitchell (1990).

56. Goldberg (1986).

57. On the importance of solidarity to the Marxist notion of class consciousness, see Booth (1978), esp. pp. 168–69, and Elster (1985), p. 347.

58. This conception of voice is therefore more narrow than that used by Freeman and Medhoff. Discussing the role of unions in western democracies, they include participation on the electoral arena as an exercise of voice. Here I assume, as is true in the Egyptian case, that elections and party politics do not provide workers with a genuine channel for interest articulation. On the limits to democracy in Egypt, see, inter alia, the articles by Post and by Lesch.

59. Because deliberations of this kind often occur in secret in the third world context, voice is difficult to document decisively; it may appear to outsiders that regime elites never in fact attempted the change. On the other hand, government leaders resisting external pressures for reform may cite opposition from societal forces, even when no discussion has in fact taken place.

60. Outside of the state-centered literature on the third world cited above, most of the work on "the new institutionalism" is focused on advanced industrial economies. A seminal work is Hall (1986). For an excellent compendium, see the edited volume by Steinmo, Thelen, and Longstreth.

61. This call for blending institutional with interpretive analyses takes inspiration from the papers by Locke and Thelen (1993, 1994) and Hall and Taylor (1994).

62. In the Egyptian case, a better job would generally be considered a white collar government position; these are available only to college graduates. Ironically, such positions generally pay less than blue-collar employment, but carry more prestige, besides being far less demanding physically.

63. Stepan, ch. 1.

64. See the works by Ogle and by Choi.
65. Skidmore (1988).
66. I discussed these issues with a group of Egyptian labor activists at a seminar sponsored by the al-Jeel Center for Youth and Sociological Studies, Ain el-Sera, December 28, 1994. I am grateful to Dr. Ahmed Abdalla, the Center director, for arranging this seminar. See *Al-Badil,* August 1995, pp. 8–15, for a summary of my presentation and the ensuing discussion.
67. Locke and Thelen (1993, 1995).
68. On the concept of critical junctures, see also the Colliers (1991), pp. 27–39).
69. This discussion is drawn from their 1979 article, which is incorporated without modification, into the 1991 book.
70. The importance of disaggregating the labor movement with regards to structural issues is also stressed by Roxborough (1981), esp. p. 86, and Valenzuala (1989), p. 446. The latter argues that these internal differences become especially salient in periods of democratic transition.
71. Choi (1989), pp. 146–47, makes a similar argument, linking it to Michel's "iron law of oligarchy."
72. Bianchi (1989).
73. In a different theoretical framework, Kevin Middlebrook's study of Mexican labor also argues for blending state- and society-centered approaches.
74. Thompson, p. 79.
75. Przeworski (1986), pp. 50–53.
76. One important study of the legitimacy problems confronting Arab rulers in the 1950s through 1970s is that by Hudson.
77. This point was suggested to me by Roger Owen.
78. The term "takeaways," coined by labor activists in the United States, refers to management-imposed reductions in workers' wages or benefits. It is related to, but analytically distinct from, "givebacks," which are similar reductions ceded by unions in negotiations.
79. The *Yearbook* is published in Geneva by the International Labour Office.
80. See also the chapter by Barbara Ibrahim in Tessler et al.
81. El-Issawy, pp. 2–3.
82. ETUF (1982), p. 25.
83. Interview with Ahmed Disuqi, March 1988.

1. Corporatism and Etatism Take Shape, 1952–1964

1. See the works by O'Donnell, and the critical assessments of it in the Collier volume.
2. Cf. Vitalis, pp. 196–98.
3. For more detail on the Egyptian communist movement, see note 19 below.
4. Beinin and Lockman, p. 419; Gordon, pp. 39–59.
5. Hansen and Nashashibi, pp. 38–41; Mead, pp. 49–50; O'Brien, pp. 68–72, 81–82. Cf. Vitalis (1995), esp. ch. 6.
6. Beinin and Lockman, pp. 421–42.

7. Ibid; 'Izz al-Din, pp. 801–5. The death sentence was controversial enough to warrant a confirmation vote by the RCC itself. 'Abd al-Nasir joined Khalid Muhyi al-Din and Yusuf Sadiq in opposing the decision, but they were in the minority, and the two workers were publicly hung on September 7.

8. Vatikiotis (196)1, pp. 79–80.

9. Quoted in 'Izz al-Din, p. 807.

10. Beinin and Lockman, p. 429.

11. These are documented in Beinin and Lockman, *passim.*

12. Beinin and Lockman, pp. 409–14; 'Amir, pp. 95–96; Kamil, pp. 116–17.

13. 'Amir, p. 96; Kamil, pp. 119–21; 'Izz al-Din, p. 810.

14. Beinin (1989), p. 81; Beinin and Lockman, p. 428; 'Amir, pp. 98–99.

15. Amir, pp. 99–100; 'Izz al-Din, p. 810; A. M. Sa'id (1985), p. 15.

16. 'Izz al-Din, p. 811; italics mine.

17. Beinin and Lockman, pp. 428–29; 'Amir, p. 100.

18. Beinin and Lockman, p. 435. Note, too, that the leftist presence in the FCGF had already been weakened by arrests of communists which occurred prior to the coup.

19. As detailed by Beinin (1992), pp. 74–84, Egyptian Jews played a prominent role in the revival and dissemination of Marxism in Egypt in the 1930s and 1940s. The emergent communist movement itself, however, was highly factionalized. The proper role for Jews in the Marxist movement was one point of disagreement. Although the Jews were native to Egypt, during the European colonial period many Jews acquired foreign citizenship when the colonial powers claimed the necessity of protecting religious minorities as a means to extend European influence in the region. In addition, Jewish children from wealthy families were sent to European schools where they learned to speak foreign languages rather than Arabic. These practices contributed to the popular perception of the Jews as "foreigners."

The most renowned Egyptian-Jewish communist activist was Henry Curiel, who renounced his Italian citizenship in 1935 and in 1943 founded the Egyptian Movement for National Liberation, known by its Arabic acronym HAMITU. In 1942, Hillel Schwartz, a Francophone Jew, founded the Iskra (Spark) organization, which had a high proportion of middle- and upper-class intellectuals, Jews, and foreigners among its members. Then, early in 1947, Iskra absorbed People's Liberation (Tahrir al-Sha'b) led by Marcel Israel, a Jew with Italian citizenship, and later that same year merged with HAMITU. The new group, which now formed the largest Egyptian communist organization, was HADITU. Although Curiel and two other Jews served on HADITU's central committee, Marcel Israel believed that foreign-educated Jews like himself were indeed foreigners and therefore ineligible to lead the communist movement.

A second group, known as Popular Democracy (al-Dimuqratiyya al-Sha'biyya), was founded and led by three other Jews, Yusuf Darwish, Ahmad Sadiq Sa'd, and Raymond Douek. They converted to Islam and actually established a requirement that members of this group master Arabic and prove their identification with Egyptian culture. Nevertheless, most intellectuals, even amongst the Marxists, continued to consider them as Jews.

Finally, the third prominent communist group, known as the Communist Party of Egypt (CPE: al-Hizb al-Shuyu'i al-Misri) refused to admit Jews to its ranks, accusing them of promoting sexual and moral dissolution. During the 1940s the CPE was the smallest of the three tendencies.

The 1948 Palestine War precipitated the arrest of many communists and a split in HADITU, which had followed the Soviet Union in endorsing the United Nation's plan to partition Palestine. This position rendered the communists in general, and the Jewish Marxists in particular, vulnerable to accusations of being led by foreign powers and complicit with Zionism. When Egypt invaded Israel in May, 1948, many communists were arrested and some, including Curiel and Schwartz, were subsequently deported.

When the 1952 coup happened, HADITU, under Curiel's influence from his exile in France, was the only communist organization to support the Free Officers. The other factions, again following the Soviet Union's position, considered the new regime a military dictatorship with right-wing leanings.

For more on the early history of the Egyptian Marxists see, in English, the works by Botman, Beinin, and Lockman, and Beinin (1987 and 1992). In Arabic see Rif'at Sa'id.

20. The Labor Department was originally established in 1930 as a Labor Office, a division of the Ministry of the Interior. It was responsible for trying to resolve collective labor disputes, and oversaw Conciliation Boards that were initiated in 1919 by the British colonial government. The name was changed in 1936, reflecting an elevation in status of the division, and the conciliation apparatus was refurbished at that time. The system was reformed again by the Wafdist government in 1942, and the conciliation boards were renamed Arbitration and Conciliation Comittees at that time. Their powers were expanded in 1948. The Ministry of Social Affairs was established in 1942, and the Labor Department, including this arbitration apparatus, was transferred to its auspices sometime in the ensuing decade. Beinin and Lockman, pp. 116–17, 194; Beinin, personal correspondence, March 1996.

21. Kamil, pp. 121–22; 'Izz al-Din, pp. 813–14.

22. 'Izz al-Din, pp. 814–15; Kamil, p. 121.

23. See, e.g., Sa'id (1968), pp. 64–66; and Tomiche, p. 42.

24. Beinin and Lockman, pp. 432; Bianchi, esp. pp. 28, 138; Sa'id (1968), pp. 64–68. Beinin implicitly repudiates much of his earlier assessment in his 1989 article.

25. Sa'id (1968), p. 66; Beinin and Lockman, pp. 432–33; 'Izz al-Din, p. 820.

26. Sa'id (1968), pp. 65–66; *al-'Amal* (hereafter *AM*) March 1968, p. 25; Beinin and Lockman, p. 432.

27. Sayyid Fa'id, interview, July 1988.

28. 'Izz al-Din, p. 815.

29. Beinin (1989), p. 74; 'Izz al-Din, pp. 816–20; Audsley, p. 103. Beinin sees the modification to Law 317 as a ban on *all* arbitrary firing, including mass layoffs, and this analysis is picked up by Goldberg. I believe the interpretation I present here is more accurate. It is corroborated by Harbison (p. 182), and repeated in the 1959 labor law (discussed later), suggesting that there were no further modifications after 1953.

30. Mabro and Radwan, pp. 43–44, 85–87; Hansen and Mashashabi, pp. 39–41; Mead, pp. 46–49, 103. The economic upturn beginning in mid-54, which I report later in the text, is also based on these sources, who attribute it to the end of the cyclical downturn caused by the international economic environment.

31. For evidence, see Audsley, p. 103; Harbison, pp. 182–83; and Vitalis (1995), pp. 205–6.

32. 'Izz al-Din, pp. 816–17; Beinin (1989), pp. 76–78. Noting that workers' complaints could be heard either by special labor tribunals or by the arbitration and conciliation boards, Beinin implies that it was the workers themselves who chose between the two. However, it is my understanding that where and how a case was heard depended on the number of workers and the size of the establishment involved, rather than workers' discretion. This was definitely the case after the 1959 labor law, and my sense is that it was true in the 1953–8 period as well.

The 1959 law set the procedure for handling grievances as follows: All complaints were to be filed first with the Labor Offices, the local branch of the Labor Department. From there, however, *individual* grievances were transferred immediately to the courts; the Labor Offices continued to process only collective complaints. If my assumption is correct then Beinin's data, which shows many more cases heard by the labor tribunals than the arbitration boards, indicates that individual grievances greatly outnumbered collective complaints.

The labor tribunals predated the military coup, but their precise origins are not specified in Beinin's work.

33. The discrepancy arises in part because the cost-of-living deflator provided by the Egyptian authorities, which the economists must use to derive real wages from the nominal figures, was outdated and downwardly biased (Mabro, p. 334). Moreover, Mead (pp. 312–13) points to changes in 1953 in the statistical techniques CAPMAS used to determine the nominal wages figures, and argues that this resulted in an upward bias in the 1953 figures.

Ignoring these issues, and carefully omitting the 1953 drop in the index, Goldberg (1992) draws on Abdel Fadil's data to support his contention that workers' livelihoods were improving during this period. Neither author questions the apparent contradiction between this data and the declining employment that Fadil also reports.

34. 'Amir, p. 91; Fa'id interview; FO 371/102931/JE2183/19. The Labour attaché also reported frequent instances of labor unrest, although he provides specifics on only a few. It should be noted that, according to Beinin and Lockman, pp. 415–16, in the 1950s the British' ties with trade union leaders suffered, and they relied increasingly on the American embassy to provide or confirm information about labor affairs.

35. Vitalis (1995), p. 12, italics mine; see also pp. 195–202, 215.

36. Beinin (1989), p. 74.

37. See also Goldberg (1994). As he notes there (p. 1), he, I, and Bob Vitalis debated these issues extensively over the summer of 1994. These e-mail "conversations," along with other exchanges I had with Joey Beinin, contributed to the formulation of my argument here.

38. 'Izz al-Din, p. 814.

39. For a detailed account of this period, see Gordon (esp. pp. 58–126).

40. Beinin and Lockman, p. 435; 'Izz al-Din, p. 814, 821; Kamil, pp. 52–53. Tu'ayma's rise to predominance in labor affairs coincided with the RCC's growing disenchantment with Amin, whose wealth and ostentatious lifestyle were aggravating growing disenchantment with the junta among some second-ranking officers, especially in the artillery corps. Amin was sent abroad on a diplomatic assignment in October 1953. See Gordon, pp. 110–17.

41. Beinin and Lockman, p. 433; Sayyid Fa'id interview, June 1988; 'Izz al-Din, p. 860.

42. 'Izz al-Din, pp. 821–23; A. M. Sa'id (1985), p. 15.

43. Ibid.; 'Amir, pp. 102–4. There are some discrepancies in the historical accounts, including Sa'id's own fragmented memoirs, on various details about this group, such as the actual date of its formation and the origin of its name. This paragraph and subsequent discussion about the group represent my effort to reconcile the different versions of events.

44. For more details, see Gordon, pp. 125–57.

45. See, e.g., Harbison, p. 181, and Bianchi, p. 78.

46. My account of the March '54 events is drawn from 'Izz al-Din, pp. 843–56; Kamil, pp. 130–33; and Beinin and Lockman, pp. 440–43.

47. Fathi Kamil, who was chosen to chair the meeting, describes it in his memoirs as an informal gathering, but 'Izz al-Din reports that it was called by the Permanent Congress.

48. Beinin and Lockman, pp. 440–43, 'Izz al-Din, pp. 853–56; Beattie, p. 98.

49. O'Brien, p. 75, claims that a doubling of the minimum wage occurred sometime in the first years of the regime. However, I could find no evidence to corroborate this and remain skeptical that it occurred. O'Brien himself dates the identical increase to 1962 later in his book, p. 206, as do Abdel Fadil, p. 28, and others.

50. Sa'id 1983, p. 24.

51. Fa'id interview; 'Amir, pp. 109–10.

52. 'Izz al-Din, pp. 859–60; Kamil, pp. 134–35.

53. Fa'id interview; 'Amir, pp. 106–9.

54. The British Labour Attaché approved of the RCC's judgment, and of their general efforts to weed out communism from the labor movement. See FO 371/102931/JE2183/19. I am grateful to Joel Gordon for supplying me with the text of this memo.

55. Beinin and Lockman, p. 437; Fa'id interview; al-Sha'rawi interview, March 1988.

56. Gordon, p. 189.

57. O'Brien, pp. 68–72, 81–103; Vitalis, pp. 202–14. The reasons for Nasir's preliminary turn toward etatism have been debated elsewhere and are not central to this presentation. O'Brien, pp. 311–20, has a lengthy discussion. For a broader argument on the factors propelling postcolonial states to adopt etatist policies, see Chaudhry.

58. Audsley, pp. 105–6; O'Brien, pp. 75–76. Audsley claims that the bill was later applied to all workers, but I am skeptical; the minimum of 50 workers has been consistently applied in subsequent legislation concerning both union structure and various benefits for workers.

59. Kamel, p. 46.

60. The text of the law in English is available in the International Labour Organization (ILO) *Legislative Series*, 1959.

61. Amir, pp. 79–80, 162–64; Mabro and Radwan, pp. 134–36.

62. The deficiencies in the data, along with contradictory sources, make it difficult for me to be more precise and decisive about wage trends. Hansen, p. 183, provides a graph which shows real wages climbing only in the first three-four years after the coup, and flattening thereafter until about 1961–62; he does not specify which CAPMAS statistics were used to plot this graph.

63. 'Amir, pp. 85–93.

64. See the summary of, and debate over, O'Donnell's arguments in Collier (1979). For a more recent and comprehensive study of the incorporation process in eight Latin American countries, see the Colliers (1991).

65. Kamil, pp. 139–41. Kamil reports that it was he, at this meeting, who had the inspiration to call the group "Permanent Congress," and 'Amir notes that after its creation the CEW changed its name (to the Federation of Free Workers) to avoid confusion. Both these accounts suggest that the organization which existed previously, referred to as the "Permanent Congress" by 'Izz al-Din and Sa'id, was not commonly known by that name at the time. It is clear, though, that there was a continuity between the two groups in the eyes of the Labor Bureau, which initiated and supervised both bodies.

66. Among those he added were al-Sawi, 'Uqayli, al-Sha'rawi, and 'Abd al-'Aziz Mustafa, a transit local leader who had opposed the March 1954 general strike.

67. Kamil, pp. 141–42, 153; 'Izz al-Din, p. 823.

68. Kamil, pp. 145–46, 153–57.

69. The name was officially changed in the mid-1950s to the Ministry of Social Affairs and Labor. However, for simplicity's sake, I continue to use the MSA abbreviation for the remainder of this chapter.

70. Kamil, pp. 146–47; 'Amir, pp. 110, 134–35; 'Izz al-Din, pp. 889–91.

71. 'Izz al-Din, pp. 891–93.

72. As a consequence of the cold war, there were two rival confederations on the international scene, each with international industrial federations as affiliates. Because Israel's Histadrut belonged to the Western-backed organizations, most Arab federations joined the rival Soviet-sponsored groups, or remained unattached. Salama's federation's affiliation was probably motivated by the fact that Egypt's oil concerns were Western-owned, hence belonging to the Western-backed federation could put them in touch with European employees of the same multinational oil companies. Nevertheless, the affiliation allowed rival union leaders to use Arab nationalism as a basis to attack Salama. This issue of international union affiliations becomes important again in the 1970s; see chapter 2.

73. 'Izz al-Din, pp. 391–93; *al-'Ummal* (hereafter *U*) April 8, 1971, p. 6. The oil sector was primarily foreign-owned and, according to Harbison (p. 174), large foreign-owned enterprises such as petroleum had the most sophisticated collective relations in this period; expatriate managers who did not speak Arabic "had much to gain by promoting a strong, responsible union leadership with whom it could communicate."

74. Hakim interview, February 1988; Salama interview, June 1988.

75. 'Izz al-Din, p. 893; Kamil, pp. 147–49.

76. The Western corollary to the organization represented by the Arabic words "*ittihad 'amm*" is a confederation. I adopt the word "federation," in the title here in deference to the Egyptian organization, which officially uses that translation. However, I will continue to refer to the organization in text as a confederation. The Arabic *niqabah 'ammah* is best translated as "general union," but the organizations it refers to are actually the occupationally-based federations which comprise the confederation. Since these do not have common English names, I refer to them in both translation and text as federations.

77. 'Amir, pp. 136–37, 'Izz al-Din, p. 895.

78. Bianchi (1989), pp. 134–38.

79. 'Amir, p. 137.

80. ETUF (1977), p. 84.

81. ETUF (1977, 1982).

82. For an argument to this effect see, e.g., Trimberger, pp. 159–62.

83. See, e.g., Sa'id (1983), pp. 13–14.

84. On the nature and functioning the National Union, see Vatikiotis, pp. 100–12.

85. 'Izz al-Din, pp. 897–903; Kamil, p. 193; Turk, pp. 119–20; Fa'id interview.

86. 'Amir, pp. 145–46; 'Izz al-Din, p. 904.

87. 'Amir, pp. 140–41, 148.

88. 'Izz al-Din, p. 898.

89. 'Amir, p. 141; 'Izz al-Din, p. 896.

90. 'Amir, p. 139, claims that Kamil and his associates then appealed to the Labor Bureau to remove Salama as president. Given Kamil's history and general independence from the regime, I consider this unlikely on his part. 'Uqayli may well have done so; however, it is also possible that 'Amir was confusing this period with the months just before the ETUF's founding convention.

91. 'Amir, pp. 139–41; Kamil, pp. 178–83. Kamil declines to name the individuals directly responsible for this turn of events, but my understanding is that the move to oust him was actually initiated by 'Abd al-Mughni Sa'id and Amin 'Izz al-Din. The latter (interview, July 1988) was in Iraq during 1958, and told me he informed the regime of his concerns over Kamil's behavior there. He also acknowledges having advised Kamil at the time not to contest the government's decision to remove him.

92. Yusuf Darwish told me about a strike at the Shabaha plant in Alexandria in 1959 which prompted Nasir to go to speak to the workers; they complained about the arrest of the leftists from their plant. Sayyid Fa'id was one of the very few communists who escaped imprisonment during this period. For more on the Egyptian communist movement under Nasir, see Beinin (1987).

93. 'Amir, pp. 146–47; 'Izz al-Din, pp. 905–7.

94. Sa'id himself says (1983, p. 23), that his meeting with Fawzi was to agree *not* to intervene in the elections, but 'Amir claims that elsewhere Sa'id acknowledged making the choices for the top two posts. The new board included Kamil 'Uqayli and 'Abd al-'Aziz Mustafa, both close to Fathi Kamil. Among the new faces were

Sa'd Muhammad Ahmad and Salah Gharib, both of whom would later become presidents of the ETUF.

95. 'Izz al-Din, p. 907; 'Amir, pp. 147–48.

96. 'Abd al-Mun'im, pp. 249–50; Beinin and Lockman, p. 433; Fa'id interview; Hasan, pp. 118–9.

97. Clarke, pp. 55–56.

98. There was, however, a general resolution passed by a congress of the Textile Workers' Federation in January 1958, which called on the government to guarantee freedom of trade union activity. 'Abd al-Mun'im, pp. 284–5; Beinin (1989), p. 81.

99. 'Amir, pp. 123–25; Beinin (1989), p. 81. It is noteworthy in this regard that Sayyid Fa'id dated the repeal of the six-month extension of probation to the end of 1956, rather than 1958. While this may simply reflect a memory lapse, it suggests the possibility that the activism of the Textile Federation succeeded in winning reform in that industry before the national legislation.

100. Abu 'Alam, p. 45; 'Amir, pp. 76, 90, 114–18; 'Izz al-Din, p. 908–12.

101. 'Amir, pp. 114–15.

102. This is based on the data collected by Beinin (1989), p. 77. Grievance data is not available after 1958.

103. Goldberg (1992).

104. If the decline were registered in 1959 only, it could be traced to the removal of the closed shop provision. Since it preceded the change in the law, I discount this factor.

105. The details of the nationalizations are in Waterbury (1983), pp. 68–73.

106. The news laws and the rationale behind them are discussed in a wide variety of sources. I relied on Mabro and Radwan, pp. 135–37; Sa'id (1968), pp. 79–92; Khalid (1971), pp. 44–54; and Shaaban, pp. 72–75.

107. Law 262 excluded workers in private commercial establishments, nonprofit services, and small factories. Also, the minimum wage was set lower for workers under 18 years of age. Note that this is the increase that O'Brien at first wrongly attributes to the early 1950s, as discussed above.

108. See also Waterbury (1983), pp. 73–79 and Baker (1978), pp. 174–78.

109. Based on Mabro and Radwan, p. 42, for GDP figures, which are base year 1959/60, p. 87 for industrial output, base year 1952/53. Average growth rates computed by me.

110. Employment figures from Abdel Fadil, p. 8, drawn from CAPMAS's Census of Industrial Production. Average growth rate computed by me. Mabro and Radwan (p. 103) cite slightly different employment figures drawn from the same source; there is no apparent reason for the discrepancy.

111. Waterbury (1983), pp. 71–77, 307–22; Baker, pp. 184–86, 190–92; Khalid (1975), p. 43; 'Izz al-Din, pp. 1953–55.

112. Clarke, p. 38, and interview with Nabil al-Hilali, a labor lawyer, January 6, 1995.

113. Under the terms of decree 3536 of 1962, workers who felt they were unfairly dismissed would take their complaints to the Labor Office, which would attempt

to resolve the case amicably. If that failed, the case would be submitted to the labor tribunals. Here, if the court ruled in favor of the worker, the company was bound to reinstate him or pay compensation, unless it was found that the firing was related to trade union activity, in which case reinstatement was mandatory. Most company managers chose to pay compensation rather than reinstate the workers in these cases. Enforcement of the court ruling was the province of the Ministry of Labor. These provisions applied to both private sector workers and those in the public sector who had exhausted the tripartite committee appeals process.

In 1964 an additional legal channel for workers to appeal dismissals was created with the establishment of tripartite committees, consisting of representatives from management, unions, and the ASU, to review dismissals. Their decisions were purely advisory, however, and were often rejected by management. In these instances, or in plants where no committee of three existed, the worker could then proceed to take the case to the Labor Office and it would follow the rules explained above. *AM* 55 (December 1967), pp. 10–12; see Baker, pp. 79–80, for a discussion of managers' views on these restrictions on firing.

114. The private sector was responsible for 80% or more of value-added in such industries as leather, furniture, wood, wearing apparel, and printing. Mabro and Radwan, pp. 96–97.

115. Chaudhry argues that weaknesses in the ability of new states to regulate and tax the private sector were a major reason for the adoption of socialist schemes in late developers.

116. *AM* 3 (August 1963), p. 7; al-Ghazzali (1968), p. 61. Workers won 41% of the cases handled by the Labor Offices in 1962/63. Data on the resolution of grievances in other years was not available. The figures for 1962/63 indicate that the proportion of cases handled by the Labor offices, relative to the courts, increased considerably over the 1950s. This implies that there were more collective as opposed to individual complaints filed, which suggests that the problem of arbitrary dismissals was now reduced in scope.

117. The same stipulation also applied to joint stock companies, even if smaller than 50 employees; the joint stock companies formed with public capital operated under the more stringent public sector WRM laws. See Mahmoud, pp. 51–55, 120, and Davies, pp. 59–60, 90, and 97.

118. These practices were reported in periodic exposes in *al-'Ummal*, the ETUF's newspaper, which began publication in 1965.

119. For more on the ASU, see Waterbury (1983), esp. pp. 314–32, and Harik (1973).

120. *Al-Ahram al-Iqtisadi* (December 1, 1961), quoted in Issawi, p. 196.

121. Cited in Elsabbagh, p. 49.

122. 'Izz al-Din, pp. 943–44.

123. 'Izz al-Din, pp. 938–39.

124. Salama interview, July 1988.

125. Ibid. I was first informed about this incident in an interview with Yusuf and Eqbal Darwish, December 1987.

126. 'Izz al-Din, p. 943 (parenthetical note mine).

127. Sa'id (1983), p. 33, states that Salama was the first to head the new ministry. However, Khalid and other sources indicate that Kamal al-Din Rif'at, a leftist Free Officer, held the post briefly before Salama's appointment.

128. 'Izz al-Din, pp. 957–58; Imam (1987d), p. 162.

129. Waterbury (1983), p. 319, argues that Nasir deliberately sought to balance left and right among his subordinates. When some of his closest associates questioned the wisdom of the new "socialist" orientation Nasir kept them in positions of some authority, rather than dismiss them, as a hedge against allowing more radical elements to accumulate too much power.

130. ETUF (1977), p. 84.

131. Upon his release he was under political "quarantine" (unable to join the ASU) for several years. Sa'id (1983), pp. 32–33, blames his arrest on Zakariyya Muhyi al-Din, Minister of the Interior, with whom he had clashed in the 1950s over the issue of the confederation. Given Sa'id's prominent position in the Labor Ministry, his close ties with al-Shafa'i and Rifa't, and the fact that Nasir himself had sent Sa'id to Yugoslavia in the late 1950s to report on the structure and management of labor/state relations there, it is difficult to imagine that Nasir was unaware of his incarceration; why he went along with it is unclear.

132. ETUF (1977), p. 27; Sa'id (1983), p. 32.

133. ILO *Legislative Series*, March–April 1965; *AM*, March 1969, pp. 11–12.

134. *Al-Ahram* (hereafter A), July 25, 1963.

135. The Liberation Rally existed for all practical purposes only as a national agency; it had neither the organization nor the personnel to reach into the hundreds of union locals. The National Union had more of a mass life than its predecessor, including branches in various localities, but these did not seem to involve themselves in local union affairs.

136. Hasan, p. 165; Abu 'Alam (1966), p. 68; Kamel, p. 129.

137. Fa'id interview; Imam (1987d), p. 162

138. 'Izz al-Din, p. 979; ILO *Legislative Series*, March–April, 1965; Clarke, pp. 55–56; Hasan, pp. 141–43.

139. A, April 8, 1964; *AM* 66, November 1968, p. 15.

2. *The Continuity and Conflicts of Corporatism*

1. Imam (1987d), p. 162; Sa'id (1983), p. 35; 'Izz al-Din, pp. 960–61.

2. Waterbury (1983), pp. 93–97, 409 on the IMF talks; pp. 321–23, 356–59 on the political changes.

3. 'Izz al-Din, p, 962; Sa'id, p. 35. Neither 'Izz al-Din nor Sa'id offer an explanation for Salama's removal, but the context in which it occurred suggests that his ideological leanings and closeness to al-Shafi'i were the probable factors.

4. Waterbury (1983), p. 323; 'Izz al-Din, p. 962.

5. Bianchi (1989), p. 136.

6. 'Ali Sayyid 'Ali, for example, besides his aforementioned union positions within Egypt, also served as general secretary of the Arab Petroleum Federation, a member of the executive committee of the Arab Confederation of Trade Unions,

and vice-president of the International Petroleum Workers Federation. *Al-'Amal* (hereafter *AM*) July 1964, p. 12.

7. For more information on the development of the power centers within the ASU, see Waterbury (1983), pp. 307–42. On the subordination of the ASU to the regime, see Harik (1973).

8. The highest body of the party issued a decision in June 1965 creating a new supervisory body for the WEI. Although the new rules allowed the confederation to choose seven of the eleven members of the committee from among union leaders, the chair would be chosen by the ASU's general secretariat; in addition, the director of the WEI would now also be chosen by the ASU's leadership, rather than the supervisory committee. 'Abd al-Mughni Sa'id was returned to the directorship shortly after this. ETUF (1977), p. 28.

9. 'Izz al-Din, pp. 962–63, 980, 1011; Hasan, p. 153; Sa'id (1983), pp. 35–36; *AM*, March 1966, p. 33; *Al-'Ummal* (hereafter *U*), April 1967, p. 3.

10. *U*, May 23 1968, p. 4; Sayyid, p, 71; and *U*, 1968, various issues.

11. The ETUF finally received permission to publish the paper as a monthly in 1965.

12. Interviews, *inter alia*, with Muhammad Muhammad 'Ali, January 1988, and Amin 'Izz al-Din, June 1988. For more on the March 30 declaration, see Waterbury (1983), pp. 330–31, 355.

13. *U*, March 1968, p. 3; *Al-Ahram* (hereafter A), March 4, 1968; Khalid (1975), p. 62; Waterbury (1983), p. 330; *U*, October 18, 1968, p. 5.

14. 'Izz al-Din interview, June 1988; *AM*, December 1968, p. 5; December 1969, pp. 7–8; and July 1971, p. 6; interview with 'Abd al-Mughni Sa'id, February 1988; Imam (1987d). The WS, in conjunction with the "power centers" close to 'Ali Sabri, was also able to prevent Ahmad Fahim and Kamal Rif'at from obtaining seats on the Supreme Executive Committee of the ASU when elections to that body from among the central committee were held in the latter half of 1968. Fahim had been chosen by the ETUF to run for a seat as a representative of the workers and 'Ali, against the wishes of the confederation, decided to run against him. Sa'id (1983), pp. 37–38, was told by friends on the central committee that Shar'awi Gum'a, Minister of the Interior at the time and an ally of Sabri, pressured members of the central committee to vote for 'Ali, along with a list of candidates for other seats, which excluded Rif'at. He sees the declared victory of certain highly unpopular individuals in those elections as an indication that Fahim and Rif'at may have actually won in spite of this pressure, implying that Gum'a and others falsified the true results.

15. Khalid (1971), pp. 83–86; A, April 18, 1967; *Al-Ahram Al-Iqtisadi* (hereafter *AI*), November 15, 1968, pp. 58–59; Sa'id (1983), p. 33; *AM*, March 1969, pp. 11–12 and December 1970, pp. 7–8.

16. *U*, 1968 and 1969, various issues; Khalid (1975), pp. 66–68; 'Izz al-Din, p. 1008, and interview, June 1988.

17. This section draws heavily on interviews with rank-and-file workers, along with lower-level union officials, published by *al-'Amal* and *al-Tali'ah*. Both magazines ran series evaluating the trade union movement in the period 1964–66 and

then again in 1968–69. The earlier articles have been annotated by Taha, and will be referenced here through his book. 'Izz al Din, pp. 1003–4, claims that the second *al-'Amal* series, called "Cracks in the Wall of Silence," was actually commissioned by the WS group as part of their efforts to weaken Ahmad Fahim. While this may be true, I believe the individuals interviewed were voicing genuine concerns, since their comments accord with those found in *al-Tali'ah*, and especially since one of the most vocal critics in these articles is 'Izz al-Din himself!

18. Hasan, p. 125. An article in *al-'Ummal*, February 1967, p. 13, states that 16 federations were not in attendance.

19. Hasan, pp. 129–30, 137, 152–53; *U*, January 1967, pp. 10–11, February 1967, p. 8, September 1967, pp. 10–11, and March 1968, pp. 4–7; *AM*, September 1964, p. 47; February 1966, pp. 3–4; and November 1966, p. 7; Taha, *passim*.

20. A, April 8, 1964; *AM*, November 1968, p. 15; ILO, *Legislative Series*, March–April 1965; Hasan, pp. 124, 129–33; Taha, *passim*; *AM*, December 1968, pp. 10–12.

21. The president, vice president, secretaries, and treasurer are collectively referred to in Egypt as the "office committee" of the union organization.

22. A, October 13, 1964; *AM*, August 1964, pp. 39–40 and December 1968, p. 12.

23. For details, see Beinin (1987).

24. Taha, pp. 12–14, 69–71.

25. Taha, p. 30, originally cited in *AM*, November 1966.

26. U January 7, 1971, p. 3

27. Hasan, p. 123; Taha, *passim* ; *AM*, December 1968, pp.-10–12; *al-Tali'ah* (hereafter *T*), January 1968, pp. 22–23.

28. *AM*, July 1964, p. 17 and September 1970, p. 8; *AI*, April 15, 1968; Fa'id interview, June 1988.

29. 'Izz al-Din, pp. 1003–6; *AI*, April 15, 1968.

30. Fa'id interview; Taha, pp. 36–46.

31. Imam (1987d).

32. Sa'id (1983), p, 36; 'Izz al-Din, p. 963.

33. Elsabbagh, pp. 121–23; Baker, pp. 181–86; Imam (1987d); Fa'id interview.

34. *AM*, October 1967, pp. 24–26; *AM*, October 1968, p. 34 and December 1968, p. 13; Baker, pp. 189, 268.

35. 'Izz al-Din, pp. 954–57; A, July 21, 1964; Kamel, pp. 90–93, 122–23; *AM*, December 1968, p. 13.

36. A, April 8, 1964; *AM*, November 1968, p. 15; Khalid (1971); A December 9, 1967; *U*, December 1966, pp. 24–25, May 23, 1968, p. 4, and January 16, 1969, p. 5.

37. An additional rival to the trade union movement, especially toward the end of the decade, were the organizations representing graduates of the technical schools. These groups had their origins in the 1940s when the engineering syndicates refused membership to the graduates. Some eighty such groups were formed, and organized informally into a federation. In 1969 the federation became official, and took the name "the union for applied professions" (*niqabat al-mihan al-tatbiqiyah*). Because many technicians work in plants where there were unions locals, the existence of a separate organization for skilled workers acted as competition to the former. This was not an issue for the confederation during the 1960s,

but would become a concern later. *U*, October 10, 1968, p. 7; *Arab Strategic Report*, 1989, p. 467.

38. Hasan, p. 129; *T*, October 1966, p. 67; Imam (1987d); *U*, December 10, 1970, p. 3; *AM*, December 1968, p. 14.

39. For more details, see Waterbury (1983), pp. 349–53, and Hinnebusch, pp. 40–46.

40. See, e.g., Hinnebusch, p. 44.

41. Fa'id interview, June 1988.

42. *U*, May 20, 1971; A, May 22, 1971.

43. *U*, June 17, 1971; A, June 12, 1971.

44. However, Muhammad Mutawalli al-Shar'awi, the labor activist originally from Kafr al-Dawwar, reports (interview, March 1988) that his electoral victory at a plant in Alexandria was stolen from him by local authorities, who lied about the election results.

45. *U*, June 24, 1971; A, June 18—25, 1971.

46. The results of the federation elections were published in *U*, July 15, 1971, and July 22, 1971. Information about the role of the leftists, Sadat, and the other individuals who figure prominently in the 1971 federation and confederation elections is based on interviews with Ahmad Taha, Fathi Mahmud, Ahmad al-Rifa'i, Amina Shafiq, 'Abd al-Mun'im al-Ghazzali, Amin 'Izz al-Din, Kamil 'Uqayli, Muhammad Gamal Imam, 'Aisha 'Abd al-Hadi, 'Abd al-Latif Bultiya, Ahmad al-'Amawi, Ga'far 'Abd al-Mun'im Ga'far, and Muhammad Muhammad Ali.

47. Sadat brought two communists onto his cabinet at the end of 1971. Waterbury (1983), p. 352, sees this as a move to reassure the Soviets, but my sources indicate that the communists were also being rewarded for their assistance in smashing the pro-Sabri cliques.

48. For more on the vanguard, see Waterbury (1983), pp. 333–36.

49. One of these, the Federation of Workers in Military Production, was not allowed to hold elections at all, presumably because of the role of military plant workers in the 1968 events in Helwan. Hasan, p. 154.

50. Bianchi (1989), pp. 84, 129, 141.

51. See also Imam (1987d), pp. 162–63. The reader should be aware that the communist movement was factionalized; those who allied with Sadat did not represent the entire movement. For details, see Beinin (1987).

52. On average, there were about three candidates per seat on the federation boards; only in two federations did a candidate list win by default. *AM*, August 1971, pp. 11–12.

53. Interview with 'Aisha 'Abd al-Hadi, March 1988.

54. It was this argument that persuaded 'Abd al-Hadi to support Gharib in the end. There are veiled references to these negotiations in *U*, July 22, 1971, p. 3.

55. Sadat had also moved quickly to rectify his ignorance of the trade union movement by convening a meeting of the new ETUF board in August 1971. He was particularly concerned about the confederation's position on developments in the Sudan, where Egyptian forces had recently helped to reinstate his friend Ga'far Numeiri in power after a coup attempt in which leftists had been implicated. Numeiri had ordered the execution of several communist union leaders

as collaborators, and the International Confederation of Arab Trade Unions, with the support of Egypt's confederation, had issued a sharp denunciation of these events and sent a delegation to the Sudan to investigate. The details of the ICATU actions were published in *al-'Ummal*, August 19, 1971. Rather than offering any of his own opinions to the unionists, Sadat encouraged them to speak freely, especially on the Sudan issue, as a means of identifying their political leanings. Fathi Mahmud, then a Nasirist in the opposition camp, believes (Mahmud interview, June 1988) that the Egyptian communists who had obtained seats on the ETUF board were marked for removal by Sadat as a result of this meeting.

56. 'Izz al-Din believes that this move was also motivated by Sadat's residual distrust of 'Abd al-Latif Bultiya because of his former associations.

57. Interviews with Sa'id Gum'a, Sabir Barakat, and 'Abd al-Rahman Khayr, as well as the unionists, labor journalists, and leftist labor activists cited above (note 46) also contributed to the information contained in the remainder of this section; only additional references are noted.

58. Al-Sirafi (1972), pp. 124–5; *AM*, March 1972, pp. 14–55.

59. This outcome suggests that al-Maghrabi and Khalifa had gone along with Rifa'i's support of Gharib back in 1971, but no one said that to me explicitly.

60. A, May 21, 1974; Khalid (1976), p. 84; *AM*, July 1973, p. 18.

61. *AM*, July 1975, p. 2; *Rose al-Youssef* (hereafter *RY*), No. 2470, October 13, 1975, pp. 34–37; Fa'id interview, June 1988.

62. An exception is al-Ghazzali, who apparently supported the move out of his bitterness toward al-Rifa'i.

63. A, February 3, 1975, June 9, 1975, August 18, 1975; *RY*, No. 2478, January 13, 1976, pp. 7–8; *AM*, October 1975, p. 22.

64. Beinin (1987).

65. Cf. Beinin (1989).

66. *U*, July 14, 1975; A, June 2, 1975, July 13, 1975.

67. *RY*, No. 2458, July 21, 1975, p. 6, and No. 2461, August 11, 1975, pp. 22–23; A, July 21, 1975 and August 11, 1975; *U*, August 4, 1975.

68. *RY*, No. 2466, September 15, 1975, pp. 18–19, and No. 2470, October 13, 1975, pp. 34–37; A, October 12, 1975; Imam (1987a, 1987c) and interview, October 1987.

69. *AM*, September 1973, p. 50; A, November 18, 1974, July 21, 1975, September 17, 1975, October 7, 1975; Khalid (1976), pp. 83–84.

70. *AM*, November 1975, p. 37; A, September 15, 27, 29. 1975, June 28, 1976.

71. *RY*, No. 2474–2478, November 10–December 8, 1975.

72. Ibid.

73. The full text of the 1976 law is available in Arabic in Hasan, pp. 195–203, and in English in the ILO *Legislative Series*, January 1977. All subsequent references to the law are based on these works.

74. Bianchi (1989), pp. 84, 129.

75. Relatedly, I found no mention of direct involvement by Sadat in the preparation of this legislation. His role and intent with regard to it remain to me ambiguous.

76. Samples of these criticisms can be found, *inter alia*, in *RY*, January 13, 1976, pp. 7–8, and June 28, 1976, pp. 22–25; *AM*, July 7, 1975, p. 2; and *Al-Ahali* (hereafter *AH*), August 24, 1983. See also the series on workers' concerns published by *al-Tali'ah* during 1976–77.

77. Interview with Sabir Barakat, October 1987. "The New Dawn" was the name of a newspaper published by one of the communist groups active in the 1940s and '50s. See Beinin and Lockman, *passim*.

78. *RY*, No. 2507, June 28, 1976, pp. 22–25.

79. Ibid., and *AM*, July 1976, p. 2.

80. Ibid., and Barakat interview.

81. Prior to the elections, Bultiya had decreed an increase in the number of federations to 21, and in the ETUF board to 25 members, while retaining the provision that each federation have at least one representative on the EC. Eleven of the federations nominated only one candidate, so these won by default. The elections were to choose the remaining 14 board members from among 29 candidates nominated by the other 10 federations. *AM*, August 1976, pp. 4–5.

82. *RY*, No. 2512, August 12, 1976, p. 4; *U*, July 26, 1976, p. 1.

83. For more details see Waterbury (1983), pp. 364–73.

84. *AM*, December 12, 1976, pp. 14–16; *U*, January 16 and July 26, 1976.

85. Information about the affiliation of unionists with the platforms comes mainly from interviews with Amina Shafiq, Muhammad Gamal Imam, Sabir Barakat, 'Abd al-Rahman Khayr, Niyazi 'Abd al-'Aziz, 'Abd al-Hamid al-Shaykh, 'Abd al-Majid Ahmad, and Fathi Mahmud; references in print are cited.

86. *AM*, December 1976, pp. 14–16; *U*, November 29, 1976, p. 1.

87. How the committees were to be constituted after the demise of the ASU was not specified.

88. *U*, January 24 and 31, 1977; *AM*, March 1977, p. 20.

89. *U*, February 28, March 7 and 28, May 16, 1977; *AM*, March 1977, p. 20.

90. Sa'id (1983), p. 39; *U*, May 16, 1977. The name of the ministry was changed around this time to the Ministry of Manpower and Vocational Training. For continuity and writing ease, I shall continue to refer to it as the Ministry of Labor.

91. 'Abd al-'Aziz clung to his Tagammu' membership through much of 1978, but ultimately left because of differences with the party's leadership.

92. My argument here resembles that of al-Sayyid (1983), Qandil (1985), and Bianchi (1989).

93. *U*, November 20, December 5 and 19, 1977; *AM*, March 1978. Some senior unionists nevertheless harbored personal disagreements with Sadat's actions; Sa'id Gum'a (interview, June 1988) indicated to me that the Camp David trip was done reluctantly and under pressure, and his associates said he was still embarrassed by it. Afterward the ETUF refused to establish relations with the Israeli Histadrut despite pressures in that direction from Israel. In 1981 the EEMWF challenged Sadat and Israel by refusing to lead Israeli President Benjamin Navon on a tour of metal factories in Helwan, and encouraged its Helwan members to organize work stoppages in protest of Navon's visit.

94. It was while in prison during these years that Barakat and others decided to produce the independent workers' magazine *Sawt al-'Amil*, which became a voice

for leftists both within and outside of the Tagammu' who were critical of the party's leadership. Barakat was carrying the first issue to the printer on the day of Sadat's assassination, which marked his last arrest during this period. Those lay-outs were confiscated by police, and the magazine did not appear until 1985.

95. Khayr and Barakat interviews, and interview with Niyazi 'Abd al-'Aziz, December 1987.

96. On the general tenor of this period and the events leading up to Sadat's 1981 assassination, see Waterbury (1983), pp. 359–88.

97. Heikel, pp. 117–18; Ahmad, p. 165. Hendriks (1983), p. 268, cites a higher figure of 78 for the number of initial OSP rejections, but I believe this to be in error since the figure of 56 was confirmed by numerous official and unofficial sources. Sabir Barakat was among the five who lost their court appeals.

98. *AM*, October 1979, pp. 12–13, 36; A, September–November 1979, various issues.

99. Some of the details are available in the 1978 issues of *al-Ahali*, the Tagammu' newspaper, before it folded in the middle of that year.

100. Ahmad, p. 165–67; *AM*, October 1979, p. 36; A, August 20 and October 16, 1979; *AH*, May 11 and August 31, 1983, September 16, 1987; Barakat interview.

101. This was Law 3 of 1977. *AM*, June 1987, p. 46; 'Abd al-Magid, p. 165.

102. The text of the 1981 law is available in English in the ILO *Legislative Series*, February 1982. In Arabic see Hasan, pp. 204–16. See also Ahmad's statements about the draft law in A, February 9, 1979.

103. See the critique of the law by Ahmad Sharaf al-Din, a member of the *Sawt al-'Amil* group, in *AH*, July 20, 1983, p. 5. Significantly, Sharaf al-Din's arguments, which challenged the very notion of a hierarchically functioning confederation, were criticized as syndicalist by some unionists affiliated with the party.

104. For more on the changing domestic political climate under Mubarak, see the works by Lesch, Post, and Hendriks.

105. See chapter 1 on the limitations of the raw data reflected in these tables. Although there were no changes in the legal provisions regarding union membership, several laws affecting specific occupations, when combined with the economic opening, did lead to a dramatic growth in the membership of certain federations, although the reported numbers may be inflated. One of these was the Land Transport Workers' Federation, whose membership more than tripled, from 130,875 in 1971–73 to 423,232 in 1983–87. This was mainly the result of the expansion of privately owned forms of mass transit, such as taxis and minivans; obtaining a license to operate one of these vehicles requires a certificate from the union which drivers must pay dues to obtain. The Building and Construction Workers' Federation grew almost as rapidly, from 85,577 members in 1971–73 to 249,971 in 1983–87. In this case the surge was reportedly due to the fact that occupational certificates from the union were required for construction workers to obtain permission to leave the country for lucrative jobs in the Gulf. Thanks largely to these sectors, nonagricultural union membership grew more rapidly than the total nonagricultural workforce between 1973 and 1985, and the overall proportion of unionization increased somewhat. Information about the reasons for these spurts in union membership is based on discussions with workers in these sectors, and

with industrial unionists. On employment and unionization of the construction industry in Egypt, see also Assaad (1990).

106. *RY*, October 26, 1987, n.p.

107. Lists of federation officers elected in 1983 are from the appendices to Turk, pp. 157–202. The 1979 lists are from ETUF (1982), pp. 93–136.

108. AM 246 (November, 1983), pp. 54–5; A November 20, 1983.

109. See, *inter alia*, *AH*, June 2 and August 11, 1982, February 23, 1983, and June 27, 1984.

110. Ahmad, p. 167; *AH*, March 3, June 2, June 15, 1983.

111. *AM*, November 1983, pp. 54–55.

112. Barakat, for example, is from the same plant as Sa'id Gum'a and maintains that the latter supported his exclusion because otherwise the two would compete for local office. Relatedly, leftists complained that the ETUF also did little to seek the release from prison of those local leaders still incarcerated as a result of the 1977 riots. See Ahmad, p. 167, and *AH*, May 19, 1982.

113. *AH*, August 31, October 21, 1983; June 27, 1984.

114. Interviews with Sabir Barakat and 'Abd al-Rahman Khayr.

115. Sharaf al-Din (1986a), p. 16; *AM*, June 1987, pp. 46–48. Nasir had signed a similar accord during the 1960s, but because it was never published in the *Gazette*, it did not constitutionally have the force of local law.

116. *RY*, October 26, 1987.

117. After the ETUF was excluded from the Arab Confederation of Trade Unions because of Camp David, numerous federations did move to affiliate with the ICFTU. These included some of the historically more progressive federations such as the PCWF. Unionists suggested to me that in most cases the actions were not a political statement of support for Camp David but rather an effort to find a replacement for the perks, in the form of overseas travel and acquaintances, that had previously come to the unions through their Arab identities. Sa'd Muhammad Ahmad, under pressure from opponents of the affiliations, put a freeze on them in the mid-80s.

118. While Bianchi (1989), pp. 140–41, confidently attributes Gharib's disfavor with Sadat to the 1975–76 strike wave, he ignores the 1985–86 wildcats completely as a possible factor in Mubarak's disenchantment with Ahmad, attributing it instead to Ahmad's opposition to public sector employment cuts in 1986. I prefer to acknowledge the ambiguities surrounding Mubarak's actions. Bianchi was also evidently unaware that Gharib had been removed from the Ministry more than a year before he was ousted from the ETUF presidency, hence his erroneous claim that Ahmad's removal was "the first time since Ahmad Fahim" that the posts of ETUF president and Minister of Labor were separated.

119. Known in Arabic as *Majlis al-Shura*, the Consultative Council was a body created by Sadat which functioned like an upper house of parliament. See Waterbury (1983), pp. 382–83, and Springborg, p. 139.

120. *U*, November 17, 1986; January 14, 1987.

121. 'Uqayli at the time was vice-president of the ETUF.

122. The five men supported by Ahmad were Abu Bakr Gadd al-Muwali (Health Services), Anwar 'Ashmawi (federation unknown), Hassan 'Id (Postal),

'Abbas Mahmud (Mining and Minerals), and Ibrahim Shalabi (Textiles). Two of them left the NDP and joined the Wafd as a result.

123. 'Abd al-Hamid may have played a key role in the selection of 'Asim for the Ministry, and in helping the regime choose a new group of loyal unionists. As president of the Agricultural Workers' Federation, he had strong ties with Yusuf Wali, who became minister of Agriculture in 1982.

124. *AH*, March 11, 1987; Imam (1987c). Khidr, ironically, had been the unionist considered closest to Salah Gharib; he was particularly keenly disliked by the Nasirists and leftists in the trade union movement.

125. Information about the government's actions against Ahmad and his associates is drawn from *AH*, April–June 1987, various issues.

126. Interview with Mahmud Dabur, March 1988.

127. A, November 11–19, 1987; Imam (1987c).

128. Results of the 1987 ETUF elections were published in the official dailies and the December issue of *al-'Amal*. The members of the 1983 board were taken from Turk, pp. 153–54; for 1979 the names were taken from ETUF (1982), p. 27.

129. *AH*, November 11, 1987; A, November 25–26, 1987; Imam (1987c).

130. Gum'a married the daughter of one of the foremost communist leaders of Egyptian labor in the 1940s and 1950s, whose sister remained active in leftist politics. Also, the EEMWF had continued its relations with federations in the socialist bloc countries, while the PCWF was one of those that affiliated with the ICFTU.

131. In the main, the decree merely amounted to enforcing the language of the 1976 and 1981 union laws, which specified that union membership, and hence the ability to hold union office, ended upon retirement; this provision had been benignly overlooked during Ahmad's tenure. However, 'Abd al-Haqq also sought to apply it against individuals who turned 60 but continued working.

132. Interview with Amin 'Izz al-Din, January 4, 1995. 'Izz al-Din maintains that 'Abd al-Haqq started on this path in the 1980s, currying the favor of Boutrus Ghali, then a senior figure in the regime's foreign policy establishment (and in 1995 Secretary-General of the United Nations) and Mustafa Khalil, an NDP leader in charge of the party's office for normalization with Israel.

133. The 1976 union law had maintained the right of the Ministry of Labor to inspect the financial records of the unions. The leading unionists at that time continued to object to this provision, and in the 1981 law it was modified to specify joint supervision of union finances by the confederation and the Ministry. The concern was largely dropped by the confederation after this modification, until 'Abd al-Haqq's leadership at the Ministry brought it to the fore again, prominently.

134. Sharaf al-Din (1986a), p. 16; *AM*, June 1987, pp. 46–48.

135. During the 1987 election campaign, I was privileged to attend the monthly meetings of the Tagammu's labor bureau. In addition to the cited references, information about the elections contained in this section is based on notes from these meetings.

136. The party also objected to the fact that no campaign literature was permitted to be distributed before the final candidate lists were issued. They charged that this stipulation favored candidates who had the backing of management or the government and could therefore have campaign materials ready on such short

notice, while workers running independently and lacking financial and organizational resources were put at a disadvantage. *AH*, October 21, 1987. Actually, however, those candidates with official backing from the Tagammu' would not be as hurt by this restriction as independents, since the party did provide its members with campaign literature.

137. A, October 22, 1987; *AH*, September 17, October 21, 1987. The other main opposition parties, the Wafd and the Socialist Labor Party, also fielded a larger number of candidates and gave the campaign much greater coverage than in 1983. See *al-Sha'b*, September 14–15 and October 19–23, 1987, and *al-Wafd*, September 19–27 and October 22–29, 1987.

138. The Tagammu' had no representatives in the parliament at that time, so the bill was introduced by Ahmad Taha, a former communist labor activist who, although remaining close to the Tagammu', had obtained a parliamentary seat by running on a Wafd ticket. *Al-Yasar*, December 1992.

139. El-Shafei, pp. 65–70.

140. *AH*, December 3, 1986.

141. The defense against this, mostly implicit but sometimes stated, was that Mubarak was the best option among the plausible alternatives. In the mid-1980s certain Tagammu' leaders were particularly worried about the developing closeness between then-Minister of Defense 'Abd al-Hakim Abu Ghazzala and the United States. They feared that if the Mubarak government was not more forthcoming on economic reforms, the United States might sponsor a coup to put Abu Ghazzala in power. For more on Abu Ghazzala, see Springborg, pp. 118–23.

Mubarak was also seen as a bulwark against the rise of Islamic fundamentalists, who were periodically subjecting leftists to brutal physical attacks. Although the Abu Ghazzala threat was effectively eclipsed when Mubarak removed him from the Defense Ministry, the fundamentalist threat, and with it the accommodation of the Tagammu' to the regime, intensified in the 1990s. See Beinin (1993), pp. 25–26.

142. The Tagammu' approach closely resembles that of the Arab Nasirist Party (ANP). Although not officially recognized by the regime until the early 1990s, the party began publishing its newspaper, *al-Arabi*, shortly after its formation in the mid-1980s. Some Nasirists from the Tagammu', including most prominently Fathi Mahmud, became leaders of the ANP.

143. Based on interviews with leftists working both within and outside of the Tagammu' who asked for anonymity. See also el-Shafei, pp. 52–53, and Beinin (1994).

144. Sabir Barakat, whose views and experiences are featured prominently in this and the next two chapters, was a founding member of this group. In addition to the cited references, the comments about VOW below are based on my interviews with him.

145. See the article by Ahmad Sharaf al-Din in *AH*, July 20, 1983, and the response by Fathi Mahmud, *AH*, August 10, 1983.

146. As we have seen, the leagues were technically outlawed as rivals to the union locals in 1964, but were able to survive by redefining themselves as private voluntary organizations. The ETUF objected vehemently to their existence and

won a promise from Sadat to abolish them in 1971, but no official action was taken, and in 1973 there were some 350 leagues of workers and professionals registered with the Ministry of Social Affairs. The 1976 union law reaffirmed that only one organization in any plant could perform the functions assigned to the union locals, implicitly outlawing the leagues for the second time, but once again no action was actually taken. The reason, apparently, is that the associations have strong patrons in the Ministry of Social Affairs, for whom they provide a source of power as well as wealth. *RY*, August 16, 1971, p. 4; *AM*, July 1973, p. 20; A, December 16, 1974; *U*, February 10, 1975; *RY*, No. 2745, January 19, 1981, pp. 8–10; Springborg, pp. 171–73; *Arab Strategic Report* (1989), p. 467.

3. *Workers at the Point of Production: Moral Economy and Labor Protest*

Portions of chapter 3 appeared previously in "Irrational Workers: The Moral Economy of Labor Protest in Egypt," *World Politics* 46, No. 1 (October 1993): 83–120, and in "Collective Action and the Consciousness of Egyptian Workers, 1952–87," in Zachary Lockman, ed., *The Historiography of Middle East Labor* (Albany: SUNY Press, 1993).

1. Earlier versions of the empirical sections that follow appeared in Posusney 1993. I have somewhat revised the formulations here, and incorporated some new information including, especially, the 1989 factory occupation at the Helwan ironworks. A section on the spontaneous and informal nature of rank-and-file protests has also been added.

2. For an overview of these arguments, see Kennan.

3. On the importance of solidarity to the Marxist notion of class consciousness, see Booth (especially pp. 168–69), and Elster (1985), p. 347.

4. 'Amir, p. 91; Sayyid Fa'id interview, June 1988; FO 371/102931/JE2183/19.

5. Beinin (1989), pp. 76–78.

6. As noted in chapter 1, the Labor Offices immediately transferred individual grievances to the courts, and continued to process only collective complaints. Of these, those emanating from plants employing more than 50 workers went directly to conciliation committees. If conciliation failed, the cases would then go to an arbitration board, whose decisions were considered final. However, either party could appeal the verdict to the Court of Cassation. For plants employing less than 50 workers, the conciliation committee stage was replaced by mediation efforts on the part of Labor Office officials, with arbitration in order only when those efforts failed (see Girgis). For a contrasting interpretation of the grievance data, see Goldberg (1994).

7. The deficiencies in the data, along with contradictory sources, make it difficult for me to be more precise and decisive about wage trends. Hansen, p. 183, provides a graph which shows real wages climbing only in the first three to four years after the coup, and flattening thereafter until about 1961–62; he does not specify which CAPMAS statistics were used to plot this graph.

8. This is based on the data collected by Beinin (1989), p. 77. Grievance data is not available after 1958.

9. Interview with Muhammad Mutawalli al-Sha'rawi, March 1988; Tomiche, p. 44. The only reported job action after this upsurge was a politically motivated strike in 1959 at the Shabaha factory in Alexandria, protesting the arrest of communist labor leaders; Nasir went there personally in an effort to calm the workers. Sayyid Fa'id interview, July 1988.

10. Mabro and Radwan, pp. 135–37; Sa'id (1968), pp. 79–92; Khalid (1971), pp. 44–54.

11. Cited in Dekmejian, p. 140.

12. Waterbury (1983), pp. 93–97, 409; Abdel-Fadil, pp. 33–34.

13. *al-'Ummal* (hereafter *U*), August 1967, pp. 4–5, 18–19; *al-Ahali* (hereafter *AH*), October 24, 1984.

14. Hussein (1973), pp. 234–37.

15. Fa'id interview.

16. The rise in real wages shown in 1972 followed an increase in the minimum wage ordered by the government in March of that year. See Starr, table 2, pp. 13–14.

17. Goldberg's argument should also imply different patterns of protest between public and private sector workers, although how they would differ according to his logic is unclear. Since private sector workers did not gain the same protections as those in the public sector, job security should have remained their dominant concern, and continued to be an incentive for collective action. Yet Goldberg could also argue that collective activity in the private sector would be weaker, precisely because the risks were higher—which is the dominant effect? Minimally, his logic should lead him to expect, as in the 1940s, variations in the frequency of private sector protest according to macroeconomic conditions, with interindustry differentials according to specific labor market conditions. My data is too imprecise to test the interindustry hypothesis.

18. Beinin (1989).

19. Buroway (1984), pp. 41–42.

20. See, e.g., the application of Buroway's arguments to Egypt in Henley and Ereisha.

21. Brown (1990).

22. Sabel (1982), pp. 128–36.

23. See Posusney (1996).

24. Interview with 'Abd al-Rahman Khayr, October 1987

25. Hinnebusch, p. 71; Baker, p. 165; 'Abd Al-Raziq, pp. 80–84; *MERIP* No. 56 (April 1977), p. 6; see also Shoukri (1981), p. 323.

26. For more on the economics of this period, see Waterbury (1985).

27. *Rose al-Youssef* (hereafter *RY*), No. 2551, May 2, 1977, p. 4; *U*, March 7 and 28, and May 16, 1977; Starr, p. 3; Salmi, p. 32; Waterbury (1983), p. 228.

28. 'Abd al-'Aziz interview, February 1988; Starr, p. 4. Union leaders say that the maximum possible raise under this law is actually greater than that normally available to public sector workers.

29. The trend shown in table 3.3 is confirmed by interviews with numerous workers from both the public and private sectors.

30. *AH*, April 12, 1978.

31. MEED, 1982 and 1983 volumes.

32. The leftist weekly *Al-Ahali* was my primary source of information on the frequency, causes, and nature of labor protest in the 1980s. Hereafter only supplemental references are cited.

33. MEED, 1984–87 volumes.

34. See also works by Adli, El-Shafei, and Badawi.

35. Goldberg (1992), p. 157; italics mine.

36. This and all subsequent material on incidents in 1952–53 is drawn from Audsley, pp. 99–100; Gordon, p. 83; Beinin and Lockman, pp. 429, 434; and FO 371/102931/JE2183/19&20.

37. *al-Tali'a* (hereafter *T*), August 1976, pp. 55, 58. Subsequent references to the 1960s incidents at this plant are from the same source.

38. Barakat interview, October 1987. The incident described occurred in the Delta ironworks plant; all subsequent information concerning actions at this plant is from the same interview.

39. Fa'id interview; *U*, February 3, 1975; *AH*, October 24, 1984.

40. *T*, September 1976, p. 55; *al-'Amal* (hereafter *AM*), No. 141 (February 1975), pp. 10–11; *al-Ahram* (hereafter A), September 21–2, 1976; *U*, September 27 and October 18, 1976; *RY*, No. 2520, September 27, 1976, pp. 4–5; Khayr interview.

41. For a critical evaluation of leftist interpretations of the riots, see Beinin (1994).

42. *Al-Sha'b*, April 27, 1984.

43. *MEED*, October 19, 1984.

44. *Awraq 'Ummaliyah*, No. 5, January 1986, pp. 6–7.

45. *al-Akhbar*, February 10, 1986.

46. Based on interviews with lawyers who represented workers detained during this event; see also Adli, p. 186. All subsequent information on this incident is also based on these sources.

47. A compromise agreement was finally negotiated between the Ministry of Labor, company officials, and workers' representatives in October, 1987. *Al-Ahram al-Iqtisadi*, October 26, 1987.

48. See Abbas et al., and Al-Shafei, pp. 65–71. All subsequent references to the Helwan incident in 1989 are based on these sources.

49. It should follow logically that, if there are certain industries which are experiencing more takeaways than others, the level of protest there should be higher. This would therefore be an important test of moral economy. However, it would require documenting all instances of both takeaways and protests against them on an industry by industry basis; such detailed and accurate data was unavailable to me.

50. For workers younger than 18, the minimum wage was set lower.

51. *U*, October 16 and November 6, 1972. Subsequent references to the 1972 incidents are from the same sources.

52. Fa'id interview.

53. *AM*, No. 144, May 1975, pp. 10–14; A, March 22–23, 1975.

54. Shoukri (1981); Barakat interview.

55. A, September 21–22, 1976; *U*, September 27 and October 18, 1976; *RY*, No. 2520, September 27, 1976, pp. 4–5; Khayr interview.

56. Khalid (1975), pp. 130–31; interviews with Khayr and 'Abd al-'Aziz Higazi, March 1988.

57. *AM*, No. 144, May 1975, pp. 10–14; A, March 22–23, 1975; Shoukri, pp. 240–41.

58. Interviews with al-Sha'rawi and M. Gamal Imam, October 1987; Tomiche (1974), p. 80. All subsequent references to this incident are based on the same sources.

59. *Sawt al-'Amal* (hereafter SA), No. 5, August 1986, pp. 27–34.

60. Sharaf al-Din et al (1986).

61. *T*, September 1976, p. 5; *AM*, No. 141, February 1975, pp. 10–11. The factory is unnamed in these sources.

62. A, September 21–22, 1976; *U*, September 27, 1976; *RY*, No. 2520, September 27, 1976, pp. 4–5, and No. 2450, May 26, 1975, p. 9.

63. On these incidents, see also El-Shafei, pp. 43–64.

64. See Olson.

65. El-Shafei, p. 52.

66. For an argument to this effect, see El-Shafei.

67. For more on the community solidarity of the Egyptian *sha'b*, see the study by Singerman.

68. See the volume edited by Cook and Levi, especially the chapters by Michael Taylor and Russell Hardin; Putnam, pp. 167–71, cites the community savings associations common in developing countries as an example of an institution which survives based on norms of trust. See Singerman, pp. 165–66, for more on these associations in Egypt.

69. See Frank for a persuasive case that behaviors which facilitate trust are in fact biologically selected.

70. Scott (1976) suggested much the same for peasants in his discussion of agrarian rebellions. See also Scott (1985), pp. 242–47, 320–26.

71. Investigating such differences was in fact my original goal in Egypt. I found that research conditions there rendered this impossible.

72. *MEED*, October 19, 1984.

73. *Al-Ahram al-Iqtisadi*, October 26, 1987.

74. Barakat interview.

75. Sharaf al-Din et al., pp. 28–30.

76. SA, No. 3, October 1985, p. 5.

77. El-Shafei, p. 56.

78. *U*, March 23, 1968.

79. Fa'id interview.

80. *RY*, March 17, 1986, p. 5.

81. Shoukri, pp. 240–41; Khayr interview.

82. Barakat interview.

83. *U*, June 23, 1975.

84. A, September 21–22, 1976; *U*, September 27 and October 18, 1976; *RY*, No. 2520, September 27, 1976, pp. 4–5.

85. *U*, January 24 and 31, 1977; *AM*, March 1977, p. 20.

86. SA, No. 3, October 1985, p. 5.
87. *MEED*, October 19, 1984.
88. *al-Akhbar*, February 10, 1986.
89. SA, No. 5, August 1986, pp. 27–34.
90. See also *al-Akhbar* and Reuters from July 9, 1986.

4. The Union Movement With and Against Reform

Portions of chapter 4 appeared previously in "Labor as an Obstacle to Privatization: The Case of Egypt, 1974–87," in Iliya Harik and Denis Sullivan, eds., *Privatization and Liberalization in the Middle East* (Bloomington: Indiana University Press, 1992).

1. Harbison, p. 182; Goldberg 1992.
2. Mabro and Radwan, pp. 96–97.
3. The 1959 law specified that all collective grievances emanating from plants where a union local existed must be submitted by the local president. Girgis, p. 18. Hence the involvement of local officials in many of these collective grievances can be assumed.
4. It negotiating 16 collective agreements between 1961 and 1967, generally winning yearly raises, production bonuses, compensations, paid holidays, and better medical care. The federation's president at the time was Sa'd Muhammad Ahmad. *al-'Ummal* (hereafter *U*), December 1966, p. 3, January 1967, p. 17, and September 1967, p. 5.
5. Khalid (1971), pp. 83–86.
6. Devaluations can have an immediate inflationary effect, but that is only when consumers are very dependent on imports, and when import prices are vulnerable to movements in the exchange rate. Where subsidies and price controls are in operation, however, as was the case in Egypt, they can cushion consumers from the effects of devaluation.
7. For details on all the nationalizations, see Waterbury (1983), pp. 68–76, and the additional references he cites.
8. Baker, p. 181; Khalid (1975), p. 43; 'Izz al-Din, p. 956; Waterbury (1983), p. 93–97, 409.
9. Mabro and Radwan, p. 42, for GDP figures, base year 1959/60, average annual growth rate computed by author; p. 87 for rates of industrial growth.
10. Abdel-Fadil, p. 8, for employment figures, based on CAPMAS, Census of Industrial Production. Average annual growth rate computed by author.
11. Waterbury (1983), pp. 93–97, 409; Abdel-Fadil, pp. 33–34; *U*, December 1966, pp. 3–5.
12. *U*, August 1967, pp. 4–5, 18–19; *al-Ahali* (hereafter *AH*), October 24, 1984.
13. See *U*, February 11, 1971, p. 3, August 14, 1969, p. 5, and September 3, 1976, p. 3, Balal et al. (December 1968), pp. 13–14, and *AM*, December 1967, pp. 10–12. For most workers the first line of defense against dismissals in this period was the tripartite committees mentioned previously, but these more often than not proved ineffective. In the first ten years after their creation, the tripartite committees successfully returned 13,000 workers to their jobs, but there were 19,000 cases where

the company rejected a committee ruling in favor of the workers. In 1968 the committees heard 4,817 cases, and agreed with the firing in 2,997 cases, or about 64%. However, of the 1695 appeals which the committee upheld, workers were actually reinstated in only slightly more than half (911) of the cases.

Under the new law, decree 3309 of 1967, cases where no tripartite committee existed, as well as appeals of committee rulings in favor of management or stemming from those instances where management refused to accept a committee recommendation for reinstatement, no longer went to the Labor Office or the Labor Tribunals. Instead they were handled by the legal apparatus of the parliament. Workers had to raise money in advance to pay for a parliamentary hearing, and there was no mechanism for enforcing the rulings in their favor. Furthermore the law released managers of personal responsibility for paying fines stemming from violations of the labor code.

The genesis of Law 3309 is unclear, but it clearly represented an eclipse of the power of the Labor Ministry, which made its displeasure with the new law known in the *al-'Amal* articles cited above. It seems likely that the Workers' Secretariat group would also have objected, given the suspicions of the 'Ali Sabri clique toward public sector managers; the authorship of one of the exposes of this problem by 'Abd al-Hamid Balal, an affiliate of the WS group, is also indicative of this. Opposition to decree 3309 thus appears to have been an issue around which the ETUF and WS groups were on the same side. Which forces in the regime were pushing the law remains unclear.

14. 'Izz al-Din reports that there were 289 union leaders in attendance at the conference. The one dissenter was apparently Mutawalli al-Shar'awi, who told me (interview, March 1988) that he refused to support the call for sacrifices.

15. 'Izz al-Din, pp. 979–81; Khalid (1971), pp, 75–77.

16. *U*, May 8, 1968, March 27, 1969; *al-'Amal* (hereafter *AM*), February 1970, p. 9.

17. Moore, p. 451.

18. Interview with Muhammad Gamal Imam, October 1987.

19. Kamel, pp. 132–47.

20. After peaking at 135,644 in 1964–65, the number of complaints fell back to 119,736 in 1965–66. *al-Tali'ah* (hereafter *T*), January 1868, pp. 21–59, November 1968, pp. 21–47; al-Ghazzali, 1968, p. 61. There was no information on the disposition of these cases or on the number of grievances filed in subsequent years.

21. Since most large industrial and service firms were now under public ownership, I believe the data in this table can be considered a more reliable reflection of wage trends *in the formal sector* than the 1950s data, as public sector firms would be more likely to respond to questionnaires from other government agencies, and to provide more accurate information. It must be remembered, though, that administrative employees' salaries are included in these figures, while shops of less than 10 employees, where pay was generally lower, are excluded from the statistics.

22. Salama interview, July 1988; Balal (1969). I interviewed Balal in March 1988, when he was secretary to the Minister of Labor. Although he reaffirmed his assertions in the article, he would not provide any further details, and was in general reluctant to discuss this period with me.

23. The selection of Kafr al-Dawwar as a cite for the 1968 Mayday celebration may itself have been a symbolic statement by the ETUF, although no mention of the 1952 incident was made at the event.

24. *U*, 1970, various; 'Izz al-Din, p. 1009.

25. Waterbury (1983), p. 139; Khalid (1975), pp. 95–103. The committee report was reviewed in a lengthy article by Sa'id Sanbal in the daily newspaper *Akhbar al-Yawm*. The full text of the article, which appeared February 9, 1974, is reprinted in Khalid.

26. Khalid (1975), pp. 95–193; *AM*, No. 143, April 1975, p. 16.

27. 'Umar, pp. 249–50.

28. *Rose el-Yusef* (hereafter *RY*), No. 2279, February 14, 1972, pp. 14–18; *U*, January 27, 1972, pp. 4–5, February 10, 1972, p. 6.

29. Equivalent minimum wage protection was extended to cover the remaining private concerns in 1974, but its effects may not be fully reflected in these data since they cover only establishments employing more than 10 workers.

30. The data used in this survey were collected at the end of October. At that time many public sector manufacturing workers who were not called up to fight were earning overtime pay, while soldiers continued to earn only their basic wage. Most private sector establishments reportedly did not offer any salary compensation to workers who were drafted. See *RY*, No. 2290, April 1, 1974, p. 6.

31. Mahmoud, pp. 51–55, 120, and Davies, pp. 59–60, 90, 97.

32. Based on my interviews with workers during 1987–88.

33. Mustafa, pp. 18–19, and interview with Fathi Mahmud (Mustafa), July 1988; *T*, May 1974, pp. 113–14.

34. 'Abd al-Rahman, pp. 16–17; *al-Ahram* (hereafter A), November 11, 1974.

35. 'Abd al-Rahman, pp. 16–17; *RY*, No. 2457, July 14, 1975, p. 3; A, December 2, 1974.

36. A, June 2, 1975; Mahmud interview; *RY*, No. 2450, April 26, 1975, pp. 3–6, and 2451, May 2, 1975, pp. 14–17.

37. Mahmud interview.

38. Though their origins date back to the 1950s, the joint companies became part of the public sector in the 1960s and were thus subject to its laws.

39. Mursi (1976), pp. 151–59; al-Maraghi (1983), pp. 88–92; Rif'at, pp. 6–7; *RY*, No. 2706 (May 5, 1980), pp. 8–11.

40. One likely source of extra-union opposition was public sector managers. In the late 1980s, the Minister of Industry himself was a staunch defender of the public sector. However, preliminary research by Waterbury (1988), p. 21, suggests that state managers were not in active opposition to privatization in Egypt.

41. Rif'at, pp. 6–7; Qandil (1985), p. 460; A, June 2 and June 9, 1975.

42. *AM*, No. 146, July 1975, p. 14; Qandil (1985), p. 460; Rif'at, pp. 6–7.

43. *RY*, No. 2458, July 21, 1975, p. 6; 2474, November 10, 1975; 2475, November 17, 1975; and 2576, November 24, 1975; Mahmud interview.

44. The figures may also indicate the lagged effect of the equalizing legislation of the early 1970s whose implementation, as suggested above, was somewhat interrupted by the October 1973 war.

45. During the *infitah* years blue-collar employment in the private sector grew more rapidly than in the public sector. According to the SEWHW statistics, the blue-collar public sector workforce included 548,720 workers in 1970. This climbed to 627,813 in 1982, for an average annual growth rate of 1.12%. In the formal (establishments of more than 10 workers) private sector, blue-collar employment was 133,744 in 1970, and rose to 177,214 in 1982. This means an average annual growth rate of 2.37%, more than double that in the public sector.

46. Some private company owners admitted to evading the labor laws in a study conducted for the U.S. Agency for International Development (hereafter AID) in mid-1985. See the study by the Partnership for Productivity.

47. Waterbury (1983), pp. 140–42; Maraghi (1983), p. 92; *AH*, September 2, 1987.

48. Waterbury (1993), p. 241; Zaytoun, pp. 224–25; Salmi, pp. 36–40.

49. ETUF (1980), p. 84.

50. Waterbury (1983), p. 141; *RY*, No. 2707, April 28, 1980, p. 11, 2730, October 6, 1980, pp. 6–7, and 2731, October 13, 1980, p. 6.

51. *RY*, No. 2650, March 26, 1979, p. 4, and 2735, November 10, 1979, p. 7; *AM*, No. 211, December 1980, p. 18; *AH*, November 3, 1982. A detailed report prepared by the Federation of Bankers and Insurance Workers on how the proposal would affect workers and the public sector generally was reprinted in *al-Ahram al-Iqtisadi* (hereafter AI) 615, October 17, 1980; for summaries of the research papers presented at the conference see *RY*, No. 2734, November 3, 1980, pp. 12–13, and *AM*, No.211, December 1980.

52. These meetings are summarized in the memo by Lubell.

53. *AH*, November 3, 1982; *AM*, No. 218, July 1981, pp. 16–19.

54. *AH*, November 3, 1982. The article does not specify the precise mechanism for privatization in this plan.

55. *AH*, November 3, 10, and 17, 1982, December 1 and 8, 1982.

56. The growth of these associations and their interactions with the government are discussed in Qandil (1985a and 1986a). See also the chapter on interest groups in the 1986 *Arab Strategic Report*.

57. Qandil (1986b), p. 19; *AH*, January 8, 1986, February 26, 1986; 'Umar, p. 246.

58. *AH*, December 24 and 31, 1986, January 14 and 21, 1987.

59. *AH*, January 21 and March 4, 1987.

60. Ibid.

61. A single firm entering Egypt under Law 43 is called an "investment company" (*sharikah istithmariyah*), or sometimes just a "Law 43" company. A joint venture under Law 43 is referred to as an "investment project" (*mashru' istithmari*) or a "joint project" (*mashru' mushtarik*), as distinguished from the "joint stock company" (*sharikah mushtarikah*) referred to above.

62. Waterbury (1983), pp. 142–43, 147; Maraghi (1983), pp. 92–95; Mursi (1987), p. 57; Sa'id and Harun, p. 65.

63. *Rose al-Youssef* ran a five-part critique of joint ventures beginning May 5, 1980. *Al-Ahali* also ran periodic exposes; see especially July 24, 1985, October 2, 1985, and June 11, 1986. See also Waterbury (1983), pp. 141–42, and the works by Mursi, al-Maraghi (1983), and Sa'id and Harun, cited above.

64. *AM*, No. 198, November 1979, p. 16, and 211, December 1980, p. 15.

65. Waterbury (1983), pp. 153–54; *RY*, No. 2641, January 22, 1979, pp. 12–15.

66. Ibid.; see also Hinnebusch, p. 141.

67. The generally assembly is literally a meeting of shareholders. In wholly publicly owned companies, the shareholders consist of the company's management and various other officials appointed by the Minister of Industry. The assembly normally meets once a year to approve the production plan for the factory. Youssef, pp. 128–29.

68. *RY*, No. 2706 May 5, 1980, pp. 8–11; *AM*, No. 198, November 1979, pp. 16–17; Maraghi (1983), p. 95.

69. *RY*, No. 2739, December 8, 1980.

70. The Chloride story is compiled from the following sources: interviews with 'Abd al-'Aziz Higazi, president of the Chloride-Egypt management council, January 30, 1988, and Fayyiz Mahmud al-Karta, president of the GBC local, February 7, 1988; *Sawt al-'Amal* 4, January 1986, and 5, May 1986; *Awraq 'Umaliyya* 2, May 1985; *RY*, No. 2715, June 23, 1980; *AI*, No. 617, November 10, 1980; and *Al-Ahali* from September 29, 1982. through May 19, 1983, and February 6, 1985, February 12, 1986, and November 27, 1986. Some discrepancies in these accounts are noted.

71. Higazi said that workers' salaries were doubled when they transferred to the joint venture, and claimed that it was only leftists among the workforce who opposed the project.

72. MEED, August 17, 1985.

73. *AH*, March 19, June 16, and July 2, 1986. Another leftist critique of the project can be found in Sa'id and Harun. The authors noted that GM was on the Arab boycott list because of its relations with Israel, and charged that the company hoped to use the project to circumvent the boycott by exporting cars from Egypt to the Arab world.

74. *AH*, June 16, 25, and July 2, 1986.

75. *AH*, July 22 and October 15 and 29, 1986; *Tali'at al Sina'a* (hereafter *TS*) 21, July 1986, pp. 4–5, and 22, December 1986, p. 17.

76. *AH*, July 2, 1986.

77. Personal interview, January 1987. Mr. Rahman's remarks were made in the context of comparing Nasco's workers to those at the GM Egypt factory in 6th October city. The latter, a strictly private sector joint venture, had recently opened and was producing vans and light trucks. Workers at that plant received relatively high wages and an attractive benefits package, including mortgage loans for nearby new housing, in part to lure them to the remote location of the factory. Rahman stated that GM Egypt workers were more loyal to the company than to the union.

78. *AH*, October 29 and December 10 and 31, 1986, and January 14, 1987.

79. *AH*, October 22, 1986.

80. Maraghi (1983), p. 94; Waterbury (1983), p. 163.

81. 'Aziz' departure from the Tagammu' was caused by disagreements with other leaders over internal functioning. During most of the 1980s he was affiliated with a Nasirist party that the government refused to recognize.

82. An exception is the new Semiramis hotel, which opened in Cairo in 1987, and is strictly privately owned.

83. *Business Monthly,* December 1987, p. 46.
84. Ibid.
85. Ibid.
86. Interview with Mustafa Ibrahim Mustafa, July 1988.
87. Khair interview, October 1987; *U,* January 3 and 10, 1977; ETUF (1982), pp. 56–57; *AM,* No. 165, February 2, 1977, pp. 16–18.
88. See, e.g., Waterbury (1993), p. 240.
89. *U,* January 24, 1977. The statement is reprinted in al-Sayyid, pp. 75–76, and 'Abd al-Raziq, pp. 86–88.
90. *U,* February 7, 1977.
91. The material on government price policy in this section is drawn from annual volumes of *MEED* and *Middle East Economic Survey* from 1983 through 1988, as well as from conversations with ordinary Egyptians in the summer of 1984, and from June 1986 through August 1988.
92. See also Singerman, pp. 166–67.
93. This idea was embodied in the proposal raised by some economists to replace direct subsidies with ration cards.
94. *AH,* March 20, 1985; A, July 15, 1986; Qandil (1985b), p. 464; 'Umar, pp. 236–37.
95. *AH,* May 1, 1985, October 1, 1986.
96. See Harik (1988) for more discussion of the 1987 agreement.
97. Khalid (1975), pp. 86, 95–103; Waterbury (1983), pp. 125–39 and (1980), p. 352; Cooper, pp. 44–54, 88–89. Both Cooper and Waterbury argue that the basis for the new economic policies was actually laid under Nasir. Cooper dates the transition to the aftermath of the 1967 war, citing a number of price hikes, several new regulations aimed at encouraging the private sector, some "minor desequestrations," and new public sector management policies that were seen by some as an attack on the very concept of parastatals. Waterbury sees the winds of change starting to blow in 1966, when legislation was prepared to make Port Said into a free zone.
98. Waterbury (1983), pp. 125–26; Khalid (1975), p. 86.
99. Khalid (1975), pp. 88–91.
100. Ibid., p. 105.
101. Cooper, pp. 88–89.
102. *AM,* June 1974, pp. 38–39; *RY,* No. 2290, April 1, 1974, p. 6, 2293, April 15, 1974, pp. 5–7, and 2294, April 22, 1974, p. 12.
103. Projects with foreign capital which began under the auspices of the law were legally considered to be "joint-stock companies," under Law 26 of 1954, the private companies law. Law 26 had been amended in the 1960's to require that joint-stock companies implement systems of workers' representation in management and workers' share of profits similar to those in the public sector. However, Law 43 exempted the new joint ventures from these provisions. Law 26 was repealed and replaced by Law 159 of 1981; the exemptions for Law 43 companies either remained in effect or were obviated by the new law. Mahmoud, pp. 51–55, 120; Davies, pp. 59–60, 90, 97.
104. These are the only published indications of workers' views. Only a small number of workers were interviewed, and they were not selected randomly, but the

exchanges have all appearances of being genuine, since the workers often expressed opinions at variance with the journal, especially on foreign affairs. The interviews seemed aimed mainly at exposing the abysmal poverty of the masses. The journal was shut down by the government several months after the January 1977 riots, an action some blame on the publication of these interviews.

105. 'Umar, pp. 249–61.

106. *U*, April 14, 1975, pp. 5–7.

107. ETUF (1977), pp. 93–94.

108. Khayr interview; Qandil (1985b), p. 459; al-Sayyid, p. 81.

109. Qandil (1985b), pp. 459–62; 'Umar, p. 262; Waterbury (1985), pp. 72–73.

110. The laws applied only to joint-stock companies. The World Bank [1983), pp. 214–16, and John Waterbury (1985), p. 80, suggest that all joint stock companies were made subject to the same profit-sharing requirements as the public sector in the 1960s. This would imply that 25% of net profits had to be distributed to workers. However, the ETUF position before the law was passed stated that the private sector was required to share only 10% of profits, and this is the figure that Davies says Law 43 companies were exempted from, although they were required to have some form of profit sharing. This would imply that the WSP rules for private sector joint ventures before Law 159 was passed were less stringent than those incumbent on the public sector. Similarly, it appears that there were requirements for WRM in the private sector, before law 159, but these applied only to joint-stock firms or those with more than 50 employees; before the *infitah,* most joint ventures involved partial government ownership and were therefore considered part of the public sector, and most private sector firms had fewer than 50 employees. As with profit sharing, the new law made some system of WRM incumbent on the new joint ventures, but both the percentage of workers' participation and the areas subject to discussion in WRM committees were reduced.

111. Qandil (1985b), pp. 459–62; 'Umar, p. 262; Waterbury (1985), pp. 72–73; Davies, pp. 59–60, 90, 97; *AM*, No. 211 (December 1980), p. 14.

112. In this case the domestic private sector bore much of the suffering, and factory owners also called for import curbs. For description and analysis of the *infitah's* effect on the textile industry, see *AH*, August 21, 1985, November 20, 1985, and March 26, 1986.

113. *RY*, No. 2734, December 8, 1980; A, May 16, 1985; *AH*, September 10, 1986.

5. *Toward a New Era in Labor/State Relations*

Portions of chapter 5 appeared in "Egypt's New Labor Law Removes Worker Provisions," *Middle East Report*, No. 194/5 (May–August 1995): 52–53, 64.

1. One Western businessmen complained that, "at best, privatization [in Egypt] will be a lengthy process, at worst a perpetual one." Cited in Middle East Economic Survey (hereafter *MEES*), May 17, 1993, p. B3.

2. *MEED*, 1987 volume. All subsequent information in this section and the next is derived from the MEED volume for the relevant years; only additional references are cited.

3. Traditionally, the IMF has restricted its focus to stabilization programs, while structural adjustment has been the province of the World Bank. In the case of Egypt, there was a clear coalescence of the two agencies around privatization beginning at this time. On the development of the economic programs of these agencies, see Kahler (1989 and 1990). Al-Sayyid (1990) also discusses the policy approach of the United States Agency for International Development (USAID), a major lender in Egypt.

4. *MEES*, October 9, 1989, p. B1. The article is based on a report in *al-Ahali* (hereafter *AH*), September 20, 1989.

5. Information in this paragraph and the one following is drawn from al-Sayyid (1990), pp. 53–8).

6. However, some small furniture-making companies were also included, and some government owned vehicles were also slated for sale. *AED* (Africa Economic Digest), February 19, 1990, p. 19.

7. *AED*, February 26, 1990, p. 16.

8. *MEES*, April 9, 1990, p. B4.

9. Al-Sayyid, p. 15. According to Sullivan (1992), p. 32, however, AID was ultimately prevented from pursuing this project, and instead attempted to finance workers' equity in an entirely new, wholly private sector tire company; that project was stillborn two years later.

10. He excluded, in particular, the textile plans at Kafr al-Dawwar and Mahalla al-Kubra, and the iron and steel complex at Helwan, all of which had been scenes of labor militancy in the 1970s and 1980s; the Kima fertilizer plant and the Naga Hammadi aluminum smelting operation were also marked for preservation.

11. See also *The Banker*, July 1990, pp. 62–66, and *Euromoney*, special supplement on Egypt, September 1990, pp. 42–48.

12. The impact of this on the workforce in these establishments is not clear, but the absence of information to the contrary suggests that these workers either kept their jobs, were transferred to other public sector employment, or remained on the government payroll even while idle, like the workers in the tourist sector.

13. See the pamphlet by the Popular Committee for Constitutional Reform (in Arabic).

14. Interview with Niyazi 'Abd al-'Aziz, then president of the EEMWF, December 31, 1994.

15. The selection process had been delayed by the Gulf War. When the award was finally given to a consortium that included the Coopers and Lybrand accounting firm and the Bechtel Corporation, some disappointed competitors charged that regional politics played a heavy role in the decision. Bechtel has vast interests in the Gulf region and enjoyed strong ties to the Bush administration; the company also received lucrative contracts for rebuilding Kuwait's oil facilities after the war.

16. See *AH*, December 2, 1992, *al-Ahram* (hereafter A), March 5, 1993, *al-Gumhuriyya*, October 21, 1993, and *al-Tadamun* (a newsletter of opposition unionists), No. 1, December 1993, pp. 14–16. Although the ILO had historically served as a liberal counterweight to the orthodox multilaterals, some Egyptian leftists cite

evidence that it had now allied with these agencies in promoting structural adjustment. See the works by Freeman (1993) and Hilali.

17. Bromley and Bush, p. 204.

18. *MEES*, July 20 and August 17, 1992.

19. *al-Wafd* (hereafter *W*), December 10, 1992, and October 18, 1993.

20. Fu'ad 'Abd al-Wahab should be distinguished from Muhammad 'Abd al-Wahab, the Minister of Industry. After the PEO's creation, Abu Ghazzala apparently ceased to be involved publicly in the privatization effort.

21. *MEES*, May 17, 1993.

22. The anchor for the Al-Nasr Bottling Company, which holds the Coca Cola franchise, is Mak for Investments, a firm of private Arab investors. Pepsi Cola won the bidding to be anchor in the purchase of the Egyptian Bottling Company, which holds the Pepsi franchise.

23. See, e.g., the statements from Mubarak in *October*, January 24, 1993, and 'Atif Sidqi in A, January 10, 1993.

24. See *al-Sha'b* (hereafter *SH*) from January 5, January 26, and March 30, 1993; *AH*, January 22, 1993, and *al-Haqiqa*, April 3, 1993.

25. Wickham (1996).

26. I am grateful to Ragui Assaad for calling this to my attention. On employers' objections to the strike/layoff trade-off, see *Al-'Alam al-Yawm*, June 5, 1993.

27. Cited in *MEED*, June 18, 1993.

28. The term "ritual dances" comes from Callaghy, 1989, p. 129.

29. A schedule of firms slated for privatization was published in *Al-Ahram al-Iqtisadi*, December 26, 1994, p. 33.

30. The 1976 union law had maintained the right of the Ministry of Labor to inspect the financial records of the unions. The leading unionists at that time continued to object to this provision, and in the 1981 law it was modified to specify joint supervision of union finances by the confederation and the ministry. The concern was largely dropped by the confederation after this modification, until 'Abd al-Haqq's leadership at the Ministry brought it to the fore again, prominently.

31. *Al-Akhbar*, February 19, 1991; *Al-Yasar*, December 1992; *AH*, January 29, 1992.

32. The election campaign and results were covered regularly in *al-Ahali*, *al-Sha'b*, *al-Wafd*, and the official press from October through December. This paragraph draws especially on the summary in *AH*, December 4, 1991. See also the study prepared by the Arab Research Center. The discussion of the elections in the following section also draws on these sources.

33. *Al-Yasar*, December 1992; *AH*, January 29, 1991.

34. *Al-Yasar*, June 1992; *AH*, April 22, 1992.

35. *October*, July 18, 1993. The journalist Muhammad Khalid was a long-standing labor correspondent who authored two books on union affairs under Nasir. See also *al-Yasar*, July 1992.

36. W, July 12, 1993; *Rose al-Youssef* (RY), October 18, 1993; 'Izz al-din interview, January 4, 1995.

37. *Al-Ahram al-Masa'i*, January 1, 1993; *Al-Yasar*, December 1992; *SH*, December 18, 1992 and January 22, 1993.

38. This paragraph and the next draw on the extensive coverage of the controversy in the opposition press from late April through December of that year, and occasional articles in the government-owned dailies. See especially *AH*, April 28 and August 4, 1993; *Al-Yasar*, September 1993; and *al-Haqiqa*, October 22, 1993.

39. W, August 7, 1993.

40. Interview with 'Abd al-Rahman Khayr, December 31, 1994.

41. *Al-Arabi* (hereafter *AR*), July 19, 1993.

42. Interview with Fathi Mahmud, January 4, 1995. Mahmud was then general secretary of the Arab Nasirist Party.

43. See, e.g., 'Uthman (1991).

44. Based on discussions with VOW members at a forum sponsored by the Al-Jeel Center in 'Ain al-Sera, December 29, 1994.

45. National Committee to Fight Privatization, 1994 (in Arabic).

46. Interview, January 4, 1995.

47. Originally created in the 1970s by Sadat as a counterweight to the Tagammu', the SLP formed an electoral alliance with the Muslim Brotherhood, legally precluded from contesting elections in its own name, during the 1987 parliamentary elections.

48. These appeared periodically in *al-Sha'b*; see also al-Banna (1994).

49. Al-Sayyid (1990), p. 39.

50. Qandil (1986), p. 19.

51. *AH*, June 7 and June 21, 1989; *SH*, June 6 and July 4, 1989.

52. Al-Sayyid, p. 42–43; *AH*, December 13, 1989.

53. *AH*, April 18, 1990; *SH*, April 24, 1990.

54. *AH*, June 12 and September 11, 1991.

55. *AH*, June 19, 1991.

56. Interviews with Khayr and 'Abd al-'Aziz.

57. Egyptian Trade Union Federation, *al-Masirah al-tarikhiyah lil-ittihad al-'amm li-niqabat 'ummal misr fi 35 'aman, 1957–1992*. Cited in Samya Sa'id Imam (1994), p. 211.

58. Some of the mechanisms which protect and entrench senior union personnel are described in *Al-Ahram al-Iqtisadi*, December 3, 1991, *Al-Ahali*, December 4, 1991, and *Al-Sha'b*, November 26 and December 3, 1991.

59. Interviews with leftist workers in Helwan, January 1995.

60. *AH*, January 22 and February 8, 1992.

61. *SH*, August 7, 1992.

62. *SH*, July 21 and 28, 1992; *AH*, July 22 and 29, 1992; *AR*, November 22, 1993.

63. *AH*, September 12, 1992.

64. *SH*, January 1 and 5, 1993; *AH*, January 6, 1993.

65. *AR*, November 22 and December 20, 1993; W, November 21, 24 and December 2, 1993.

66. 'Abd al-'Aziz interview.

67. W, October 18 and 29, 1993; *SH*, October 30, 1993; *Al-Ahram Weekly*, December 29, 1994.

68. The full text of the statement appears in the newsletter *al-Tadamun* (Solidarity), December 1993, pp. 17–18.

69. *AH*, April 17, 1996.
70. *AH*, September 12, 1992.
71. *SH*, April 2, 1996.
72. Mahmud interview.
73. *AH*, June 7 and 21, 1989; *SH*, June 13, 1989; meeting with Helwan workers at Dar al-Khidamat al-Niqabiyah (Center for Trade Union Services), January 4, 1995.
74. *AH*, March 27, 1991; meeting with Helwan workers.
75. On the performance of Islamic forces in the 1987 and 1991 trade union elections, see Posusney (1996).
76. Several months later, a similar committee was formally established at the national level by leaders of the Tagammu' and the Nasirist party; it declared itself in a state of "ongoing meeting." *Al-Arabi*, August 9, 1993.
77. *AH*, March 10, 1992, May 12, 1992, April 7 and 21, 1993.
78. *Al-Tadamun*, No 1, December 1993; interview with 'Abd al-Rahman Khayr, an activist with the committee, December 31, 1994.
79. *Al-Ahram*, September 9–14, 1992; see also Assaad (1993), pp. 40–41. I am grateful to Ahmad Abdalla for first calling my attention to this incident.
80. *Hurriyati*, December 13, 1992; *AH*, January 22, 1993.
81. *AH*, May 12, 1992 and December 8, 1993.
82. *AH*, June 14, 1993; Interviews with Amin 'Izz al-Din and Taha Sa'd 'Uthman, January 1995.
83. The total population of the city at the time was about 500,000.
84. The Kafr al-Dawwar story is based primarily on information provided by human rights activists in Egypt, including lawyers who defended the workers arrested. Because of the recency of the incident and the fact that court cases are still pending, I am protecting the identity of these sources. See also *Al-Badil*, No. 3 (January 1995), pp. 20–26.
85. These are penalties in addition to the days originally absent and mark a new reinterpretation of existing law which allows a 1 day penalty per day of excess absence with a maximum of 5 days loss of wages.
86. Based on an interview with the lawyers representing the detained workers.
87. This discussion of the law is based on the complete draft published in *Al-Ahali*, November 30, 1994. A slightly altered version was published in *al-'Amal*, March 1995, but was not available to me at the time of this writing.
88. 'Abd al-'Aziz interview.
89. *Al-Ahram Weekly*, May 23–29, 1996.
90. The courts were empowered to order reinstatement of the worker only if it was established that the dismissal was not punishment for trade union organizing; otherwise, they could mandate only financial compensation. However, the burden of proof that the firing was not union-related fell on the employer. The new law maintains this provision.
91. Employers complained that the tripartite committees were an obstacle to firing because both government and union representatives usually sided with the worker. In contrast, the courts have upheld dismissals in most cases. See al-Muhami, p. 20.
92. Mahmud interview.

93. See also Assaad (1993), pp. 2, 41–42.
94. Mahmud interview.
95. Ibid.
96. This paragraph and the one following based on personal correspondence from Sabir Barakat, June 1995.
97. Written communique from Egyptian labor activists, Spring 1995.

Conclusions

1. Thus this study resides in what Peter Evans (1995), p. 2, has proclaimed as the "eclectic, messy center" of comparative politics.
2. For an excellent overview, see the article by Armijo, Biersteker, and Loewenthal.
3. On the general political economy of the East Asian miracle countries, see the works by Haggard (1990) and Wade (1990); on labor in particular, see titles by Deyo, Luther, Ogle, and Choi. The contrasts made here are also based on discussions with economists who work on the Asian cases.
4. By the same logic, it should also be politically easier to "sell" privatization in the former Soviet republics where private sector enterprise was previously nonexistent, so workers lack a standard of comparison.
5. Armijo et al., pp. 22–23.
6. See Introduction, note 7.
7. See the paper by Epstein.
8. See the papers by Nikoi and Pastor.

Bibliography

Books, Articles, and Dissertations in English

Abdalla, Ahmed. *The Student Movement and National Politics in Egypt*. London: Al-Saqi Books, 1985.

Abdel-Fadil, Mahmoud. *The Political Economy of Nasserism*. Cambridge: Cambridge University Press, 1980.

Abdel-Malek, Anouar. *Egypt: Military Society*. New York: Random House, 1968.

Alexander, Chris. Between Accommodation and Confrontation: State, Labor, and Development in Algeria and Tunisia. Ph.D. dissertation, Duke University, 1996.

Almond, Gabriel. *A Discipline Divided: Schools and Sects in Political Science*. Newbury Park, Calif.: Sage, 1990.

Armijo, Leslie, Thomas Biersteker, and Albert Loewenthal. "It's Fine in Practice, But Does it Work in Theory?" Paper presented at the Workshop on Political and Economic Reform, Watson Institute for International Affairs, Brown University, September 1994.

Arrighi, G. and J. Saul. *Essays of the Political Economy of Africa*. New York: Monthly Review Press, 1973.

Arthur R. Little International. "Review and Evaluation of Small Scale Enterprises in Egypt." Mimeograph, March 1982. U.S. A.I.D. Library, Cairo.

Assaad, Ragui. "The Employment Crisis in Egypt: Trends and Issues." *Working Papers in Planning*, Cornell University Dept. of City and Regional Planning, No. 86 (January 1990).

——. "Structural Adjustment and Labor Reform in Egypt." Paper presented at the 1993 annual meeting of the Middle East Studies Association, Raleigh/Durham, N.C., November 11–14, 1993.

——. "Compliance with Labor Market Regulations in Egypt." Paper presented at the 1995 annual meeting of the Middle East Studies Association, Washington, D.C., December 6–10, 1995a.

——. "The Effects of Public Sector Hiring and Compensation Policies on the Egyptian Labor Market." Paper prepared for the World Bank as a background document for the 1995 *World Development Report.* (1995b)

——. "An Analysis of Compensation Programs for Redundant Workers in Egyptian Public Enterprise." Paper presented at the workshop on "The Changing Size and Role of the State-Owned Enterprise Sector." Amman, Jordan, May 15–17, 1996.

Assaad, Ragui and Simon Commander. "Egypt: The Labour Market Through Boom and Recession." World Bank, May 1990.

Audsley, M. T. "Labour and Social Affairs in Egypt." *St. Antony's Papers* 4 (1958): 95–106.

Awad, Ibrahim. "Socio-Political Aspects of Economic Reform: A Study of Domestic Actor's Attitudes towards Adjustment Policies in Egypt." In Handoussa and Potter, eds., *Employment and Structural Adjustment*, pp. 275–94.

Baker, Raymond. *Egypt's Uncertain Revolution Under Nasser and Sadat*. Cambridge: Harvard University Press, 1978.

Baldwin, David A. "Interdependence and Power: A Conceptual Analysis." *International Organization* 34, no. 2 (Autumn 1980): 471–506.

Bayat, Assef. *Workers and Revolution in Iran.* London: Zed Books, 1987.

Beattie, Kirk J. *Egypt During the Nasser Years: Ideology, Politics, and Civil Society*. Boulder: Westview Press, 1994.

Beinin, Joel. "The Communist Movement and Nationalist Political Discourse in Nasirist Egypt." *Middle East Journal* 41 (August 1987): 568–84.

——. "Labor, Capital, and the State in Nasirist Egypt, 1952–61." *International Journal of Middle East Studies* 21 (February 1989): 71–90.

——. "Exile and Political Activism: The Egyptian-Jewish Communists in Paris, 1950–59." *Diaspora* 2, no. 1 (Spring 1992): 73–94.

——. "The Egyptian Regime and the Left: Between Islamism and Secularism." *Middle East Report*, No. 185 (November-December 1993): 25–26.

——. "Will the Real Egyptian Working Class Please Stand Up?" In Lockman, ed., *Workers and Working Classes in the Middle East*, pp. 247–70.

Beinin, Joel and Zachary Lockman. *Workers on the Nile*. Princeton: Princeton University Press, 1987.

Bellin, Eva. The Social Origins of Accountability: Industrialization and Democratization in Tunisia, the Middle East, and Beyond. Forthcoming manuscript, Harvard University.

Berger, Suzanne, ed. *Organizing Interests in Western Europe*. Cambridge: Cambridge University Press, 1981.

Bergquist, Charles, ed. *Labor in the Capitalist World-Economy.* Beverly Hills: Sage, 1984.

Bianchi, Robert. "The Corporatization of the Egyptian Labor Movement." *Middle East Journal* 40, no. 3 (Summer 1986): 429–44.

——. *Unruly Corporatism: Associational Life in Twentieth Century Egypt.* Oxford: Oxford University Press, 1989.

Bienefeld, Manfred, Martin Godfrey, and Hubert Schmitz. "Trade Unions and the 'New' Internationalization of Production." *Development and Change*, No. 8 (1977), pp. 417–39.

Biersteker, Thomas J., "Appropriating the Concept of Privatization for Development." Paper presented to the 1989 annual meeting of the International Studies Association, London, England, March 1989.

Booth, Douglas E. "Collective Action, Marx's Class Theory, and the Union Movement." *Journal of Economic Issues* 12, no. 1 (March 1978): 163–85.

Botman, Selma. *The Rise of Egyptian Communism, 1939–70*. Syracuse: Syracuse University Press, 1988.

Boyd, Rosalind E., Robin Cohen, and Peter C. W. Gutkind. *International Labor and the Third World*. Aldershort: Gower, 1987.

Bromley, Simon and Ray Bush. "Adjustment in Egypt? The Political Economy of Reform." *Review of African Political Economy* 60 (1994): 201–13.

Brown, Nathan. *Peasant Politics in Modern Egypt: The Struggle vs. the State.* New Haven: Yale University Press, 1990.

Buroway, Michael. "The Contours of Production Politics." In Charles Bergquist, ed., *Labor in the Capitalist World Economy*, pp. 23–47.

Callaghy, Thomas M. "Toward State Capability and Embedded Liberalism in the Third World: Lessons for Adjustment." In Nelson et al., *Fragile Coalitions*, pp. 123–38.

Callaghy, Thomas M. and Ernest James Wilson II. "Africa: Policy, Reality, or Ritual?" In Vernon, ed., *Prospects for Privatization*, pp. 179–230.

Cardoso, Fernando Henrique and Enzo Faletto. *Dependency and Development in Latin America*. Berkeley: University of California Press, 1979.

Chaudhry, Kirin Aziz. "The Myth of the Market and the Common History of Late Developers." *Politics and Society* 21, no. 3 (September 1993): 245–74.

Choi, Jang Jip. *Labor and the Authoritarian State: Labor Unions in South Korean Manufacturing Industries 1961–1980*. Seoul: Korean University Press, 1989.

Clarke, Joan. *Labor Law and Practice in Egypt*. Washington, D.C.: Labor Bureau, 1965.

Cohen, Robin. "Review of Frobel at al." *Labour, Capital, and Society* 14, no. 1 (April 1981).

——. "Resistance and Hidden Forms of Consciousness Amongst African Workers." *Review of African Political Economy*, No. 19 (September-December 1980), pp. 8–22.

Cohen, Youssef. " 'The Benevolent Leviathan': Political Consciousness Among Urban Workers Under State Corporatism." *American Political Science Review* 76, no. 1 (March 1982): 46–59.

Collier, David, ed. *The New Authoritarianism in Latin America*. Princeton: Princeton University Press, 1979.

Collier, David and Ruth Berins Collier. "Inducements Versus Constraints: Disaggregating Corporatism." *American Political Science Review* 73 (December 1979): 967–86.

——. *Shaping the Political Arena: Critical Junctures, the Labor Movements, and Regime Dynamics in Latin America*. Princeton: Princeton University Press, 1991.

Cook, Karen Schweers and Margaret Levi, eds. *The Limits of Rationality*. Chicago: University of Chicago Press, 1990.

Cook, Paul and Colin H. Kirkpatrick, eds. *Privatization in Less Developed Countries*. New York: St. Martin's Press, 1988.
Cooper, Mark N. *The Transformation of Egypt*. Baltimore: Johns Hopkins University Press, 1982.
Coronis, Susan Dee. The Impact of Trade Unions on Policy Making in Egypt, 1952–84. Ph.D. dissertation, Northwestern University, 1985.
Davies, Michael H. *Business Law in Egypt*. Deventer, Netherlands: Kluwer Law and Taxation Publishers, 1984.
de Almeida, Angela Mendes, and Michael Lowy. "Union Structure and Labor Organization in the Recent History of Brazil." *Latin American Perspectives* 3, no. 1 (Winter 1976): 98–119.
Dekmejian, Hrair. *Egypt Under Nasir: A Study in Political Dynamics*. Albany: SUNY Press, 1971.
Delarbe, Raul Trejo. "The Mexican Labor Movement, 1912–75." *Latin American Perspectives* 3, no. 1 (Winter 1976): 133–56.
Deyo, Frederic C. *Beneath the Miracle: Labor Subordination in the New Asian Industrialism*. Berkeley: University of California Press, 1989.
——. "Export Manufacturing and Labor: The Asian Case." In Bergquist, ed., *Labor in the Capitalist World-Economy*, pp. 267–88.
——. "State and Labor: Modes of Political Exclusion in East Asian Development." In Deyo, ed., *The Political Economy of the New Asian Industrialism*, pp. 182–202.
Deyo, Frederic C., ed. *The Political Economy of the New Asian Industrialism*. Ithaca: Cornell University Press, 1987.
DiTella, Torcuato S. "Working Class Organization and Politics in Argentina." *Latin America Research Review* 16, no. 2 (1981): 33–56.
Elsabbagh, Zoheir Naim. An Analysis of the Impact of the Political Changes on Labor Unions in Egypt. Ph.D. dissertation, North Texas State University, 1977.
Elster, Jon. "Marxism, Functionalism, and Game Theory." *Theory and Society* 11 (July 1982): 453–82.
——. *Sour Grapes*. Cambridge: Cambridge University Press, 1983.
——. *Making Sense of Marx*. Cambridge: Cambridge University Press, 1985.
——. "When Rationality Fails." In Cook and Levi, eds., *The Limits of Rationality*, pp. 19–51.
Epstein, Jerry. "MNC's, Stagnation, and Inequality." Paper presented at the PAWSS seminar on "Global Economic Inequity." Amherst, Mass., June 12, 1996.
Erickson, Kenneth Paul and Patrick V. Peppe. "Dependent Capitalist Development: U.S. Foreign Policy and Reperssion of the Working Class in Chile and Brazil." *Latin American Perspectives* 3, no. 1 (Winter 1976): 19–44.
Erickson, Kenneth Paul, Patrick V. Peppe, and Hobart A. Spalding, Jr. "Dependency vs. Working Class History: A False Contradiction." *Latin American Research Review* 15, no. 1 (1980): 177–81.
Evans, Peter. *Dependent Development: The Alliance of Multinational, State, and Local Capital in Brazil*. Princeton: Princeton University Press, 1979.
——. Comments in "Symposium: The Role of Theory in Comparative Politics," *World Politics* 48 (October 1995): 1–49.

Evans, Peter, Dietrick Rueschemeyer, and Theda Skocpol. *Bringing the State Back In*. Cambridge: Cambridge University Press, 1985.
Fernando, L. "The State and Class Struggle in Sri Lanka." *Labour, Capital, and Society* 16, no. 2 (1983).
Frank, Robert H. *Passions Within Reason: The Strategic Role of the Emotions*. New York: Norton, 1988.
Freeman, Richard B. "Labor Market Institutions and Policies: Help or Hindrance to Economic Development?" In *Proceedings of the World Bank Annual Conference on Development Economics*, 1992, pp. 117–44. Washington, D.C,: World Bank, 1993,
Freeman, Richard B. and James L. Medoff. *What Do Unions Do?* New York: Basic Book, 1984.
Frey, Frederick W. "Comment: On Issues and Non-Issues in the Study of Power." *American Political Science Review* 65, no. 4 (December 1971): 1081–1104.
——. "The Problem of Actor Designation in Political Analysis." *Comparative Politics* (January 1985): 127–52.
——. "Analytical Foci for Power Analysis: Issues Regarding Issues." Mimeograph, University of Pennsylvania, 1991.
Frobel, F., J. Heinrichs, and O. Kreye. *The New International Division of Labor*. London: Cambridge University Press, 1980.
Galenson, Walter, ed. *Labor in Developing Countries*. Berkeley: University of California Press, 1963.
Girgis, Fakmy K. "The System of Preventing and Settling Industrial Disputes in the Arab Republic of Egypt." *Labour* 10 (1974): 17–25.
Goldberg, Ellis. *Tinker, Tailor, and Textile Worker: Class and Politics in Egypt, 1930–52*. Berkeley: University of California Press, 1986.
——. "The Foundations of State-Labor Relations in Contemporary Egypt." *Comparative Politics* 24, no. 2 (January 1992): 147–61.
——. "Including Capital Out: Perverse Incentives and Mixed Messages in Nasser's Egypt." Mimeo, University of Washington, 1994.
Golden, Miriam. "Conclusion: Current Trends in Trade Union Politics." In Miriam Golden and Jonus Pontusson, eds., *Bargaining for Change: Union Politics in North America and Europe*. Ithaca: Cornell University Press, 1992.
——. "The Dynamics of Trade Unionism and National Economic Performance." *American Political Science Review* 87, no. 2 (June 1993): 439–54.
Gordon, Joel. *Nasser's Blessed Movement: Egypt's Free Officers and the July Revolution*. New York: Oxford University Press, 1992.
Greenstone, J. David. "Group Theories." In Fred I. Greenstein and Nelson W. Polsby, eds., *Micropolitical Theory*. Reading, Mass.: Addison-Wesley, 1975.
Haggard, Stephan. "The Politics of Adjustment: Lessons from the IMF's Extended Fund Facility." In Miles Kahler, ed., *The Politics of International Debt*, pp. 157–86. Ithaca: Cornell University Press, 1986.
——. *Pathways from the Periphery: The Politics of Growth in the Newly Industrializing Countries*. Ithaca: Cornell University Press, 1990.
Haggard, Stephan and Robert R. Kaufman. "The Politics of Stabilization and Economic Adjustment." In Jeffrey D. Sachs, ed., *Developing Country Debt*

and Economic Performance, pp. 209–43. Chicago: University of Chicago, 1989.
——. "Economic Adjustment in New Democracies." In Joan M. Nelson et al., *Fragile Coalitions*, pp. 57–78.
——. "The Challenges of Consolidation." *Journal of Democracy* 5, no. 4 (October 1994): 5–16.
Haggard, Stephan and Robert R. Kaufman, eds. *The Politics of Stabilization and Economic Adjustment*. Princeton: Princeton University Press, 1992.
Hall, Peter. *Governing the Economy*. Oxford: Oxford University Press, 1986.
Hall, Peter and Rosemary Taylor. "Political Science and the Four New Institutionalisms." Paper presented at the annual conference of the American Political Science Association. New York, September 1994.
Hansen, Bent. *The Political Economy of Poverty, Equity, and Growth: Egypt and Turkey*. New York: Oxford University Press, 1991.
Handoussa, Heba and Gillian Potter, eds. *Employment and Structural Adjustment: Egypt in the 1990s*. Cairo: American University in Cairo Press, 1991.
Handoussa, Heba and Karim Nashashibi. "Egypt." In National Bureau of Economic Research, *Foreign Trade Regimes and Economic Development*, vol. 4. New York: Columbia University Press, 1975.
Harbison, Frederick H. "Egypt." In Walter Galenson, ed., *Labor and Economic Development*, pp. 146–85. New York: Wiley, 1959.
Hardin, Russell. *Collective Action*. Baltimore: Johns Hopkins University Press, 1982.
Hardin, T. F., and H. A. Spalding. "The Struggle Sharpens: Workers, Imperialism, and the State in Latin America." *Latin American Perspectives* 3, no. 8 (1976).
Harik, Iliya. "The Single Party as a Subordinate Movement: The Case of Egypt." *World Politics* 26 (1973): 80–105.
——. "The Intractable Problems of the Patron State: The Role of the Public and Private (Sectors) in Egypt." Paper presented at the JCNME/SSRC Conference on "Retreating States and Expanding Societies." Aix-en-Provence, March 1988.
Harik, Iliya and Denis Sullivan, eds. *Privatization and Liberalization in the Middle East*. Bloomington: Indiana University Press, 1992.
Heikel, Muhammed H. *Autumn of Fury*. London: Corgi Books, 1984.
Helmy, Eglal Ismael, Alienation Among Industrial Workers in Egypt: A Comparative Study of Weaving and Motorcycle Assembly Operations. Ph.D. dissertation, Pennsylvania State University, 1979.
Hendriks, Bertus. "The Legal Left in Egypt." *Arab Studies Quarterly* 5, no. 3 (Summer 1983): 260–75.
——. "Egypt's Election, Mubarak's Bind," *MERIP* 129 (January 1985): 11–18.
Henley, John S. and Mohamed M. Ereisha. "State Control and the Labor Productivity Crisis: The Egyptian Textile Industry at Work." *Economic Development and Cultural Change* 1987: 491–521.
Hindess, Barry. "Rational Choice Theory and the Analysis of Political Action." *Economy and Society* 13 (1984): 255–77.
Hinnebusch, Raymond A. *Egyptian Politics Under Sadat: The Post-Populist Development of an Authoritarian-Modernizing State*. Cambridge: Cambridge University Press, 1985.

Hudson, Michael. *Arab Politics: The Search for Legitimacy*. New Haven: Yale University Press, 1977.
Hussein, Mahmoud. *Class Conflict in Egypt: 1945–70*. New York: Monthly Review Press, 1973.
Hyman, Richard. "Third World Strikes in Comparative Perspective." *Development and Change* 10 (1979): 321–37.
Issawi, Charles. *Egypt in Revolution: An Economic Analysis*. London: Oxford University Press, 1963.
el-Issawy, Ibrahim H. "Labour Force, Employment, and Unemployment." Technical Papers, No. 4, *Employment Opportunities and Equity in Egypt*. Geneva: International Labour Office, 1983.
Jelin, Elizabeth. "Comment." *Development and Change* 8, no. 2 (1977): 256–57.
Jelin, Elizabeth. "Labor Conflict Under the Second Peronist Regime: Argentina 1973–76." *Development and Change* 10 (1979): 233–57.
Kahler, Miles. "International Financial Institutions and the Politics of Adjustment." In Nelson et al., *Fragile Coalitions*, pp. 139–60. New Brunswick, N.J.: Transaction Books, 1989.
——. "Orthodoxy and Its Alternatives: Explaining Approaches to Stabilization and Adjustment." In Nelson, ed., *Economic Crisis and Policy Choice*, pp. 33–62. Princeton: Princeton University Press, 1990.
Kahneman, Daniel, Jack L. Knetsch, and Richard Thaler. "Fairness as a Constraint on Profit Seeking: Entitlements in the Market." *American Economic Review* 76, no. 4 (September 1986): 728–41.
Kamel, Ibrahim Ahmed. The Impact of Nasser's Regime on Labor Relations in Egypt. Ph.D. dissertation, University of Michigan, 1970.
Kandil, Amany. "Labour and Business Representation in Egypt," Paper prepared for the conference on "Dynamics of States and Societies in the Middle East." Cairo University, June 17–19, 1989. (In Arabic, see Qandil, Amani.)
Kassalow, Everett M. and Ukandi G. Damachi, eds. *The Role of Trade Unions in Developing Countries*. Geneva: International Institute for Labor Studies, 1978.
Kaufman, Bruce E. "Bargaining Theory, Inflation, and Cyclical Strike Activity in Manufacturing." *Industrial and Labor Review* 34 (April 1981): 333–55.
Kaufman, Robert R. "Industrial Change and Authoritarian Rule in Latin America: A Concrete Review of the Bureaucratic-Authoritarian Model." In Collier, ed., *The New Authoritarianism*, pp. 165–254.
Kennan, John. "The Economics of Strikes." In Orley Ashenfelter and Richard Layard, eds., *Handbook of Labor Economics*, vol. 2, pp. 1091–1138. Amsterdam: North-Holland, 1986.
Klotz, Audie. *Norms in International Relations*. Ithaca: Cornell University Press, 1995.
Kraus, J. "African Trade Unions: Progress or Poverty?" *African Studies Review* 19, no. 3 (1976): 95–108.
——. "Comment" *Development and Change* 8, no. 2 (1977): 258–59.
——. "Strikes and Labor Power in Africa," *Development and Change* 10 (1979): 259–86.

Lange, Peter. "Unions, Workers, and Wage Regulation: The Rational Bases of Consent." In John H. Goldthorpe, ed., *Order and Conflict in Contemporary Capitalism*. Oxford: Oxford University Press, 1984.

Lesch, Ann "Democracy in Doses." *Arab Studies Quarterly* (1989), pp. 87–108.

Locke, Richard and Kathleen Thelen. "Apples and Oranges Revisited: Contextualized Comparisons and the Study of Comparative Labor Politics." *Politics and Society*, 1995.

——. "The Shifting Boundaries of Labor Politics: New Directions for Comparative Research and Theory." *Harvard University Center for European Studies Working Paper Series*, No. 44 (1993).

Lockman, Zachary, " 'Worker' and 'Working Class' in pre-1914 Egypt: A Rereading." In Zachary Lockman, ed., *Workers and Working Classes in the Middle East*, pp. 71–110. Albany: SUNY Press, 1994.

Lowi, Theodore. "American Business, Public Policy, Case-Studies, and Political Theory." *World Politics* 16 (1964): 677–715.

Lubell, Harold. "Views of Trade Union Officials on Public Sector Enterprises." U.S. AID, Internal Memorandum, August 2, 1982.

Luther, Hans. "Repression of Labor Protest in Singapore: Unique Case or Future Model?" *Development and Change* 10 (1979): 287–99.

Mabro, Robert. "Industrial Growth, Agricultural Under-Employment, and the Lewis Model: The Egyptian Case, 1937–65." *Journal of Development Studies* 3, no. 4 (July 1967): 322–39.

Mabro, Robert and Samir Radwan. *The Industrialization of Egypt 1939–1973*. Oxford: Claredon Press, 1976.

Mahmoud, M. Rashad. *Investment Guide in Egypt*. Cairo: Middle East Observer, 1977.

Malloy, James M, ed. *Authoritarianism and Corporatism in Latin America*. Pittsburgh: University of Pittsburgh Press, 1977.

Marshall, Gordon. "Some Remarks on the Study of Working-Class Consciousness." *Politics and Society* 12, no. 3 (1983): 263–301.

Mead, Donald C. *Growth and Structural Change in the Egyptian Economy*. Homewood, Ill.: Richard D. Irwin, 1967.

Mericle, Kenneth S. "Corporatist Control of the Working Class: Authoritarian Brazil Since 1964." In Malloy, ed., *Authoritarianism and Corporatism in Latin America*, pp. 303–38.

Middlebrook, Kevin. *The Paradox of Revolution: Labor, the State, and Authoritarianism in Mexico*. Baltimore: Johns Hopkins University Press, 1995.

Mitchell, Timothy. "Everyday Metaphors of Power." *Theory and Society* 19 (1990): 545–77.

——. "The Limits of State: Beyond Statist Approaches and their Critics." *American Political Science Review* 85, no. 1 (March 1991): 77–96.

Moore, Richard V. "The Ecology of Egyptian Labor." *Indian Journal of Industrial Relations* 1 (1966): 442–56.

Muller, Edward N. and Karl-Dieter Opp. "Rational Choice and Rebellious Collective Action." *American Political Science Review* 14 (June 1986): 471–87.

Munck, Ronaldo. "The Labour Movement in Argentina and Brazil: A Comparative Perspective." In Boyd, Cohen, and Gutkind, *International Labor*, pp. 108–33.

Nagel, Jack. "Some Questions About the Concept of Power." *Behavioral Science* 13, no. 2 (March 1968): 129–37.

Nelson, Joan M. "The Political Economy of Stabilization: Commitment, Capacity, and Public Response." In Robert H. Bates, ed., *Toward a Political Economy of Development: A Rational Choice Perspective*, pp. 80–130. Berkeley: University of California Press, 1988.

——. "Organized Labor, Politics, and Labor Market Flexibility in Developing Countries." *The World Bank Research Observer* 6, no. 1 (January 1991): 37–56.

Nelson, Joan M., ed. *Economic Crisis and Policy Choice: The Politics of Adjustment in the Third World*. Princeton: Princeton University Press, 1990.

Nelson, Joan M. et al., *Fragile Coalitions: The Politics of Economic Adjustment*. New Brunswick, N.J.: Transaction Books, 1989.

Nikoi, Kote Nikoi, "The Distributional Impact of Structural Adjustment in Sub-Saharan Africa." Paper presentation at the PAWSS seminar on "Global Economic Inequity." Amherst, Mass., June 13, 1996.

North, Douglas. *Structure and Change in Economic History*. New York: Norton, 1981.

O'Brien, Patrick. *The Revolution in Egypt's Economic System*. London: Oxford University Press, 1966.

O'Donnell, Guillermo A. *Modernization and Bureaucratic Authoritarianism: Studies in South American Politics*. Berkeley: University of California Press, 1973.

——. "Corporatism and the Question of the State." In Malloy, ed., *Authoritarianism and Corporatism in Latin America*, pp. 47–88.

——. "Tensions in the Bureaucratic-Authoritarian State and the Question of Democracy." In Collier, ed., *The New Authoritarianism*, pp. 285–318.

——. "Challenges to Democratization." *World Policy Journal*. 1988, pp. 281–300.

Offe, Claus. *Disorganized Capitalism*. Cambridge, MIT Press, 1985.

Ogle, George E. *South Korea: Dissent Within the Economic Miracle*. London: Zed Books, 1990.

Olson, Mancur. *The Logic of Collective Action*. New York: Basic Books, 1971.

Panitch, Leo. "Recent Theorizations of Corporatism: Reflections on a Growth Industry." *British Journal of Sociology* 31, no. 2 (June 1980): 159–87.

——. "Trade Unions and the Capitalist State." *New Left Review* 125 (1981): 21–43.

Partnership for Productivity. "Small Business Capacity Development: Outlook and Prospects for the Egyptian Private Sector." Report prepared for the USAID, Egypt, August 1985. (FSA No. 263–0090–3–10339—FSA 18)

Pastor, Manuel. "The Latin America." Paper presented at the PAWSS seminar on "Global Economic Inequity." Amherst, Mass., June 13, 1996.

Pizzorno, Alessandro. "Political Exchange and Collective Identity in Industrial Conflict." In Colin Crouch and Alessandro Pizzorno, eds., *The Resurgence of Class Conflict in Western Europe Since 1968*. Vol. 2, *Comparative Analyses*. New York: Macmillan, 1978.

Popkin, Samuel. *The Rational Peasant*. Berkeley: University of California Press, 1979.
Post, Erika. "Egypt's Elections." *MERIP*, No. 147 (July–August 1987), pp. 17–22.
Posusney, Marsha Pripstein. Workers Against the State: Actors, Issues and Outcomes in Egyptian Labor/State Relations. Ph.D. dissertation, University of Pennsylvania, 1991.
——. "Labor as an Obstacle to Privatization: The Case of Egypt." In Harik and Sullivan, eds., *Privatization and Liberalization in the Middle East*, pp. 81–105.
——. "Irrational Workers: The Moral Economy of Labor Protest in Egypt." *World Politics* 46, no. 1 (October 1993): 83–120.
——. "Islam, Islamists, and Labor Law in Egypt." Paper presented to the seminar on "Moral Economy of Political Islam," University of California at Berkeley, April 24–26, 1996.
Przeworski, Adam. "Material Bases of Consent: Economics and Politics in a Hegemonic System." *Political Power and Social Theory* 1 (1980): 21–66.
——. *Capitalism and Social Democracy*. Cambridge: Cambridge University Press, 1985.
——. "Marxism and Rational Choice." *Politics and Society* 14, no. 4 (1985).
——. "Some Problems in the Study of the Transition to Democracy." In Guillermo O'Donnell, Philippe C. Schmitter, and Laurence Whitehead, eds., *Transitions from Authoritarian Rule: Comparative Perspectives*. Baltimore: Johns Hopkins University Press, 1986.
Putnam, Robert. *Making Democracy Work*. Princeton: Princeton University Press, 1993.
Remmer, Karen L. and Gilbert W. Merkx. "Bureaucratic-Authoritarianism Revisited." *Latin American Research Review* 17, no. 2 (1982): 3–52.
Richards, Alan. "The Political Economy of Dilatory Reform: Egypt in the 1980s." *World Development* 19, no. 12 (December 1991): 1721–30.
——. "The Political Economy of Dilatory Reform Revisited: Egypt in the 1990s." Paper presented at the seminar on "Oil Revenues and State Strategies in the Middle East," Center for Middle East Studies, Harvard University, November 20–21, 1992. Cited with permission.
Riker, William H. "Political Science and Rational Choice." In James E. Alt and Kenneth A. Shepsle, eds., *Perspective on Positive Political Economy*. Cambridge: Cambridge University Press, 1990, pp. 163–81.
Roemer, John E. "Neoclassicism, Marxism, and Collective Action." *Journal of Economic Issues* 12 (March 1978): 147–61.
Roxborough, Ian. "The Analysis of Labor Movements in Latin America: Typologies and Theories." *Bulletin of Latin American Research* 1, no. 1 (October 1981).
——. *Unions and Politics in Mexico: The Case of the Automobile Industry*. Cambridge: Cambridge University Press, 1984.
Sabel, Charles. "The Internal Politics of Trade Unions." In Berger, ed., *Organizing Interests in Western Europe*, pp. 209–48.
——. *Work and Politics: The Division of Labor in Industry*. Cambridge: Cambridge University Press, 1982.

Salmi, Aly El. "Public sector management: An analysis of decision-making and employment policies and practices in Egypt." No. 6, Technical Papers to *Employment Opportunities and Equity in Egypt*. Geneva: International Labour Office, 1983.
el-Sayed, Salah. *Workers Participation in Management*. Cairo: AUC Press, 1978.
al-Sayyid, Mustafa Kamel. "Privatization: The Egyptian Debate." *Cairo Papers in Social Sciences* 13, Monograph 4 (Winter 1990).
Schamis, Hector E. "Reconceptualizing Latin American Authoritarianism in the 1970s." *Comparative Politics* 23, no. 2 (January 1991): 201–20.
Schmitter, Philippe C. "Still the Century of Corporatism?" In Fredrick B. Pike and Thomas Stritch, eds., *The New Corporatism*. Notre Dame: University of Notre Dame Press, 1974.
Scott, James C. *The Moral Economy of the Peasant*. New Haven: Yale University Press, 1976.
——. *Weapons of the Weak: Everyday Forms of Peasant Resistance*. New Haven: Yale University Press, 1985.
Sen, Amartya K. "Rational Fools," *Philosophy and Public Affairs* 6, no. 4 (1977): 317–44.
Shaaban, Mustafa Ibrahim. Collective Bargaining and Labor Policy Under Egyptian Socialism. Ph.D. Dissertation, Indiana University, 1975.
el-Shafei, Omar. Workers' Struggles in Mubarak's Egypt. MA thesis. The American University in Cairo, Political Science Department, 1993.
Shalev, Michael. "Industrial Relations Theory and the Comparative Study of Industrial Relations and Industrial Conflict." *British Journal of Industrial Relations* 18 (1980): 26–34.
——. "Trade Unionism and Economic Analysis: The Case of Industrial Conflict," *Journal of Labor Research* 1 (Spring 1980): 133–73.
Shoukri, Ghali. *Egypt: Portrait of a President*, 1971–81. London: Zed Press, 1981.
Simon, Herbert. "Human Nature in Politics: The Dialogue of Psychology with Political Science." *American Political Science Review* 75, no. 2 (1985): 293–304.
Singerman, Diane. *Avenues of Participation: Family, Politics, and Networks in Urban Quarters of Cairo*. Princeton: Princeton University Press, 1995.
Skidmore, Thomas. *Military Rule in Brazil* 1964–85. New York: Oxford University Press, 1988.
Solow, Robert M. *The Labor Market as a Social Institution*. Cambridge: Blackwell, 1990.
Southall, Roger, ed. *Labour and Unions in Asia and Africa*. New York: St. Martin's Press, 1988a.
——. *Trade Unions and the New Industrialization of the Third World*. London: Zed Press, 1988b.
Springborg, Robert. *Mubarak's Egypt: Fragmentation of the Political Order*. Boulder: Westview Press, 1989.
Starr, Gerald, "Wages in the Egyptian Formal Sector." No. 5, Technical Papers, *Employment Opportunities and Equity in Egypt*. Geneva: International Labour Office, 1983.

Steinmo, Sven, Kathleen Thelen, and Frank Longstreth. *Structuring Politics: Historical Institutionalism in Comparative Perspective*. Cambridge: Cambridge University Press, 1992.

Stepan, Alfred. *The State and Society: Peru in Comparative Perspective*. Princeton: Princeton University Press, 1978.

Sullivan, Denis. "Extra State Actors and Privatization in Egypt." In Harik and Sullivan, eds., *Privatization and Liberalization in the Middle East*, pp. 24–45.

Swenson, Peter. *Fair Shares: Unions, Pay, and Politics in Sweden and West Germany*. Ithaca: Cornell University Press, 1989.

Tessler, Mark A, Monte Palmer, Tawfic E. Farah, and Barbara Lethem Ibrahim. *The Evaluation and Application of Survey Research in the Arab World*. Boulder: Westview Press, 1987.

Thompson, E. P. "The Moral Economy of the English Crowd in the 18th Century." *Past and Present* 50 (February 1971): 76–136.

Tomiche, Fernand J. *Syndicalisme et certains aspects du travail en Republique Arabe Unie (Egypte)*, 1900–1967. Paris: G-P Maisonneuve and La rose, 1974.

Trimberger, Ellen. *Revolution from Above*. New Brunswick, N.J.: Transaction Books, 1978.

Tversky, Amos and Daniel Kahneman. "Loss Aversion in Riskless Choice: A Reference-Dependent Model." *Quarterly Journal of Economics* 106, no. 4 (1991): 1039–1061.

——. "Choices, Values, and Frames." In Neil J. Smelser and Dean R. Gerstein, eds., *Behavioral and Social Science: Fifty Years of Discovery*. Washington, D.C.: National Academy Press, 1986.

——. "Rational Choice and the Framing of Decisions." In Cook and Levi, eds., *The Limits of Rationality*, pp. 60–89.

U.S. Agency for International Development, "Views of Trade Union Officials on Public Sector Enterprises." Memo DMOO/Eg-AAd-096, August 2, 1982. Library, Cairo.

Valenzuala, J. Samual. "Labor Movements in Transitions to Democracy: A Framework for Analysis." *Comparative Politics*, July 1989, pp. 445–71.

Vatikiotis, P.J. *The Egyptian Army in Politics*. Bloomington: Indiana University Press, 1961.

Vermeulen, Bruce and Gustav F. Papanek. "Labor Markets and Industry in Egypt: Analysis and Recommendations for Employment-Oriented Growth." Report of the Industrial Strategy Assessment Project of the USAID, June 1982.

Vernon, Raymond, ed. *The Promise of Privatization*. New York: Council on Foreign Relations Books, 1988.

Vitalis, Robert. *When Capitalist Collide: Business Conflict and the End of Empire in Egypt*. Berkeley: University of California Press, 1995.

——. "The End of Third Worldism in Egyptian Studies." *Arab Studies Journal* 4, no. 1 (Spring 1996): 13–33.

Wade, Robert. *Governing the Market: Economic Theory and the Role of Government in East Asian Industrialization*. Princeton: Princeton University Press, 1990.

Waterbury, John. *The Egypt of Nasser and Sadat: The Political Economy of Two Regimes*. Princeton: Princeton University Press, 1983.

——. "The 'Soft State' and the Open Door: Egypt's Experiences with Economic Liberalization, 1974–1984." *Comparative Politics* 18, no. 1 (October 1985): 65–83.

——. "Twilight of the State Bourgeoisie?" Paper presented at the JCNME/SSRC Conference on "Retreating States and Expanding Societies." Aix-en-Provence, March 1988.

——. *Exposed to Innumerable Delusions: Public Enterprise and State Power in Egypt, India, Mexico, and Turkey*. Cambridge: Cambridge University Press, 1993.

Waterman, Peter. "The 'Labour Aristocracy' in Africa: Introduction to a Debate." *Development and Change* 6, no. 3 (July 1975): 57–73.

——. "Workers in the Third World." *Monthly Review* 29, no. 4 (1977): 50–65.

Weintraub, Andrew R. "Prosperity vs. Strikes: An Empirical Approach." *Industrial and Labor Review* 19 (October 1965): 231–38.

Wickham, Carrie Rosefsky. "Islamic Mobilization and Political Change: The Islamist Trend in Egypt's Professional Associations." In Joel Beinin and Joe Stork, eds., *Political Islam: Essays from Middle East Report*, pp. 120–35. Berkeley: University of California Press, 1996.

World Bank. *World Development Report 1995*. New York: Oxford University Press, 1995a.

——. "Will Arab Workers Prosper or Be Left Out in the Twenty-First Century?" *Regional Perspectives on World Development Report 1995*. Washington, D.C: World Bank, 1995b.

Youssef, Samir M. *System of Management in Egyptian Public Enterprises*. Cairo: American University in Cairo Press, 1983.

Zaytoun, Mohaya A. "Earning and the Cost of Living: An Analysis of Recent Developments in the Egyptian Economy." In Handoussa and Potter, eds., *Employment and Structural Adjustment*, pp. 219–58.

Books, Articles, and Dissertations in Arabic

Arabic periodicals (with abbreviations used in citations):

A	*al-Ahram*, main official daily newspaper	SA	*Sawt al-'Amil*, nonperiodic rank-and-file magazine
AH	*al-Ahali*, weekly newspaper of the opposition Tagammu' Party	SH	*al-Sha'b*, weekly newspaper of the Socialist Labor Party
AM	*al-'Amal*, monthly magazine of the Ministry of Labor	T	*al-Tali'ah*, leftist monthly journal
AI	*al-Ahram al-Iqtisadi*, official weekly economics magazine	TS	*Tali'at al Sina'a*, monthly magazine of the EEMWF
AR	*al-'Arabi*, weekly newspaper of the Arab Nasirist Party	U	*al-'Ummal*, weekly newspaper of the ETUF
RY	*Rose al-Youssef*, leftist weekly magazine	W	*Al-Wafd*, daily newspaper of the New Wafd Party

'Abbas, Kamal et al. *Malhamat 'ummal al-sulb* (The battle of the steel workers). Cairo: n.p., 1990.

'Abd Allah, Zaki. "Ant mafsul" (You're fired!). *AM*, No. 55 (December 1967): 10–12.
'Abd Allah, Ahmad, ed. *Humum Misr* (Egypt's concerns). Cairo: Al-Jeel Center for Youth and Social Studies, 1994.
'Abd al-'Aziz, 'Asim, 'Izat Nasr, and Isma Rashid. "al-Intikhabat al-niqabiyah" (The trade union elections). *RY*, No. 2886, October 3, 1983, pp. 19–23.
——. "Ithamat sakhina bayna al-'ummal wa al-lijan an-niqabiyah" (Hot accusations between the workers and the union locals). *RY*, No. 2887, October 10, 1983, pp. 30–33.
'Abd al-Mun'im, Ahmad Faris. "Jama'at al-masalih" (Interest groups). In 'Ali al-Din Hilal, ed., *al-Nizam al-siyasi* (The political system). Series Commemorating the 30th Anniversary of the July 1952 Revolution, No. 2. Cairo: Arab Center for Research and Publication, 1982.
'Abd al-Rahman, Ahmad. "Aghrab 'amaliyat bi' lil-qita' al-'amm" (The strangest sales operation in the public sector). *AM*, No. 143 (April 1975): 16–17.
'Abd al-Raziq, Husayn. *Misr fi 18 wa 19 Yanayir* (Egypt on the 18th and 19th of January). Cairo: Shuhdi, 1985.
Abu 'Alam, 'Abd al-Ra'uf. "al-Mu'tamar al-thalith lil-ittihad al-'amm fi mizan" (An evaluation of the third convention of the ETUF). *AM*, No. 15 (August 1964): 45.
——. "al-harakah al-'ummaliyah: min al-mafhum 'al-iqtisadi ila al-mafhum 'al-siyasi" (The workers' movement: from the economic to the political concept). *T*, No. 6 (October 1966): 47–57.
'Abud , Samih Sa'id. "al-Harakah al-'ummaliyah wa al-nidal al-qanuni" (The workers' movement and the legal struggle). *Al-Badil* 2 (August 1994): 20.
'Adli, Huwayda, "al-Harakah al-ihtijajiyah lil-tabaqah al-'amilah al-Misriyah (The protest movement of the Egyptian working class), 1982–1991." In 'Abd Allah, ed., *Humum Misr*, pp. 173–96.
Ahmad, 'Abd al-Majid. "Nizam al-mad'a al-ishtiraki" (The socialist prosecutor system). In al-'Alim, ed., *al-Tabiqah al-'Amilah al-Misriyah*, pp. 165–70.
Ahmad, Ahmad Taha (Ahmad Taha). *Dirasa wa taqdim al-qiyadat al-niqabiyah* (A study and evaluation of the trade union leaders). Cairo: Dar al-Jamahir, 1976.
al-Ahram Center for Political and Strategic Studies. *al-Taqrir al-Istiratiji al-'Arabi* (Arab Strategic Report). Cairo: al-Ahram Center, 1986, 1987, and 1989.
al-'Alim, Mahmud Amin, ed. *Al-Tabaqah al-'amilah al-misriyah* (The Egyptian working class). Cairo: Dar al-Thiqafa al-Jadida, 1987 (*Qadaya Fikriyah* 5).
'Ali, Muhammad Muhammad. "al-Qiyadat al-niqabiyah lil-'ummal al-Misriyah" (Union leaders of the Egyptian workers). *AM*, No. 247 (October 1983): 36–39, and No. 248 (January 1984): 54–56.
'Ali, Shafiq Ahmad. "Hal hunaka mukhalifat al-dustur hawla intikhabat al-'ummal fi majlis al-idarat?" (Is there a constitutional violation in the elections of workers to the management councils?) *RY*, No. 2387, March 11, 1974, pp. 15–17.
Amin, Muhammad Fakhim. *Sharh al-tanzim al-niqabi al-jadid* (An explanaation of the new trade union structure). Cairo: Alim al-Kutub, 1964.
'Amir, 'Abd al-Salam 'Abd al-Halim. *Thawrat Yulyaw wa al-tabiqah al-'amilah* (The July revolution and the working class). Cairo: Egyptian General Book Committee, 1987.

Arab Research Center. *Al-Harakah al-'ummaliyah fi ma'rakat al-tahawwal: dirasat fil-intikhabat al-niqabiyah 1991* (The workers' movement in the battle of [economic] transformation: Studies of the 1991 union elections). Helwan: Center for Trade Union Services, 1994.

Balal, 'Abd al-Hamid. "al-Tanzim al-niqabi marrah thaniyah" (The union organization one more time). *AM*, No. 74 (July 1969): 56–57.

Balal, 'Abd al-Hamid et al. "Sharkh fi jidar al-samt" (A crack in the wall of silence), 3-part series in *AM*, No. 65 (October 1968): 34–37, No. 66 (November 1968): 12–15, and No. 67 (December 1968): 10–12.

al-Banna, Jamal. *Nahu ta'adudiyah niqabiyah dun tafattut aw ihtikar* (Toward trade union pluralism without splintering or monopolization). Cairo: n.p., 1994.

Egyptian Trade Union Federation (ETUF). *al-Ittihad al-'Amm li-Niqabat 'Ummal Misr fi 'Ashrin 'Aman* (Twenty years of the Egyptian Trade Union Federation). Cairo: Al-Ahram Commercial Publishing, 1977.

——. *Injazat al-Ittihad al-'Amm (fil) Durah al-Sadisah 1976–79* (Achievements of the ETUF in the Sixth Session). Cairo: Dar Isama lil-Taba' wal-Nashr, 1979.

——. *al-'Id al-fadi lil-Ittihad al-'Amm li-Niqabat 'Ummal Misr 1957–1982* (The silver anniversary of the ETUF). Cairo: ETUF, 1982.

Fawda, Zakiyya Muhammad. al-Taghyir al-ijtima'i wa tatawwar al-harakah al-niqabiyah fil-mujtama' al-Misri (Social change and the development of the trade union movement in Egypt). Ph.D. dissertation, Ain Shams University, March 1981.

al-Ghazzali, 'Abd al-Mun'im. "al-Niqabat bayna al-'amal al-maktabi wa al-nishat al-jamahiri" (The unions: between office work and mass activity). *T*, November 1968: 58–66.

——. *Muhadarat 'an al-harakah al-niqabiyah* (Lectures on the trade union movement). Cairo: n.p., 1988.

Hamid, Ra'uf 'Abbas (Ra'uf Abbas). *al-Harakah al-'ummaliyah fi Misr 1899–1952* (The workers movement in Egypt). Cairo: 1967.

Harun, 'Adil and Samya Sa'id. "Ma zala sinariyo tasfiyat al-sina'ah al-wataniyah wa istinzaf al-mal al-'amm mustamiran" (The scenario of liquidating national industry and draining public finances continues). *AI*, No. 914, July 21, 1986, pp. 62–66.

Hasan, Ahmad 'Atif. *Tarikh al-harakah al-niqabiyah al-Misriyah* (History of the Egyptian trade union movement). Cairo: Rose al-Youssef, 1981.

al-Hilali, Ahmad Nabil. *Hadha al-mashru' lan yamurr* (This draft law will not pass). Helwan: Dar al-Khidamat al-Niqabiyah Bihilwan, 1994.

Imam, 'Abd Allah. "Naqib 'ummal al-tijarah: 'Narfud al-tafrit fi shibr min al-qita' al-'amm" (A unionist from the commerce workers: "We reject the dismemberment of the public sector"). *RY*, No. 2451, July 25, 1975, pp. 14–17.

Imam, Muhammad Jamal. "al-Niqabat al-'Ummaliyah: bayna al-wazir, al-ra'is, wa intikhabat al-ghurfah al-mughliqah" (The trade unions: between the minister, the president, and the closed door elections). *AH*, September 16, 1987a.

——. "Hal tanjah al-niqabat al-'ummaliyah fi ista'ada thiqat jamahiriyah?" (Will the trade unions succeed in regaining mass trust?). *AH*, November 11, 1987b.

——. "al-Taghayyarat fil-qa'ima wa al-istimrariyah fil-qimma" (The changes in the base, and the continuity at the top). *AH*, December 16, 1987c.

——. "Ghiyab al-dimaqratiyah al-niqabiyah," (The absence of trade union democracy). In al-Alim, ed., *al-Tabaqah al-'amilah al-Misriyyah*, pp. 161–64.

Imam, Samya Sa'id (Samya Sa'id). *Man Yamluk Misr?* (Who owns Egypt?). Cairo: Dar al-Mustaqbal, 1987.

——. "Rasmaliyat al-sab'inat bayna al-istimrar wal-taghayyar" (The capitalism of the 1970s: between continuity and change). *al-Nahj* 3, no. 12 (1987).

——. "Al-Harakah al-Niqabiyah al-'Ummaliyah" (The trade union movement). In 'Abd Allah, ed., *Humum Misr* (1994), pp. 201–20.

'Izz al-Din, Amin. *Tarikh al-Tabiqah al-'Amilah al-Misriyah Mundhu Nushu'ha hata sana 1970* (The history of the Egyptian working class from its beginnings through 1970). Cairo: House of the Arab Future, 1987.

Jirgis, Fahmy Kamil and Ra'uf Sadiq 'Ali. *Majmu'a mabadi ha'at al-tahkim fi munaza'at al-'amal* (The principles of the arbitration committees for labor disputes). *Kitab al-'Amal* 138 (August 1975).

Kamil, Fathi. *Ma'a al-harakah al-niqabiyah al-Misriyya fi nusf qarn* (With the Egyptian trade union movement for half a century). Cairo: House of the Arab Future, 1985.

Khalid, Muhammad. *'Abd al-Nasir wa al-haraka al-niqabiyah* (Abd al-Nasir and the trade union movement). Cairo: Cooperative Institute for Printing and Publishing, 1971.

——. *al-Harakah al-niqabiyah bayna al-madi wa al-mustaqbal* (The union movement between past and present). Cairo: Cooperative Institute for Printing and Publishing, 1975.

Khalil, Faruq. "Sharh qanun al-niqabat al-'ummaliyah 35 li-sanna 1976 (An explanation of trade union Law 35 of 1976)." *Kitab al-'Amal* 154 (December 1976).

Mahmud, Khayri. "al-Harakah al-niqabiyah al-'ummaliyah fi misr" (The trade union movement in Egypt). *al-Yasar al-'Arabi* 79 (December 1986): 12–15.

al-Maraghi, Mahmud. "Wazir al-tijara yatahaddath 'an al-as'ar wa al-'umulat wa bi' al-qita' al-'amm" (The Minister of Commerce discusses prices, sales commissions, and selling the public sector). *RY*, No. 2450, April 26, 1975, pp. 3–6.

——. *al-Qita' al-'amm fi mujtama' mutaghayyar: tajrubat misr* (The public sector in a changing society: Egypt's experience). Cairo: Arab Future Publishing House, 1983.

Al-Muhami, Samih Sa'id 'Abud. "Al-Harakah al-'ummaliyah wa al-nidal al-qanuni" (The workers' movement and the legal struggle). *Al-Badil* 2 (August 1994): 20.

Mursi, Amin Taha. "Mu'tamar ittihad al-'ummal tatanaqqash al-infitah fi zull al-iqtisad al-salam" (ETUF convention discusses the economic opening in the healthy economy). *RY*, No. 2734, November 3, 1980, pp. 12–13.

——. "Zahirah tuhaddid al-harakah al-niqabiyah" (A phenomenon which threatens the trade union movement). *RY*, No. 2745, January 19, 1981.

——. "Ra'is ittihad al-'ummal yajib 'ala tasawalat al-harakah al-niqabiyyah" (The ETUF president answers the questions of the trade union movement). *RY*, No. 2782, October 5, 1981, pp. 22–26.

——."Milyawn wa nisf 'amil yantakhabuna al-yawm" (A million and a half workers vote today). *RY*, No. 2507, June 28, 1986, pp. 22–25.

Mursi, Fu'ad. *Hadha al-infitah al-iqtisadi* (This is the economic opening). Cairo: New Culture House, 1984.

——. *Masir al-qita' al-'amm fi misr* (The fate of the Egyptian public sector). Cairo: Arab Research Center, 1987.

Mustafa, Fathi Mahmud (Fathi Mahmud). "Sharkh fi ittihad al-'ummal" (Fissures in the Workers' Confederation). *RY*, No. 2466, September 15, 1975, pp. 18–19.

National Committee to Fight Privatization, "Mashru' qanun ta'dil qanun al-niqabat al-'ummaliyah tadahwur khatir." (The draft law to revise the trade union law is a dangerous setback). Mimeo, Helwan, 1994.

Nur al-Din, Lutfi, "Ujur al-tabiqah al-'amilah bayna al-tadakhkhum wa 'awa'id huquq al-tamalluk," (The wages of the working class: between inflation and property taxes). In al-'Alim, ed., *al-Tabiqah al-'amilah al-misriyah*, pp. 130–140.

Popular Committee for Constitutional Reform, *al-Dustur aladhi nutlub bih* (The constitution we demand). Cairo: n.p., 1991.

Qandil, Amani. (1985a) "Su'al kabir . . . man humma rijal al-'a'mal?" (A big question . . . Who are the businessmen?). *AI*, No. 882, December 9, 1985, pp. 12–14.

——. (1985b) Sun' al-qarar fi Misr (Policy making in Egypt: A case study of economic policy making, 1974–1981). Ph.D. dissertation, Cairo University, 1985.

——. (1986a) "Wa marra ukhrah . . . man humma rijal al-'a'mal?" (And another time . . . Who are the businessmen?"). *AI*, No. 886, January 6, 1986), pp. 18–20.

——. (1986b). "Jama'at al-masalih wa tatawwur al-nizam al-siyasi al-Misri" (Interest groups and the development of the Egyptian political system). Paper presented at the Conference of the Faculty of Economics and Political Science, Cairo University, April 2–3, 1986.

Rif'at, 'Issam. "Bi' ishum al-qita' al-'amm: tanmiyah am tasfiyah?" (Selling shares in the public sector: development or liquidation?). *AI*, No. 481, September 1, 1975, pp. 6–8.

Sa'id, 'Abd al-Mughni, "Nidal al-'ummal wa thawrat 23 Yulyaw" (The workers struggle and the July revolution). *al-Silsilah al-Ummaliyah* No. 30. Cairo: Institute for Workers' Education, 1968.

——. "Safahat majhulah min Tarikh al-harakah al-niqabiyah" (Unknown pages from the history of the union movement). *AM*, No. 236 (January 1983): 11–25.

——. "Thawrat Yulyaw wa muhawalat al-tanzim al-siyasi" (The July revolution and the attempts at political organization). *Kitab al-'Amal* 257 (July 1985).

Sami, 'Azat. "Ru'yah shakhsiyah hawl qadayah 'ummaliyah" (A personal view on a workers' issue). *AM*, No. 289 (June 1987): 46–48.

Sa'udi, Fu'ad. *Sharh ihkam qanun al-niqabat al-'ummaliyah li sanna 1964* (An explanation of the trade union law of 1964). *Kitab al-'Amal* 10 (1964).

al-Sayyid, Mustafa Kamil. *al-Mujtama' wa al-siyasah fi misr: dur jama'at al-masalih fi al-nizam al-siyasi al-misri 1952–1981* (Society and politics in Egypt: The role of interest groups in the Egyptian political system). Cairo: Arab Future Publishing House, 1983.

Sha'ban, Adil. "al-Barnamij al-Iqtisadiyah wa al-Dimuqratiyah lil-Harakah al-'Ummaliyah" (The economic and democratic program of the trade union

movement). In Arab Research Center, *Al-Harakah al-'Ummaliyah*, pp. 111–38.
Shafiq, Amina. "al-tabaqah al-'amilah al-misriyah: al-nasha', al-tatawwar, al-nadalat" (The Egyptian working class: Its inception, development, and struggles). *Cultural Secretariat*, Tagammu' Party, Pamphlet No. 3, January 1987.
——. *al-Idrab* (Strikes). Pamphlet No. 2, al-maktabah al-sha'biyah. Cairo: New Culture House, 1986.
Sharaf al-Din, Ahmad. "Hawla al-haqq al-niqabi" (On trade union rights). Paper presented at the human rights festival organized by the lawyers' syndicate. Cairo, December 9–11, 1986.
——. "Hawla haqq al-idrab" (On the right to strike). Paper presented at the human rights festival organized by the lawyers' syndicate, Cairo, December 9–11, 1986.
Sharaf al-Din, Ahmad, Sabir Barakat, and Ilhami al-Mirghani. "Kifah 'ummal al-sikkah al-hadid fi thamanin 'am, 1916–1986" (The railway workers struggle over 80 years). *Sawt al-'Amil Notebooks* 1, 1986.
Sharaf al-Din, Ahmad, Sabir Barakat et al. "al-harakah al-niqabiyah al-'ummaliyah: al-ma'ziq wa al-hall" (The trade union movement: the crisis and the solution). *Sawt al-'amil Notebooks* 2, 1987.
al-Sirafi, 'Attiya. "Harakah niqabiyah jadidah yanhad fi misr" (A new union movement blossoms in Egypt). *Al-Yasar al-'Arabi* 78 (September 1986): 7.
——. "al-jam'iyah al-'umumiyah li-ittihad 'ummal misr" (The general assembly of the ETUF). *AH*, October 1, 1986.
——. "mu'tamar tashri' al-'amal: taqrir khas" (A special report on the labor law conference). *T* 8, no. 3 (1972): 124–28.
Turk, Sayyid Khalil. *al-Ittihad al-'Amm lil-Niqabat: tarikh wa mustaqbal* (The ETUF: Its History and Future). Cairo: n.p., 1987.
'Umar, Sayyid abu-Dif Ahmad. "al-Niqabat al-'ummaliyah wa al-sultah al-siyasiyah ma'a dirasah tatbiqiyah 'ala al-ittihad al-'amm li-niqabat 'ummal misr, 1957–1986" (Workers unions and political authority, with an applied study of the Egyptian Trade Union Federation). Masters' thesis, Political Science Department, Assuit University, January 1987.
'Uthman, Taha Sa'd. *al-Tanzim al-niqabi wa muham al-marhalah al-qadimah* (The union structure and the tasks of the coming period). Cairo: n.p., 1991.

Index